D0770953

Planet Earth

A PHYSICAL GEOGRAPHY

GARY BIRCHALL
Head of Geography
Barton Secondary School
Hamilton, Ontario

JOHN MCCUTCHEON
Head of Geography
Westmount Secondary School
Hamilton, Ontario

John Wiley & Sons
Toronto New York Chichester Brisbane Singapore

Canadian Cataloguing in Publication Data

Birchall, Gary, 1942–
Planet earth: a physical geography

Includes index.
ISBN 0-471-79486-4

1. Geography. I. McCutcheon, John. II. Title.

G128.B57 1993 910'.02 C92-095111-2

DEVELOPMENTAL EDITOR: Graham Draper
DESIGN: Brant Cowie/ArtPlus Limited
PAGE MAKE-UP: Heather Brunton/ArtPlus Limited
TECHNICAL ILLUSTRATION: Donna Guilfoyle, Sylvia Vanderschee/ArtPlus Limited
PHOTO REPRODUCTION SCANS: Cathy Campion/ArtPlus Limited
FILM OUTPUT: TypeLine Express Limited

Printed and Bound in Canada by Friesen Printers
4 5 DWF 99

Acknowledgments

The publishers and authors are grateful for the many useful comments and suggestions made by the following reviewers:

Fraser Cartwright
Geography Consultant
York Region Board of Education
Aurora, Ontario

Jack Davies
Geography Department Head
Fredericton High School
Fredericton, New Brunswick

Kim Evans
Social Studies Department
Moncton High School
Moncton, New Brunswick

Don Farquharson
Geography and Environmental Education Consultant
Durham Region Board of Education
Whitby, Ontario

Richard Humphrey
Geography/History Department Head
Glenlawn Collegiate Institute
Winnipeg, Manitoba

Bruce Kiloh
Geography Head
Terry Fox Senior Secondary School
Coquitlam, British Columbia

Cliff Oliver
Head of Geography
Saunders Secondary School
London, Ontario

Special thanks as well to Karen Ewing of Capilano College and Dr. Margaret North of the University of British Columbia for their input at various stages of the manuscript's development.

TABLE OF CONTENTS

UNIT 1

INTRODUCTION

CHAPTER 1 **The Nature of Physical Geography** *2*

1.1 A Framework for Physical Geography *3*
1.2 The Characteristics of Physical Geography *6*
1.3 The Development of Ideas in Physical Geography *10*

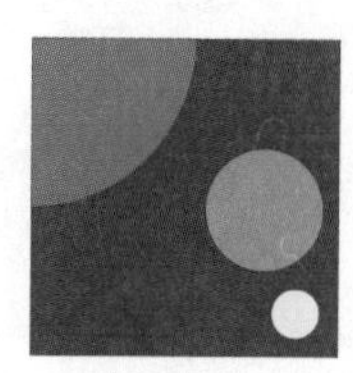

UNIT 2

THE EARTH IN SPACE

CHAPTER 2 **Earth: Its Place in the Universe** *16*

2.1 Pondering the Imponderable *17*
2.2 Our Solar System *23*
2.3 The Origin of the Solar System *25*
2.4 The Origin of the Earth *30*
2.5 Earth's Unique Place in the Solar System *34*

CHAPTER 3 **The Earth in Motion** *37*

3.1 Differing Types of Movement *38*
3.2 Earth Orbits and Tilts *44*
3.3 Rotation of the Earth *46*

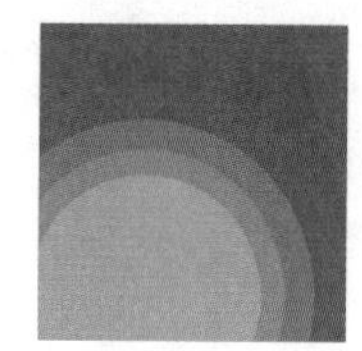

UNIT 3

ENERGY FROM BELOW

CHAPTER 4 **The Earth's Interior** *50*

4.1 The Earth's Internal Heat Sources *51*
4.2 A Layered Earth *53*
4.3 Uncovering the Earth's Interior *56*

CHAPTER 5 **The Earth's Crust** *66*

5.1 The Minerals and Rocks of the Earth's Crust *67*
5.2 Born of Fire: Igneous Rocks 70
5.3 Born of Erosion and Deposition: Sedimentary Rocks *72*
5.4 Born of Great Heat and Pressure: Metamorphic Rocks *78*
5.5 Oceans and Continents 80

CHAPTER 6 The Lithosphere in Motion: Plate Tectonics 85
6.1 A Stable Earth? 86
6.2 Lithosphere Plates: A Simple But Elegant Idea 88
6.3 Mid-Ocean Ridges: Diverging Plate Boundaries 92
6.4 Transform Fault Boundaries: Plates Sliding Past One Another 97
6.5 Collision and Subduction Zones: Converging Plate Boundaries 103
6.6 Plate Interiors: Zones of Inactivity 111

UNIT 4
ENERGY FROM ABOVE

CHAPTER 7 Solar Radiation 118
7.1 The Source 119
7.2 The Earth's Solar Radiation Balance 122
7.3 Variation in Rn Values Over Space and Time 126
7.4 The Energy Balance 128

CHAPTER 8 Climate 131
8.1 Climate and Weather 132
8.2 Climatic Classification Systems 133
8.3 Climatic Controls: The Variations in Solar Radiation Inputs 137
8.4 Climatic Controls: The Global Wind Systems 141
8.5 Climatic Controls: Ocean Currents 147
8.6 Climatic Controls: Water Bodies and Continents 149
8.7 Climatic Controls: Altitude 152
8.8 Climatic Controls: Mountains 153
8.9 Climatic Controls: Local Influences on Climate 159
8.10 Climatic Change 162

CHAPTER 9 Weather 168
9.1 Weather and Air Masses 169
9.2 Weather of the Mid- and High Latitudes 172
9.3 Equatorial Weather 177
9.4 Predicting the Weather 186
9.5 Our Complex Atmosphere as Illustrated by an El Niño Event 190

CHAPTER 10 The Hydrosphere and the Hydrologic Cycle 195
10.1 The Hydrologic Cycle 196
10.2 Saltwater Storehouses: The Earth's Oceans 198
10.3 Freshwater Storehouses 201

CHAPTER 11 Natural Vegetation and Soil Systems 210
11.1 The Ecosphere: The Home of Earth's Life Layer 211
11.2 Nutrient Cycles 213
11.3 Photosynthesis: A Vital Process 216
11.4 Ecosystems: Webs of Life 217
11.5 Vegetation Systems: The Major Biomes 218
11.6 Soil Systems: Between Living and Non-Living Matter 226
11.7 Soil Formation 230
11.8 Soil Classification and Distribution 233

UNIT 5

NATURAL LANDSCAPES

CHAPTER 12 **Denudation: Weathering and Mass Wasting** *240*

12.1 Denudational Processes and Weathering *241*
12.2 Weathering: A Key Process *243*
12.3 Weathering: Variations Over Time and Space *247*
12.4 Mass Wasting: Weathering and Gravity *248*

CHAPTER 13 **Distinctive Landscapes: Humid and Arid Environments** *254*

13.1 A World of Different Landscapes *255*
13.2 Humid Landscapes: The Work of Rivers *257*
13.3 Karst Landscapes *266*
13.4 Arid Landscapes *268*

CHAPTER 14 **Distinctive Landscapes: Glacial, Periglacial, and Coastal Environments** *275*

14.1 The Theory of Glaciation *276*
14.2 The Formation of Glaciers *278*
14.3 Continental Glaciation *280*
14.4 Alpine Glaciation *285*
14.5 Periglacial Landscapes *288*
14.6 Coastal Landscapes *291*
14.7 Erosion and Deposition by Waves *293*
14.8 Tides *297*

UNIT 6

INTEGRATIVE STUDIES

CHAPTER 15 **Natural Hazards: Disrupting Human Systems** *302*

15.1 Natural Disasters: The Environment as Hazard *303*
15.2 Atmospheric Hazards: Droughts and Floods *306*
15.3 Atmospheric Hazards: Tropical Cyclones *309*
15.4 Atmospheric Hazards: Tornadoes *311*
15.5 Geological Hazards: Earthquakes *313*
15.6 Geological Hazards: Volcanic Eruptions *316*
15.7 Natural Hazards: An Overview *318*

CHAPTER 16 **The Disruption of Natural Systems** *322*

16.1 Case Study: The Greenhouse Effect *324*
16.2 Case Study: The Depletion of the Ozone Layer *332*
16.3 Case Study: Hydro Dams — A Planned Alteration *336*

CHAPTER 17 **Fragile Environments** *344*

17.1 Environment Under Siege: The Tropical Rainforest *345*
17.2 Environment Under Stress: Boreal Forests *349*
17.3 Environment Destroyed: Desert Margins *355*
17.4 Environment at the Threshold: The Great Lakes *357*

Glossary ***366***
Index ***375***
Credits ***378***

PREFACE

The subject matter of Physical Geography is complex and diverse. This book was written to present the content of Physical Geography in a meaningful and understandable manner to our students. Meaningful in the sense that the physical environment of Planet Earth is dynamic, affecting each of our lives directly as we, simultaneously, affect it. Understandable in the sense that the varied topics within the subject are linked together in important relationships which each citizen of Planet Earth should understand. The authors believe that the unique framework presented in this text and the presentation of the material itself will allow students to better understand Planet Earth and its relationship to their personal lives.

Planet Earth is the only home we have. We are all part of a planet-wide ecosystem. In order to ensure the planet's continued health and existence, we must understand our physical world, its processes and systems, and its relationship to our human world. The health of our planet is directly related to our knowledge about, and impact on, the physical systems that operate on Planet Earth.

Our thanks are extended to our family and friends and to our many colleagues and students who gave advice and helped us in completing a long but rewarding project. Our appreciation goes to Linda Scott, Graham Draper, and Joseph Gladstone, without whose guiding hands this text would not have been completed. A special thanks is given to John Smees, the author of the Teacher's Resource Package, and a valued companion through many meetings and revisions.

Gary Birchall
John McCutcheon

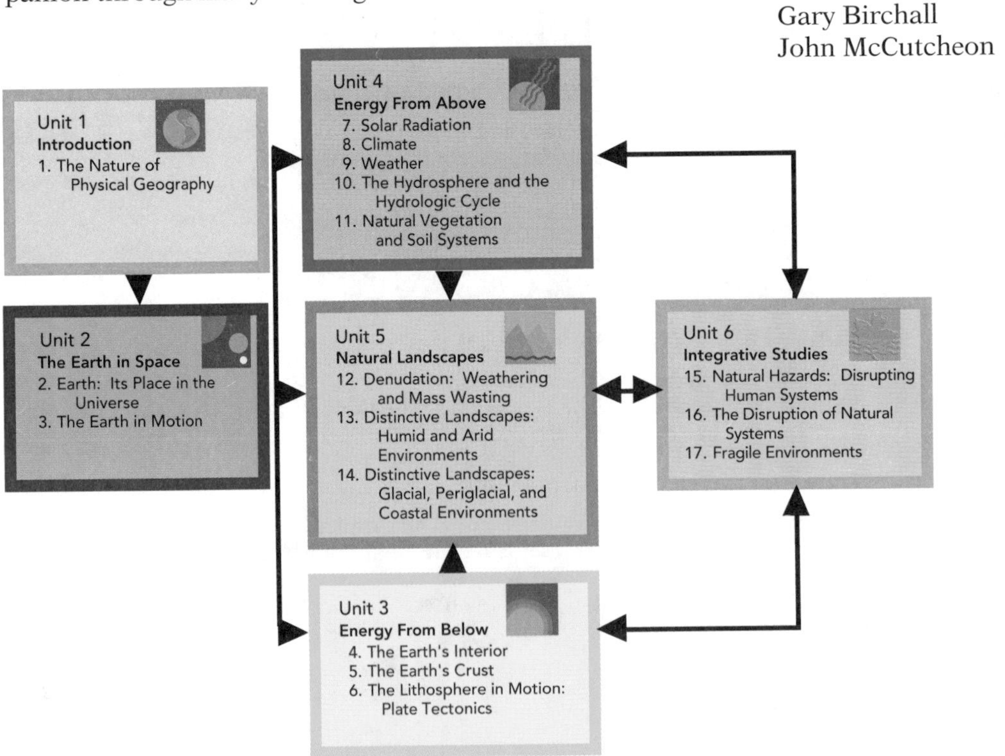

Introduction

UNIT

We all exist in place and time. Geography as a subject tries to explain our place in the world and the world itself. Physical Geography is an important subset of the broader study of geography.

This unit provides an important perspective on the study of Physical Geography and an overview of this text. It lays out the important ideas on which the subject area is built. Using these ideas, you will be better able to explore the units and chapters that follow.

As you work through this unit, consider these questions:

- What questions and issues are appropriate for the study of Physical Geography?
- What are some of the fundamental ideas in Physical Geography?
- How does Physical Geography relate to other areas of study?

CHAPTER

1

THE NATURE OF PHYSICAL GEOGRAPHY

OBJECTIVES:

By the end of this chapter, you will be able to:

- appreciate the range of topics that are included in the study of Physical Geography;
- recognize the similarities between the study of Physical Geography and the organization of this book;
- identify important concepts upon which the subject area is based.

CHAPTER 1: The Nature of Physical Geography
CHAPTER 2: Earth: Its Place in the Universe
CHAPTER 3: The Earth in Motion
CHAPTER 4: The Earth's Interior
CHAPTER 5: The Earth's Crust
CHAPTER 6: The Lithosphere in Motion: Plate Tectonics
CHAPTER 7: Solar Radiation
CHAPTER 8: Climate
CHAPTER 9: Weather
CHAPTER 10: The Hydrosphere and the Hydrologic Cycle
CHAPTER 11: Natural Vegetation and Soil Systems
CHAPTER 12: Denudation: Weathering and Mass Wasting
CHAPTER 13: Distinctive Landscapes: Humid and Arid Environments
CHAPTER 14: Distinctive Landscapes: Glacial, Periglacial, and Coastal Environments
CHAPTER 15: Natural Hazards: Disrupting Human Systems
CHAPTER 16: The Disruption of Natural Systems
CHAPTER 17: Fragile Environments

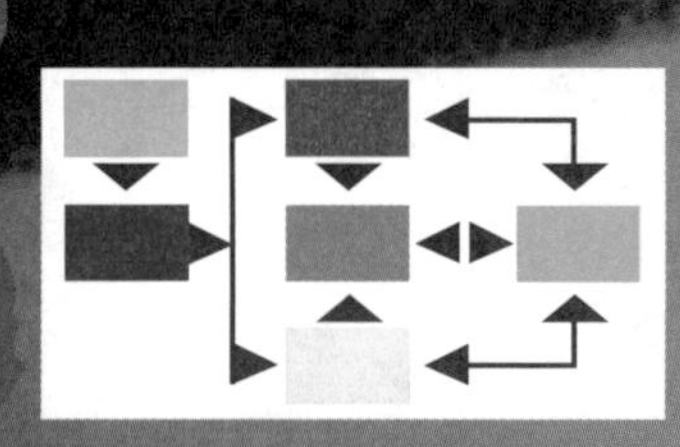

Introduction

Humans try to understand themselves and what is around them by breaking knowledge down into fields of study. The humanities, social sciences, and physical sciences are some of these fields of study. But even these cover such a large amount of knowledge that they are further broken down into subject areas. Physical Geography is one of the many subject areas that are used. Each subject area deals with a range of topics using methods or approaches that are most appropriate for its investigations.

What topics are included in the study of Physical Geography? What are the important characteristics of this subject area? Having some understanding of the range of topics that are included in the study of Physical Geography will make it easier to see the relationships among them. This first chapter provides a framework for the rest of the book.

1·1 A Framework for Physical Geography

The subject of Physical Geography is immense. It includes the study of all the processes and events that affect the appearance and form of the natural surface of our planet. Physical Geography includes the study of the tallest mountain, as well as the ripples of sand on a beach; forces that move entire continents, and forces that cause blades of grass to bend.

The earth's surface is the dividing line between two zones or spheres that physical geographers study — the **atmosphere**, which surrounds the earth, and the **lithosphere**, or outer layer, of the earth. Overlapping these two zones is the **hydrosphere**, the zone of the earth in which water exists. Although most of the planet's water is found in the oceans, the

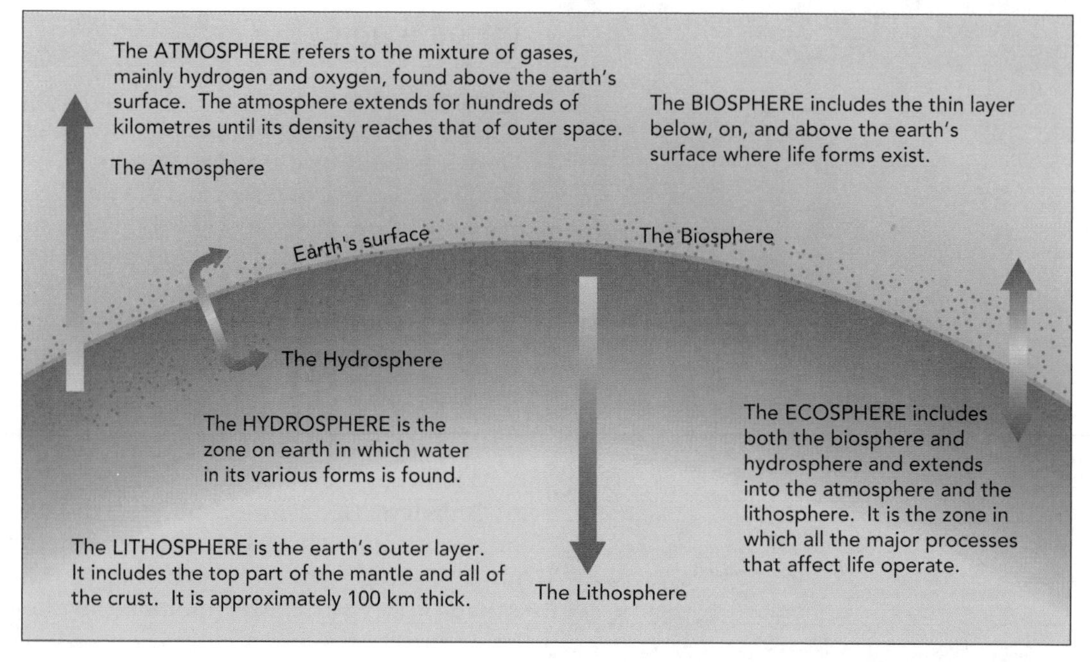

Figure 1.1 **The Spheres of Physical Geography**

hydrosphere extends into the atmosphere and beneath the earth's surface into the lithosphere. Physical geographers also deal with life on earth. The **biosphere** is a thin layer found above, on, and below the earth's surface. This is the zone in which life forms exist. The **ecosphere** is larger than the biosphere and is the zone in which the processes that influence life operate. For example, the ozone layer, which is high in the atmosphere where no life exists, can be considered part of the ecosphere because it protects life on earth from the harmful ultraviolet rays of the sun.

The immensity and diversity of Physical Geography sometimes prevent us from seeing how the different topics in the subject are interrelated and how natural systems influence one another. Some systems receive energy from above, from the sun, while others receive energy from below, which originates from the intense heat of the earth's interior. These two energy inputs or systems operate together to shape the earth's surface and to produce the many distinctive natural landscapes of which it is composed.

Figure 1.2 illustrates the framework for the study of Physical Geography upon which this text is organized. The text is divided into six major units, most of which are subdivided into chapters. The names of the chapters appear in the unit boxes. The lines and arrows connecting the boxes provide the links between the various parts of the subject. The chapters in the Integrative Studies unit examine how human beings influence and are influenced by the earth's natural systems and processes. Each unit is represented by a symbol or icon, as illustrated in Figure 1.2 The icons allow you to follow themes throughout the book and to recognize topics in one chapter that are related to topics in other parts of the book. For example, rainforests are studied in Chapter 11,

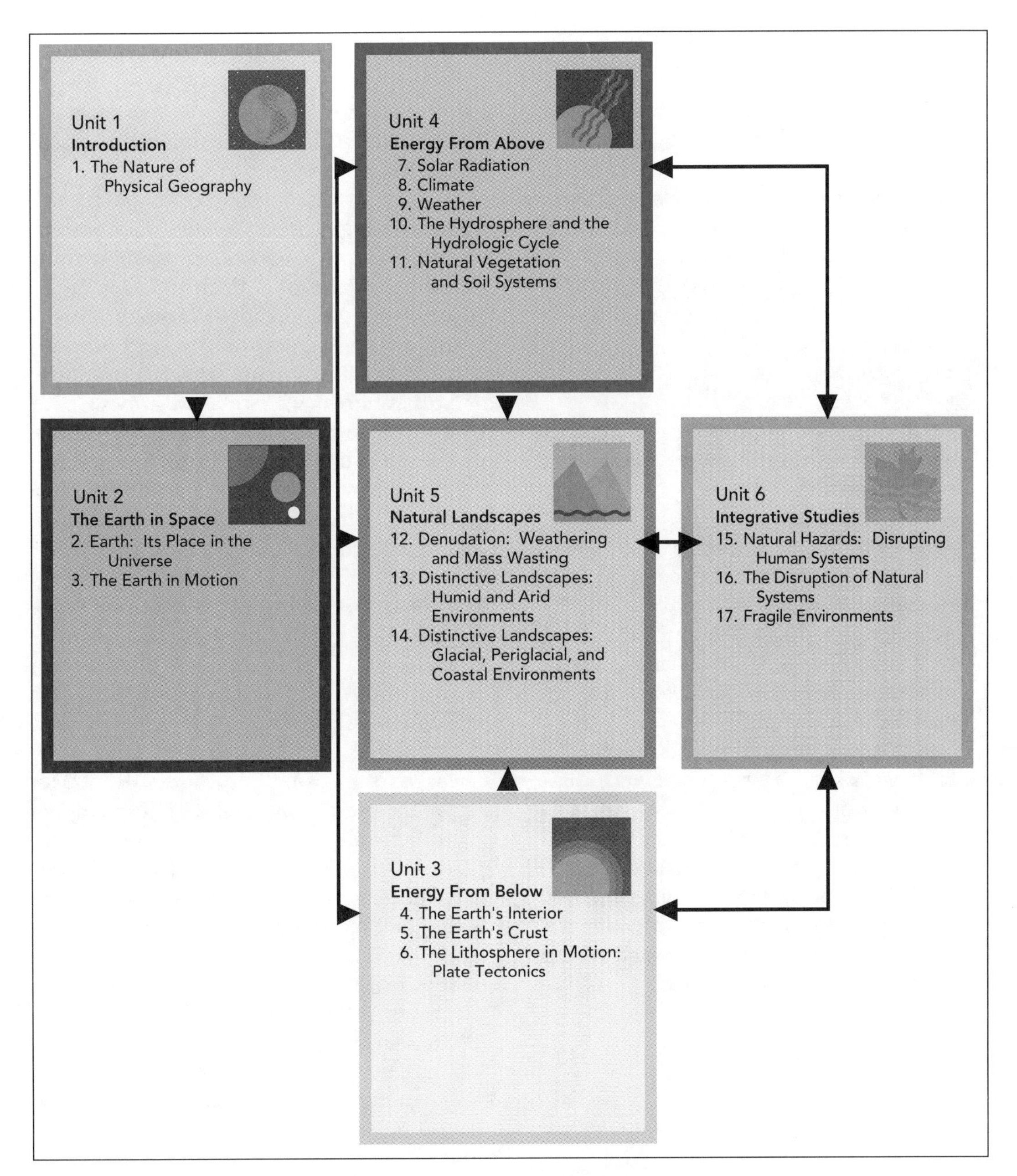

Figure 1.2 **The Organization of This Book**

while the destruction of the forests by human activity is examined in Chapter 17. The links between these two topics are indicated by the following icons.

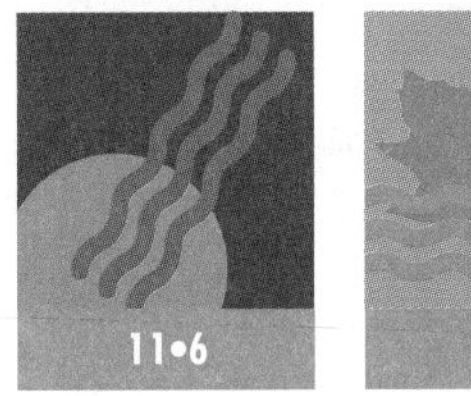

QUESTIONS

1. A few of the topics studied in this book are listed below. Use Figure 1.2 to identify the unit in the book where each topic would likely be covered to the greatest extent. For example, hurricanes would be found in the unit "Energy From Above" because the study of weather is part of this unit. Use a dictionary or encyclopedia to help you if you are having difficulty. The topics to locate are:
   - acid rain
   - hills made by glaciers
   - causes of earthquakes
   - planets in the solar system
   - revolution of the earth around the sun
   - ocean currents
   - types of rocks
   - the effect of hydro-electric dams on rivers
   - characteristics of Physical Geography
   - desert landforms
   - damage caused by earthquakes
2. Look through the book and find five topics that interest you. Which unit does each of your five topics fall into?
3. Find an icon in the book. State the section number in which the icon is found. For example, the section you just read was 1.1: the first "1" represents the chapter number and the second "1" gives the section number within Chapter 1. State the topic to which the icon is referring. Go to the section number referred to by the icon and identify the related topic. Repeat this with two other icons.

1·2 The Characteristics of Physical Geography

The characteristics that distinguish Physical Geography from other subjects are:

a) It is **integrative**. Physical Geography selects and uses information from a number of other disciplines, such as geology, meteorology, biology, chemistry, physics, astronomy, and oceanography. This approach shows how subjects are interrelated and gives a complete and balanced view of the topics under study. Figure 1.3 illustrates how physical geographers integrate information about acid rain from a number of other sciences.

b) It is **spatial**. Physical Geography studies the locations, distributions, and patterns of a wide variety of phenomena that influence the earth's surface. The scales on which these patterns and influences are studied vary from very large to very small, from the Milky Way galaxy to a small pond. Their studies could include the energy from the interior of the earth that moves entire continents, or a small rivulet of water flowing down a slope after a rainfall.

c) It is **holistic**. Physical Geography examines the "big picture"! It considers the interrelationships among the various elements and energy flows on the earth. The cutting down of a forest or the eruption of a volcano can influence the climate and vegetation of areas located thousands of kilometres away. Pollution caused by the burning of fossil fuels can acidify lakes, which can, in turn, cause the release of elements stored in rocks for billions of years. In order to understand our physical world and to cope with future changes in our environment, we must gain new knowledge and insight into the complex relationships existing on our planet. Physical Geography deals with the ways in which all processes and events on earth are interrelated.

d) It is concerned with **change**. Physical Geography sees the earth as a dynamic planet. Changes occur over the surface of the earth and over time. At times, the changes are fast and dramatic. Rivers overflow their banks causing floods and earthquakes devastate cities. At other times, the changes are subtle and slow in terms of a human time span. As you read this paragraph, new continents are being formed as old ones move back into the hot interior of the earth. Some mountain ranges are being thrust upwards while others are being eroded away. New rock is being created on the floors of the oceans and climates are changing across the globe.

e) It is **systems oriented**. Physical Geography uses a systems approach to organize and understand the diverse and complex processes operating on the planet. One classification of systems used in Physical Geography is described on pages 8 to 9. The systems differ from one another according to the focus or emphasis of each. The end products that result from inputs of energy or flows of matter are the key focus for some systems. In other systems, the flows of energy or matter are of prime importance. In yet others, the alteration of a physical system by human interference becomes the focus.

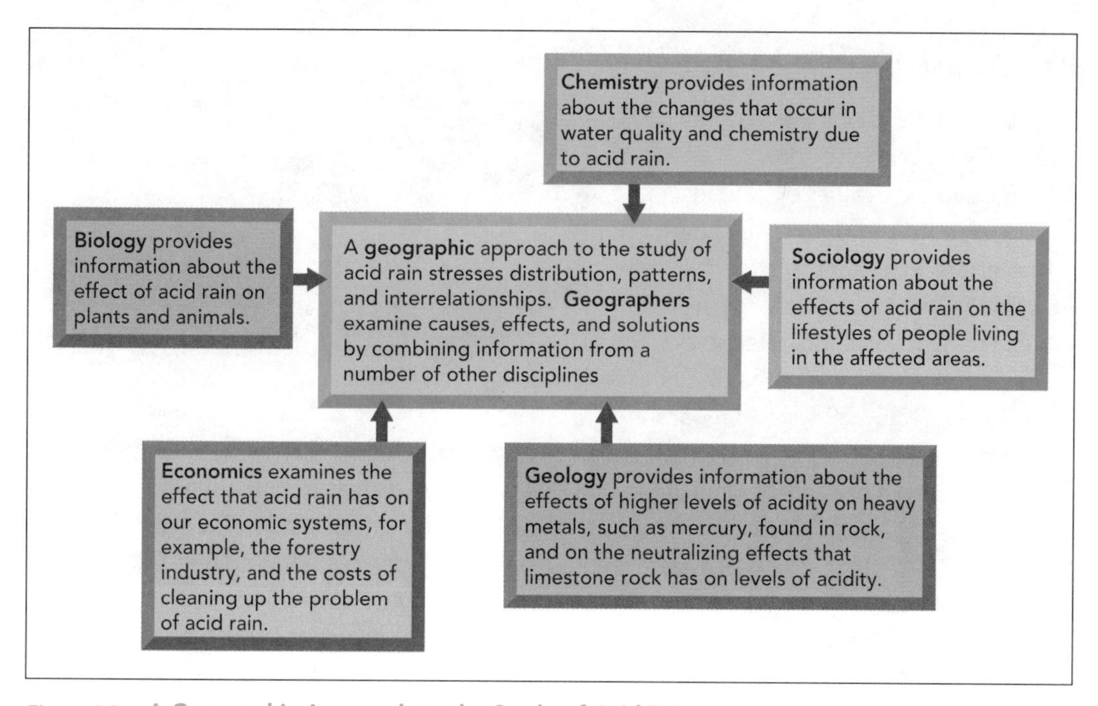

Figure 1.3 **A Geographic Approach to the Study of Acid Rain**

Systems in Physical Geography

Morphological Systems

Figure 1.4a

Morphology deals with the structure, appearance, and form of things. The focus for morphological systems is on the item, landform, pattern, or shape that results from physical processes operating on earth. A forest, a hill made by a glacier, a type of soil, and a sand dune all result from energy inputs and flows of matter. The appearance of Planet Earth itself can be thought of as the end result of the operation of a morphological system. Tiny ripples of sand on a beach are also the end result of a morphological system. When studying these systems, we are trying to understand the variables or processes that produced the feature in which we are interested. These variables or processes usually are systems themselves, known as cascading sytsems.

Cascading Systems

Figure 1.4b

The focus for cascading systems is the actual movement or flow of energy or matter and the changes of state that often occur. Inputs and outputs of energy or matter are important in studying these systems. The input of energy from our sun, the hydrologic (or water) cycle, and the flow of water are all examples of cascading systems in Physical Geography. Cascading systems can either be open-ended or closed-ended. Open cascading systems have a continuous input of energy or mass from outside the system, such as from the sun. This input is used, altered, or consumed by the elements in the system. The flow of water through a river system is an example of an open cascading system. Closed cascading systems do not have inputs of energy or mass from outside, but rather, they circulate a finite amount of mass and/or energy through the system. The water cycle is an example of a closed system, since water is never created nor destroyed, just moved in various states from one storage method to another.

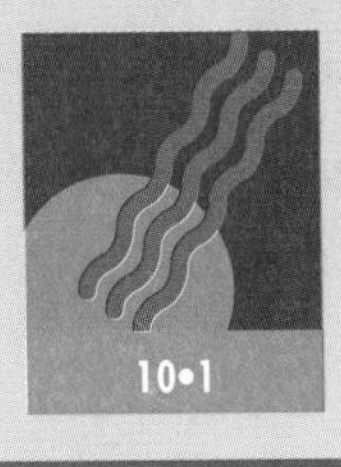

Ecosystems

Figure 1.4c

An ecosystem is a special type of system in which plant and animal life respond to and influence the components of the physical world, such as climate, soil, and relief. The interrelationships among the components of an ecosystem are of prime importance.

An ecosystem can be as small as a droplet of water or as large as the entire planet. The ecosystem on a planetary scale is called the ecosphere. A "biome" is a large ecosystem dominated by one type of vegetation, such as forests, grasslands, tundra, or deserts.

Altered Systems

Figure 1.4d

Altered systems focus on how morphological, cascading, or ecosystems have been changed by the planned or unplanned intervention of humans. The alterations may take the form of a hydro-electric dam that changes the flow of a river, higher levels of carbon dioxide that modify the earth's energy balance, or the cutting of trees in a rainforest. The planned and unplanned alteration of many of the earth's important physical systems has become a major concern to many scientists and citizens.

QUESTIONS

4. Create a collage entitled "What Is Physical Geography?" Your collage should include some of the topics studied in Physical Geography and illustrate as many of the characteristics of Physical Geography as you can.
5. Search through Chapters 2 to 17 and find four photographs or diagrams that represent each one of the four types of systems in Physical Geography. Record the figure number for each photograph or diagram and briefly explain why it represents that system.

1·3 The Development of Ideas in Physical Geography

In recent years, new discoveries have added much to our knowledge of Planet Earth. Many of these discoveries have been brought about by advances in technologies of many kinds. It is important to remember, however, that our current knowledge is an accumulation of many centuries of study and is merely a point on the long spectrum of understanding. Many ideas in Physical Geography fall into the category of "theories." Theories are explanations based on accumulated knowledge; they can be altered or proven wrong as new discoveries are made. However, as knowledge accumulates about a theory, and it is not proven to be incorrect, it gains credibility and can become widely accepted. Such an idea is the theory of **geologic time**.

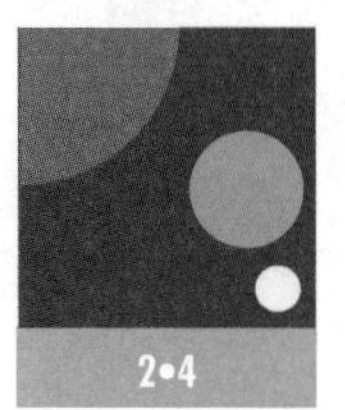

Many scientists presently believe that the earth is between 4.5 and 5 billion years old. But, it should be stressed that the estimated age of the earth has changed dramatically over the last two centuries. (See Figure 1.5.) New methods, such as measuring rates of radioactive decay, have given much support to the present view of the age of earth. However, future generations will no doubt learn more and adjust the geologic time scale accordingly. The geologic time scale, based on our present level of knowledge, is illustrated in Figure 1.6.

It's a Fact...

One of the first western scientists to suggest that the earth was much older than a few thousand years was a Scot named James Hutton. In addition to other evidence, he observed in the eighteenth century that a famous wall in Britain called Hadrian's Wall, built by the Romans, had eroded only a few centimetres in 1700 years! How much time would it take, asked Hutton, to wear down a mountain?

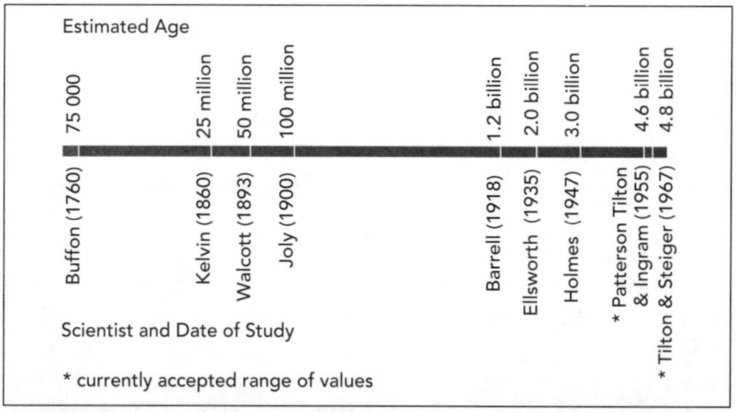

Scientist and Date of Study	Estimated Age
Buffon (1760)	75 000
Kelvin (1860)	25 million
Walcott (1893)	50 million
Joly (1900)	100 million
Barrell (1918)	1.2 billion
Ellsworth (1935)	2.0 billion
Holmes (1947)	3.0 billion
* Patterson Tilton & Ingram (1955)	4.6 billion
* Tilton & Steiger (1967)	4.8 billion

* currently accepted range of values

Figure 1.5
Changing Estimates of the Age of the Earth

Figure 1.6
Geologic Time Scale

Era	Period	Time (years before present)	Some Significant Geologic Events of Each Era	Some Significant Biologic Events of Each Era
Precambrian		4.6 to 4.8 billion	Oldest dated rock — 3.96 billion years Canadian Shield formed between 2.5 and 3.5 billion years ago.	Bacteria might have existed 4 billion years ago. Aquatic plants such as algae appeared about 3 billion years ago. Sponges and jellyfish appeared about 1 billion years ago.
		570 million		
Paleozoic (Greek for "Old Life") The Age of the Fishes and Amphibians	Cambrian Ordivician Silurian Devonian Carboniferous Permian	505 million 438 million 408 million 360 million 286 million	Sedimentary rock formed throughout this era. Major mountain building occurred in Eastern N.A., Europe, and eastern Asia. This mountain building culminated in the Appalachian Mountain System towards the end of this era. 75% of present-day coal deposits are of Carboniferous age.	Marine invertebrates developed such as snails, clams, trilobites, brachiopods. The first vertebrates appeared in the seas. The first life on land appeared while armoured fish and large coral reefs dominated the seas. First amphibians appeared. Many spiders, insects, ferns. First reptiles appeared. Fish were abundant and large trees were present on land.
		245 million		
Mesozoic (Greek for "Old Life") The Age of the Reptiles	Triassic Jurassic Cretaceous	208 million 144 million	Shallow seas covered the interior of N.A. Most of the world's oil and natural gas deposits are of Mesozoic age. Mountain building in western North and South America began.	Reptiles were the dominant life form on land and the Age of the Dinosaurs began. Great variety of fish and marine reptiles existed. First feathered birds and mammals appeared. Dinosaurs disappeared at the end of the Mesozoic era.
		66 million		
Cenozoic (Greek for "Recent Life") The Age of the Mammals	Tertiary Quaternary	2 million Present	Mountain building continued in the western parts of the Americas. The formation of the Alps and Himalayas began about 60 million years ago. Continents assumed their present shapes. Pleistocene Ice Ages began approximately 2-3 million years ago.	Mammals dominated this era as early forms of mammals developed into the species we know today. Human beings emerged and developed into their present form.

It's a Fact...

According to the Hindu Calendar, as recorded in the ancient books of that religion, the year 2000 will mark the 1 972 949 081 year since the world came into existence.

It was not until the two ideas of geologic time and an understanding of the dynamic nature of the earth came together that humans began to appreciate the nature of our planet. Before these ideas were linked, we were like ants crawling on a branch of a large oak tree; to the ants, a week seemed like an eternity and the tree branch a solid, unchanging structure. We have moved beyond the understanding of the ants to realize that a million years is but a brief moment in the long history of our ever-changing planet.

Much of our understanding of the earth is based on assumptions about past events. One such assumption is that the physical laws and processes that operate on the earth today worked in the same way in the past. For example, wave action today causes ripples in the sand at a beach; it is assumed that similar wave action in the past produced similar patterns in the sand. This assumption is called the **Law of Uniformitarianism.** The law does not state that the world has always been the same nor that the rate of

a) Ripples on a Beach Today

b) Ripples Preserved in 3-Billion-Year-Old Rock in the Canadian Shield

c) Ripples Preserved in 450-Million-Year-Old Rock on the Niagara Escarpment

Figure 1.7 **Law of Uniformitarianism** The similarity in the ripple patterns in each of these examples suggests they were formed by similar processes despite their widely differing ages.

change is always the same. But, it does maintain that the processes that influence our planet act in much the same way today as they did in the past. In other words, the effect of a wave breaking upon a shoreline hundreds of millions of years ago can be compared to a wave breaking upon a shoreline today.

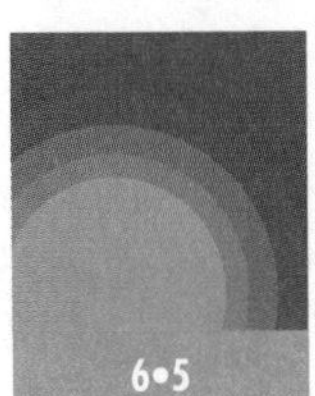

Until recently, ideas about Physical Geography assumed that geologic changes took place gradually over long periods of time. Although we know that many processes do indeed operate in this way, there is increasing evidence that sudden **catastrophic events** have also played a part in shaping our planet. Examples of such catastrophic events include the collision of large meteors with the earth and major volcanic eruptions. It is now believed that such events have influenced not only the shape of the earth's surface, but also the climate of the entire planet.

A more recent idea that is important to the study of Physical Geography is the **Gaia Hypothesis**, first proposed by James E. Lovelock in 1972 and named after the Greek goddess of the earth. The Gaia Hypothesis challenges us to go beyond the view that life exists simply as a response to the physical conditions in a place and to examine how life on earth has influenced the physical environment. The hypothesis suggests that the biosphere is like a living organism that can influence its own evolution through natural mechanisms, such as temperature changes or chemical modifications. There is increasing evidence to suggest that the evolution of the earth's atmosphere and, indeed, much of the earth's surface was profoundly influenced by life on Planet Earth.

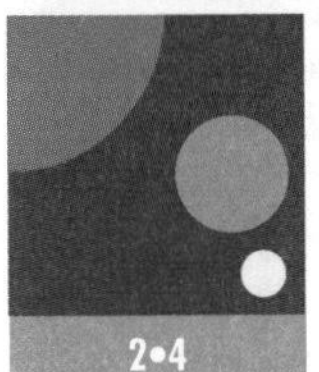

The study of Physical Geography in this book relies heavily on the types of systems outlined earlier. The text stresses how these systems interrelate and emphasizes the complexity of the planet. It looks at how human beings are influencing, and are influenced by, the physical world. The authors believe that we are reaching a critical point in the evolution of our unique planet, a point at which positive, significant action must be taken to preserve the quality of life on this planet. The oceans, the rocks, the atmosphere, the lakes, the rivers, and the soil of our planet sustain life and help define who we are as a species. We must understand our physical world if we are to protect and maintain it. The purpose of this book is to help you gain a greater understanding of the planet on which we live.

QUESTIONS

6. In your notebook, draw a line 24 cm long. Divide the line into 2 cm segments. Each segment represents one month. Refer to Figure 1.6: The Geologic Time Scale. Imagine that all of geologic time were reduced to one year. Indicate on your time line where each of the following events would appear:
   - the Canadian Shield begins to form
   - the first vertebrates develop
   - the Appalachians begin to form
   - the dinosaurs disappear
   - the Alps and Himalayas begin to form
   - human beings appear
7. a) In your own words explain how Figure 1.7 demonstrates the Law of Uniformitarianism.
   b) Why is this law so important to the study of earth history?

Review

- Physical Geography takes its subject matter from the spheres that touch on the surface of the earth.
- The subject matter of Physical Geography is very diverse; this book uses icons and flowcharts to point out the linkages between the topics.
- Physical Geography's characteristics set it apart from other subject areas. Physical Geography is integrative, spatial, holistic, concerned with change, and systems oriented.
- Certain ideas provide the basic structure of the study of Physical Geography. Some of these ideas are geologic time, the Law of Uniformitarianism, and the Gaia Hypothesis.
- A good understanding of Physical Geography can help people protect the environment on which we all depend.

Geographic Terms

atmosphere
lithosphere
hydrosphere
biosphere
ecosphere
integrative
spatial
holistic
change
systems oriented
geologic time
Law of Uniformitarianism
catastrophic event
Gaia Hypothesis

Explorations

1.a) Cut out newspaper articles for the next week that relate to Physical Geography. Classify the topics in the articles under the appropriate unit headings. (See Figure 1.2.)
   b) Choose three of the articles and explain how the topics discussed relate to your personal life.

2. a) In the classroom or library, find two other Physical Geography textbooks. Look at the chapter headings in these other textbooks and list them under the unit headings that appear in Figure 1.2.
   b) Compare the table of contents in these books with Figure 1.2. What differences exist in the way in which the books are organized? What similarities exist?

3. a) Look out the window and list fifteen items you can see. Classify these items under the following categories:
   (i) items that would be studied by physical geographers,
   (ii) items that would not be studied by physical geographers,
   (iii) items that you are not sure are part of Physical Geography.
   b) Discuss all the items you put in category (iii) with classmates or your teacher. In light of your discussions, try to fit them into one of the other two categories.

The Earth in Space

UNIT

Human knowledge and understanding of the origin of the earth and its place in the universe is constantly changing. The myths and legends of many of the earth's peoples include accounts of the creation of the earth. One of the constant themes and objectives of science has been to provide answers to questions on the origin of the universe, the earth, and of life itself. The chapters in this unit outline some of the ideas that scientists have developed to begin answering these universal but difficult questions.

The knowledge that science has accumulated about the universe, the solar system, and the earth is important to physical geographers in order to reconstruct the steps in the formation of the earth and to understand how the motions of the earth in space influence processes that directly and indirectly affect our lives. In addition, an understanding of the magnitude and beauty of the universe, the solar system, and the earth will develop an appreciation of the unique nature of earth and our need to treat it with care and affection.

As you work through this unit, consider these questions:

- What is the scientific explanation for the origin of the earth?
- How will understanding the origin of the earth help us explain processes currently observable on the earth's surface?
- In what ways do the motions of the earth in space affect the physical geography of our planet?
- How is the earth a unique planet that we need to treat with great respect and care?

CHAPTER

2

EARTH: ITS PLACE IN THE UNIVERSE

CHAPTER 1: The Nature of Physical Geography

CHAPTER 2: Earth: Its Place in the Universe

CHAPTER 3: The Earth in Motion

CHAPTER 4: The Earth's Interior

CHAPTER 5: The Earth's Crust

CHAPTER 6: The Lithosphere in Motion: Plate Tectonics

CHAPTER 7: Solar Radiation

CHAPTER 8: Climate

CHAPTER 9: Weather

CHAPTER 10: The Hydrosphere and the Hydrologic Cycle

CHAPTER 11: Natural Vegetation and Soil Systems

CHAPTER 12: Denudation: Weathering and Mass Wasting

CHAPTER 13: Distinctive Landscapes: Humid and Arid Environments

CHAPTER 14: Distinctive Landscapes: Glacial, Periglacial, and Coastal Environments

CHAPTER 15: Natural Hazards: Disrupting Human Systems

CHAPTER 16: The Disruption of Natural Systems

CHAPTER 17: Fragile Environments

OBJECTIVES:

By the end of this chapter, you will be able to:

- describe and appreciate the general structure, size, and grandeur of the universe;
- describe the scientific theory for the origin of the universe;
- describe the basic steps in the origin of the solar system and the earth;
- understand and describe the unique aspects of earth within the solar system;
- describe the conditions that allowed life to develop on the earth's surface;
- appreciate the significance of life on earth and its uniqueness in the solar system.

Introduction

The universe is so vast that no human being can fully comprehend its size and magnitude. Even the size of the Milky Way galaxy and the solar system in which earth is located is hard to grasp. Nevertheless, scientists equipped with new and powerful instruments and telescopes are continually making new discoveries about the universe and probing to its outer edges. As the information flows in, new and exciting theories emerge about the origin of the universe, the solar system, and the earth. Many of these theories are providing scientific answers to questions that have puzzled humans since the earliest times. This chapter presents the scientific theories and discoveries about the origin of the universe, the solar system, and the earth.

2·1 Pondering the Imponderable

We had spent the day visiting the sites of ancient civilizations in the Negev Desert. Darkness came shortly after we finished dinner. A group of us left the dining room to walk along the pathway near the hotel.

We looked up. The clear air of the desert and the darkness of the night gave us a spectacular panorama. The first sight to capture our attention was a thin cloud-like haze, similar to that seen in the blue skies of the day, that appeared to hang above us. But this was night. As our eyes adjusted to the scene, we realized we were gazing at millions of stars. They were so closely spaced and so visible in the clear air that they appeared to form a thin white cloud.

We were looking at the Milky Way, or rather a part of it. We had never seen it so clearly. In our field of vision, we could imagine the scale of our galaxy, a mere speck in the universe. Some of the tiny specks of light were probably other galaxies, beyond the Milky Way, each also composed of millions of stars.

This was our "window on the universe". So breathtaking was it that we just stood and gazed for many minutes, humbled by the view and its meaning.

From the diary of Gary Birchall, Israel, July 25, 1978.

This personal account gives some idea of the smallness of earth and the solar system of which it is a part. The universe is so vast that even mathematicians and astronomers have difficulty comprehending its magnitude.

In spite of its vast size, most of the universe is almost empty "space" where matter is extremely rare. The distances between the billions of galaxies and galaxy clusters are so enormous that astronomers don't use kilometres as their measurement, but rather use the **light year** (the distance light travels in a year). A light year is 9.46 trillion or 9.46×10^{12} km in length. A light minute is the distance that light travels in a minute. The light from the sun takes eight light minutes to travel the 150 million kilometres to earth.

Distances within our solar system are small by comparison with those in the universe. Yet, these distances are vast in human terms. Examine Figure 2.1. This shows the distance covered by the Voyager 2 space probe that was launched on August 20, 1977. It took four years and five days for Voyager 2 to reach Saturn travelling at a speed of about 130 000 km/h, a distance of 2.5 billion kilometres, following the circular orbit shown in Figure 2.1. Signals travelling at the speed of light took approximately 1 h 20 min to reach earth from the satellite when it was in the vicinity of Saturn.

The number of stars visible in a clear sky is overwhelming. Even so, under ideal conditions only a small fraction of the universe is visible from earth. Stars are masses of hot gases that give off light and heat. There are approximately 100 billion stars in our galaxy, the Milky Way, but most of these are hidden behind dust clouds and other stars. Beyond the Milky Way itself, the most distant object visible to the unaided eye is the Andromeda galaxy. It lies 2 million light years from earth. The largest known galaxy, M87, is located 40 million light years from earth and is thought to have anywhere from 1 to 100 trillion stars.

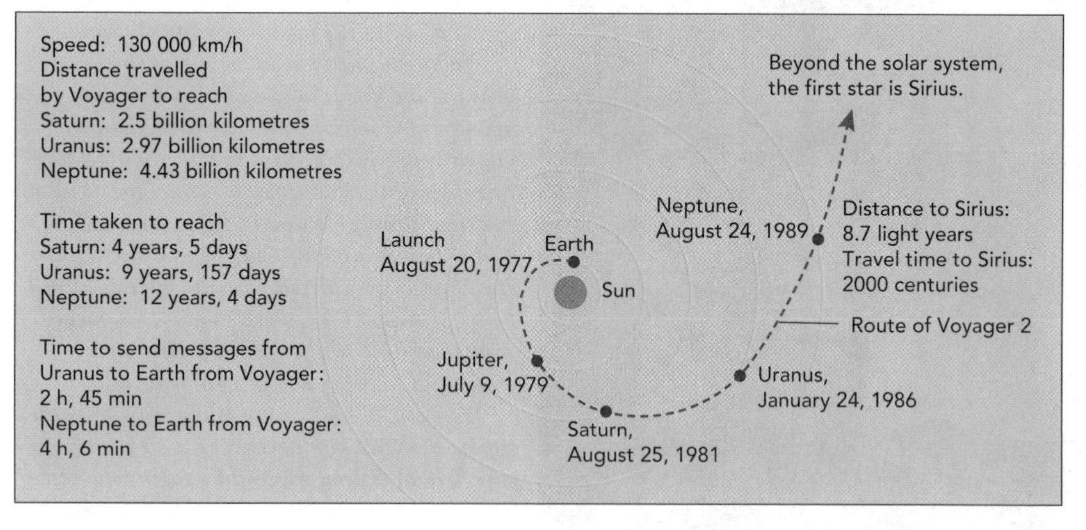

Figure 2.1 **The Journey of Voyager 2**

The most distant galaxy observed with the most sensitive telescopes is 12 billion light years away! The light from this galaxy has taken 12 billion years to reach the earth. Since our solar system and the earth are thought to be only 4.6 billion years old, this is really very ancient history!

Galaxies are the major clusters of observable matter in the universe. They form the "skeleton of the universe", as one astronomer put it. Within each galaxy, millions, billions, or even trillions of individual stars are distributed in spiral, elliptical, or irregularly shaped patterns. As Figure 2.2 shows, galaxies occur in clusters and superclusters. The Milky Way is a spiral galaxy "spinning" on its axis at a speed of 250 km/s or 900 000 km/h. Because of its immense size, it takes about 250 million years to complete one revolution.

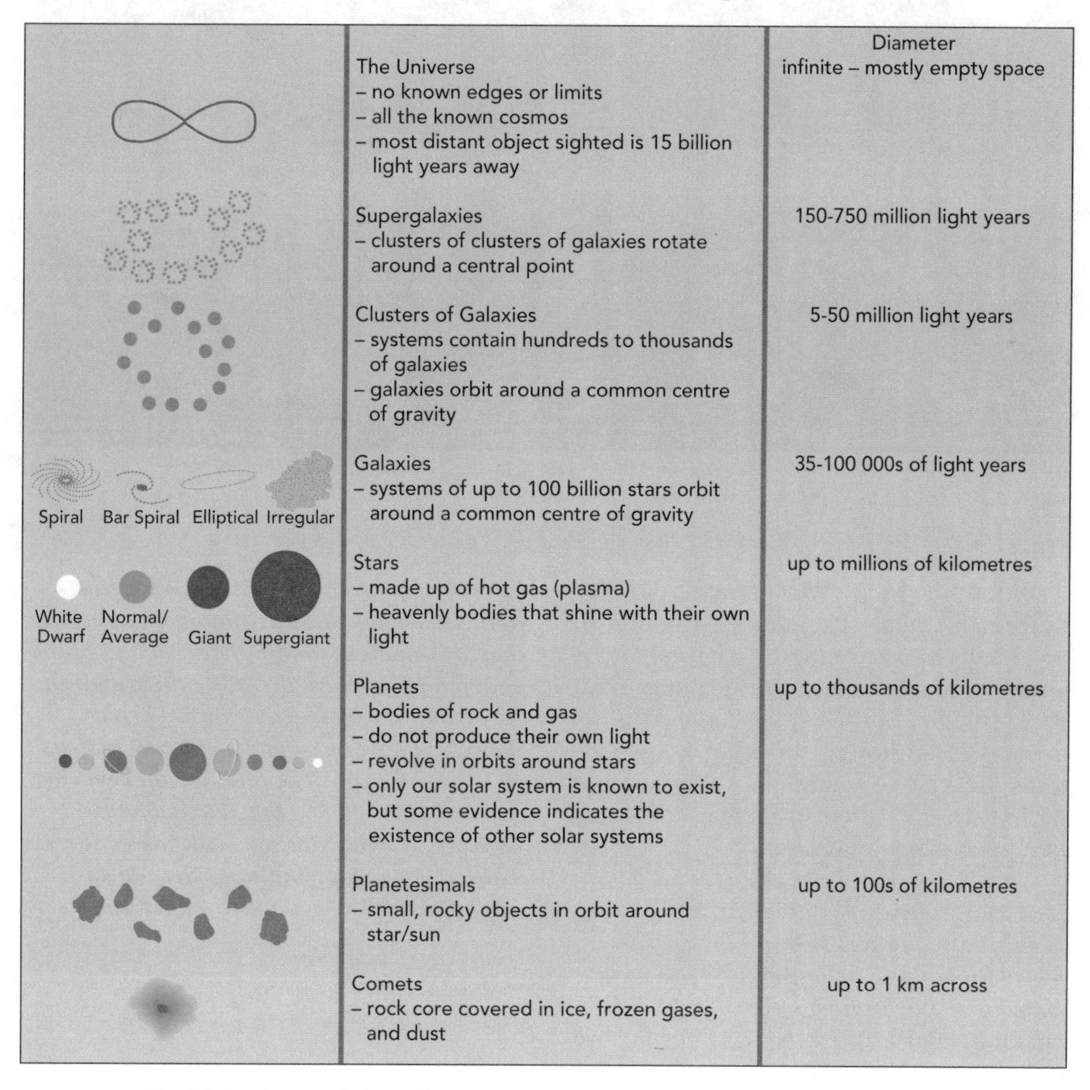

		Diameter
	The Universe – no known edges or limits – all the known cosmos – most distant object sighted is 15 billion light years away	infinite – mostly empty space
	Supergalaxies – clusters of clusters of galaxies rotate around a central point	150-750 million light years
	Clusters of Galaxies – systems contain hundreds to thousands of galaxies – galaxies orbit around a common centre of gravity	5-50 million light years
Spiral, Bar Spiral, Elliptical, Irregular	Galaxies – systems of up to 100 billion stars orbit around a common centre of gravity	35-100 000s of light years
White Dwarf, Normal/ Average, Giant, Supergiant	Stars – made up of hot gas (plasma) – heavenly bodies that shine with their own light	up to millions of kilometres
	Planets – bodies of rock and gas – do not produce their own light – revolve in orbits around stars – only our solar system is known to exist, but some evidence indicates the existence of other solar systems	up to thousands of kilometres
	Planetesimals – small, rocky objects in orbit around star/sun	up to 100s of kilometres
	Comets – rock core covered in ice, frozen gases, and dust	up to 1 km across

Figure 2.2 The Major Parts of the Solar System

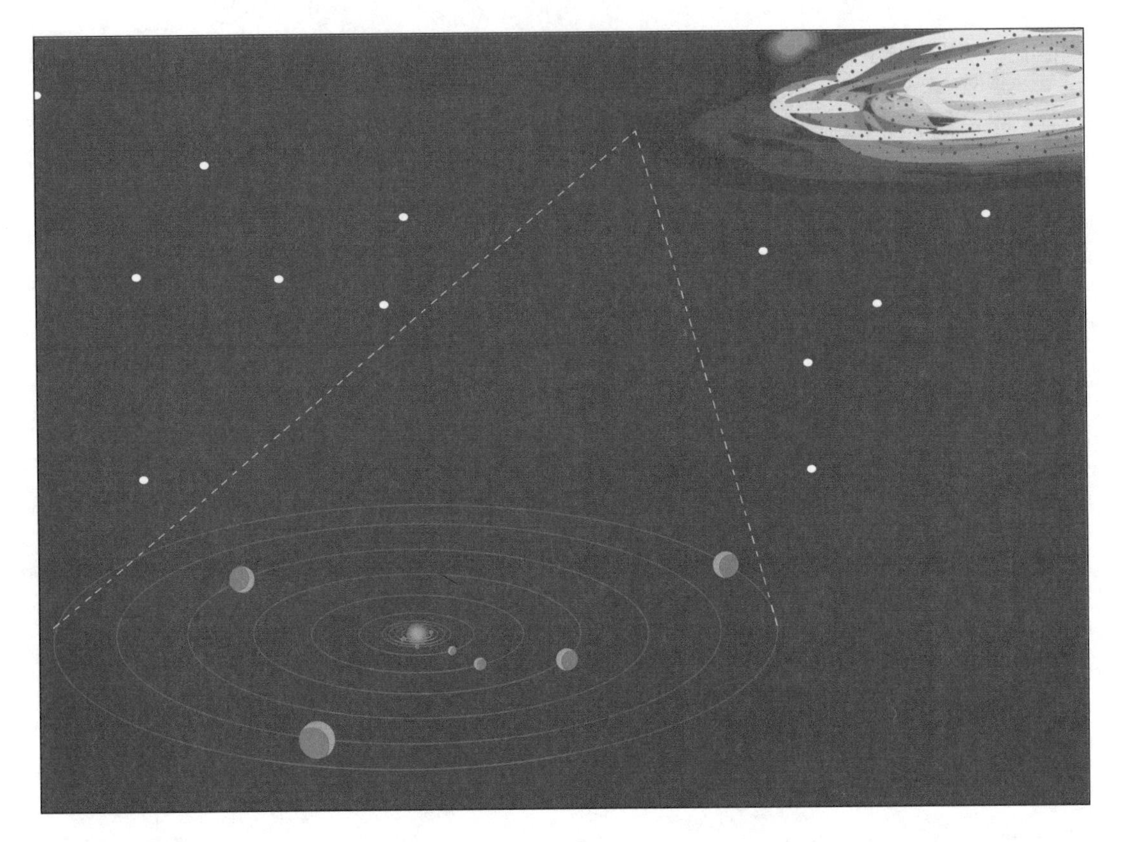

Figure 2.3 The Solar System in the Milky Way Galaxy

Within the Milky Way galaxy, our **solar system** is small. The sun is just one of the medium-sized stars of the galaxy. A solar system is made up of a star with rocky or gas bodies (planets) revolving around it. Although no other solar systems are known today, the odds seem to be in favour of their existence. It is even likely that some may contain life.

How many planets . . . are there in the universe?. . .This proves to be an awkward question, because to date even planets associated with our nearest stellar neighbours are undetectable. Since they do not glow, we cannot see them. . . . [The] nine planets and 40 or so moons . . . give every appearance of being the by-products of the sun's formation. This leads most astronomers to believe that, at least for stars in the size class of the sun, planets may be the rule rather than the exception. If this speculation is correct, then there may be as many or more planets as there are stars! The total would be a staggering 1 000 000 000 000 000 000 000 objects! . . . The evidence . . . leads us to believe that earth conditions are not so improbable as to make our planet unique.

From Wallace S. Broecker, *How to Build a Habitable Planet,* (Palisades, New York: Eldigio Press, 1985), pp. 8-9.

Big Bang and Steady-State Hypotheses

Although there are still many unanswered questions, astronomers generally agree that time and the universe began approximately 15 to 20 billion years ago when unimaginably dense matter (called a "cosmic egg" by some astronomers) ripped apart in a "big bang". This truly incredible explosion sent matter speeding outward in all directions at enormous speeds. As it hurled outwards during the birth of the universe, this matter clustered together to form up to 100 billion galaxies. To this day, these galaxies continue to speed away from one another, enlarging the universe.

The Big Bang hypothesis suggests that there are two possibilities for the future of the universe. It could continue to expand forever, or, it could eventually slow down and stop expanding. If this second possibility were to occur, the universe would begin to fall back into its centre, recreating the dense "cosmic egg" from which it began. The whole process would then start over once again.

Based on present knowledge, astronomers believe that the Big Bang hypothesis has a 90 percent certainty rating. The question of whether the universe will continue to expand or will eventually collapse is one that astronomers feel they do not have enough evidence to answer.

The Steady-State hypothesis is the only other explanation of the origin of the universe that has gathered any significant acceptance by astronomers. This hypothesis suggests that fresh hydrogen is created steadily in the voids of space from nothing. This hydrogen becomes the raw

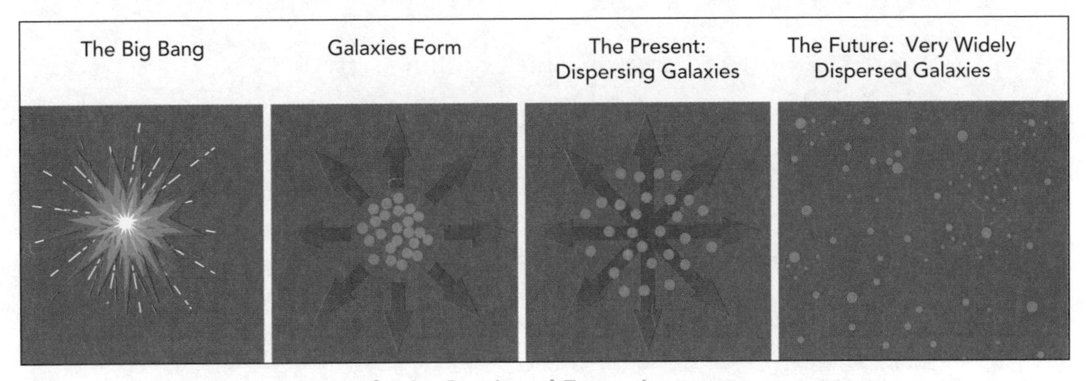

Figure 2.4a The Big Bang Hypothesis: Continued Expansion

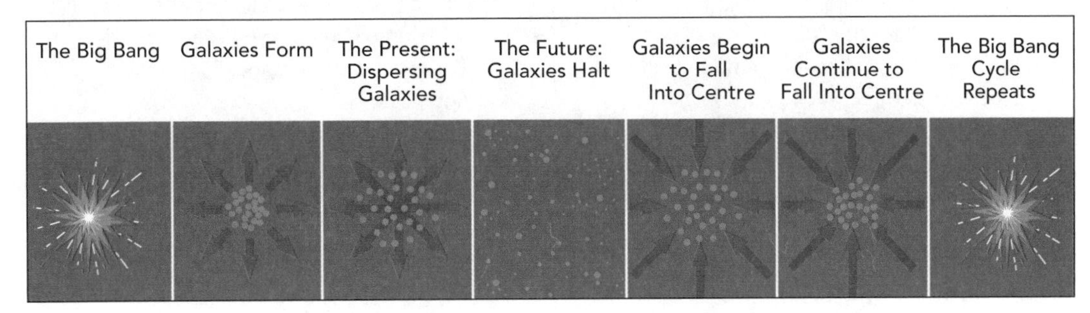

Figure 2.4b The Big Bang Hypothesis: Expansion and Contraction

material that creates new stars which replace old, dying ones. These new stars also fill in the spaces left by the steadily expanding universe. The Steady-State hypothesis suggests a universe without a beginning and without an end.

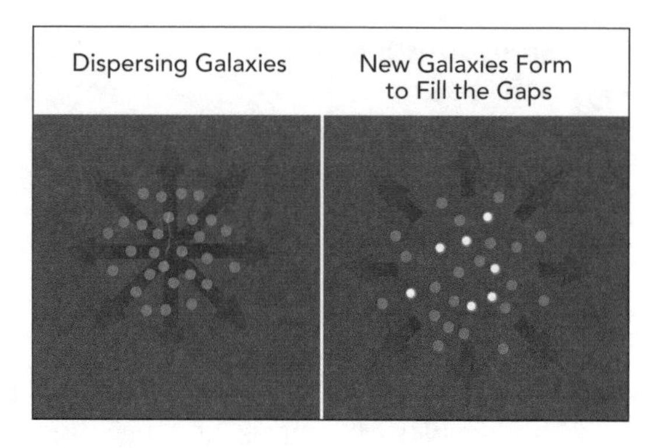

Figure 2.5 **The Steady-State Hypothesis**

It's a Fact...

- Elliptical galaxies make up about 60 percent of all known galaxies in the universe, while the various types of spiral galaxies account for 30 percent and irregular galaxies make up the remaining 10 percent.
- The nearest star to the sun is Alpha Centauri, which is 4.3 light years away from earth.

QUESTIONS

1. Describe your thoughts and feelings when you look up into the night sky and see hundreds or thousands or millions of stars.
2. a) The universe is composed of key blocks or parts. What are these parts?
   b) What are the relationships among these key blocks or parts?
3. Analyse the quotation from Wallace S. Broecker on page 20. Classify the quotation as mainly factual or mainly speculative, giving evidence to support your answer.
4. a) Explain why it is so difficult to determine if there are other planets and solar systems in the universe.
   b) What facts suggest that it is possible that other planets exist and that some might even contain life?
   c) If the odds for the occurrence of the special conditions required to produce life were 10 000 000:1, how many planets might contain life?
5. a) In a group, brainstorm methods people might use to explore for the existence of life on other planets in our galaxy.
   b) Which of these suggestions is the most practical? Explain your choice.

2·2 Our Solar System

The sun makes up about 99 percent of the total mass of our solar system. At least nine planets, 55 smaller moons, and countless **asteroids** revolve around this sun. Asteroids are small bodies of rock that orbit the sun between Mars and Jupiter. More than 2000 asteroids have diameters greater than 10 km. They are sometimes known as "minor planets". In addition, there are untold billions of **meteoroids**. Meteoroids are very small bodies of rock moving through space in orbit around the sun. The total mass of these asteroids and meteoroids is less than the mass of our moon. When meteoroids enter the earth's atmosphere, they burn up and appear as streaks of light, sometimes known as "shooting stars". **Meteorites** are meteoroids that survive the passage through earth's atmosphere and collide with the earth's surface.

The nine planets revolve around the sun, each in its own orbit. The orbit of Pluto, however, is different from the others.

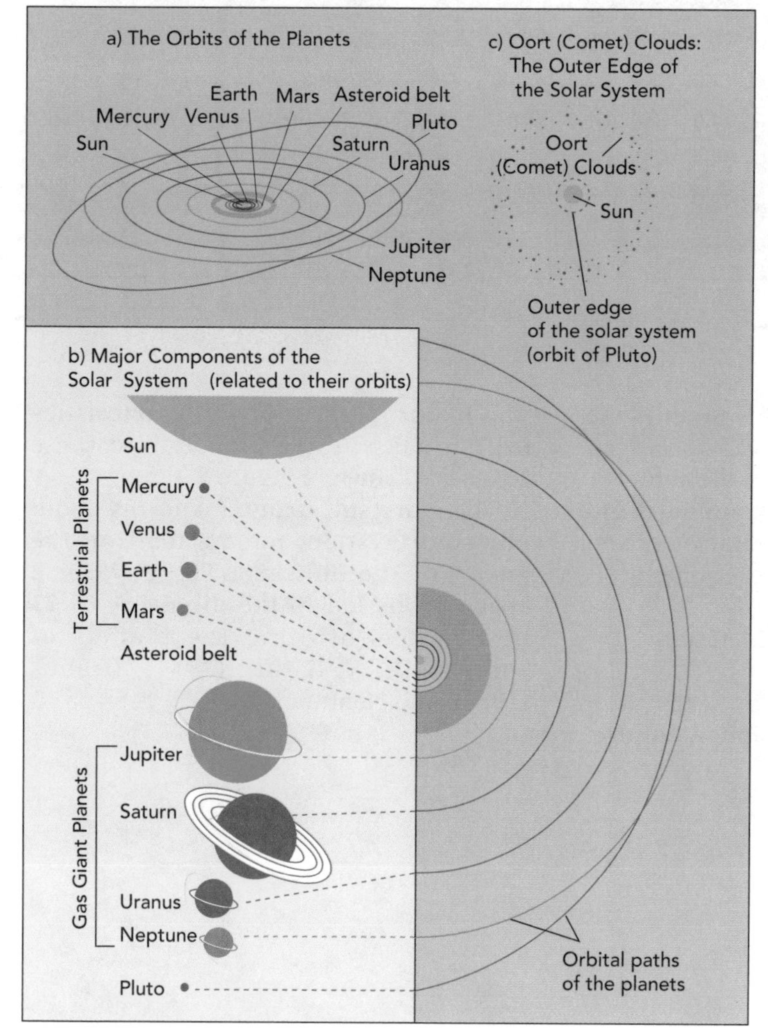

Figure 2.6
The Solar System and Its Major Components

Planet Name	Mass (in grams) 1.00E + 27	Density (gm/cm^3)	Approx. composition: Metals (Fe, Ni, etc.) percent	Oxides SiO_2, MgO, FeO percent	Ices H_2O, CH_4, NH_4, H_2S percent	Gases H_2, He... percent	Radius (in thousands of km)	Average Distance From Sun (in millions of km)	Orbital Period Around Sun (years)	Gravity (Earth =1)	Major Atmospheric Components percent
Sun	1 990 000.00	1.4	0.1	0.2	1.2	98.5	694.00				Hydrogen (74), Helium (25)
Mercury	0.33	5.4	50.0	50.0	0.0	0.0	2.42	58	0.241	0.38	Helium (Trace)
Venus	4.87	5.3	30.0	69.0	1.0	0.0	6.05	108	0.616	0.90	Carbon Dioxide (96)
Earth	5.97	5.5	29.0	69.0	2.0	0.0	6.37	150	1.000	1.00	Nitrogen (76), Oxygen (21)
Mars	0.64	3.9	10.0	90.0	0.0	0.0	3.37	228	1.880	0.38	Carbon Dioxide (95)
Jupiter	1 900.00	1.3	4.0	9.0	5.0	82.0	71.40	778	11.900	2.53	Hydrogen (90), Helium (10)
Saturn	570.00	0.7	7.0	14.0	12.0	67.0	60.50	1427	29.500	1.07	Hydrogen (94), Helium (6)
Uranus	88.00	1.3	8.0	17.0	60.0	15.0	23.60	2869	84.000	0.92	Hydrogen (85), Helium (15)
Neptune	103.00	1.7	6.0	14.0	70.0	10.0	24.80	4498	164.800	1.19	Hydrogen, Helium
Pluto	0.60	4.0	n.a.	n.a.	n.a.	n.a.	3.20	5980	247.700	0.05	Methane

Figure 2.7 **Major Characteristics of the Planets**

During part of its orbit, this ninth planet is actually closer to the sun than Neptune, the eighth planet from the sun!

Our solar system can be divided into five zones based on the characteristics of the planets themselves. These zones are:

- the **terrestrial planets** (Mercury, Venus, Earth, Mars)
- the asteroid belt
- the **gas giant planets** (Jupiter, Saturn, Uranus, Neptune)
- Pluto
- comet clouds

The major features and characteristics used to classify the planets into these zones are shown in Figure 2.7.

There are still many unknowns about our solar system, not to mention the galaxy or the universe. The Voyager 2 space probe led to the discovery of 21 new moons orbiting the planets of Jupiter (1979), Saturn (1981), Uranus (1986), and Neptune (1989).

It's a Fact...

- Just as the nine planets revolve around the sun, the solar system also revolves around the centre of the Milky Way galaxy, taking about 250 million years to complete one revolution. Even the Milky Way itself is "spinning" on its axis at a speed of 250 km/s or 900 000 km/h.
- Our knowledge of the solar system has more than doubled as a result of the data sent back from the twelve-year travels of Voyager 2. It will take scientists decades to process and make sense out of the enormous amounts of information collected by the various instruments on this one satellite.

QUESTIONS

6. List some science-fiction movies that you have seen that have dealt with extra-terrestrial life forms. How realistically do you think such encounters are depicted in these movies?
7. a) Using Figure 2.7, compare the terrestrial and the gas giant planets.
   b) Point out reasons why Pluto can be considered the "maverick" planet of the solar system.
8. If you were piloting an interplanetary space ship, which of the five zones of the solar system would be most hazardous? Explain.
9. In your own opinion, what is the value of continuing to discover more about our solar system, galaxy, and universe?

2·3 The Origin of the Solar System

From the earliest times, people have attempted to explain the origin of the solar system and the earth. Until the technological and electronic advances of the twentieth century, many of these theories were based on a limited amount of information about the sun and the planets of the solar system. More and more data are now being compiled every year, leading astronomers to revise and expand their theories. Based on present knowledge, most astronomers believe our solar system is a by-product of the origin of the sun. The most accepted explanation is known as the **nebular hypothesis**. Figure 2.8 shows the major steps in the formation of the solar system according to this hypothesis.

The nebular hypothesis — the term **nebula** means an immense cloud-like cluster of dust particles — was proposed in 1755 by Immanuel Kant, a German philosopher. This hypothesis was generally accepted until the early decades of the twentieth century when different competing hypotheses were proposed by astronomers. But, by the 1940s, new evidence and refinements in the nebular hypothesis again made it the most widely accepted explanation for the origin of the solar system.

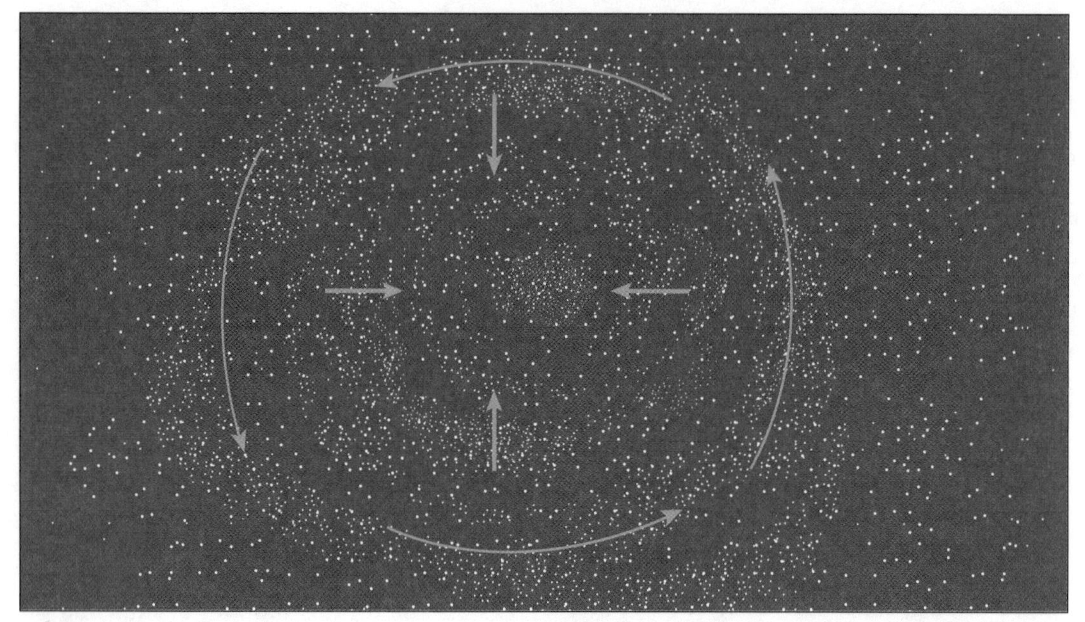

Figure 2.8a **The Formation of the Solar System** The original nebula of interstellar dust and gases, with the largest mass accumulating in the centre of the cloud, attracted even more matter due to increasing gravitational attraction. The whole nebula continued to increase its speed of rotation in order to preserve its angular momentum as its volume decreased.

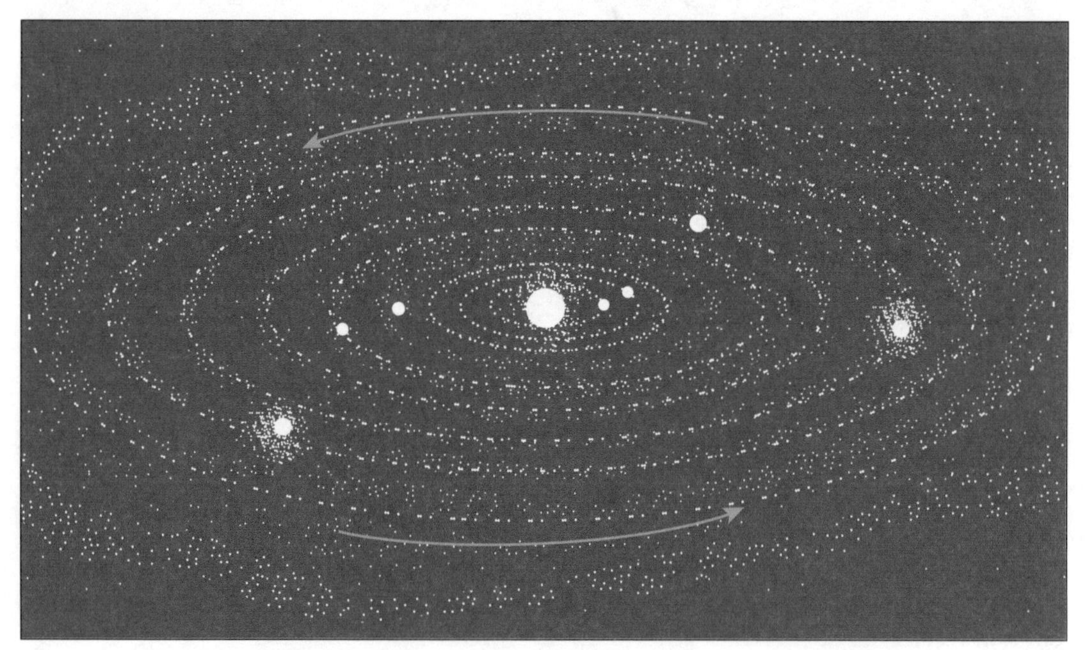

Figure 2.8b **The Formation of the Solar System** The sun and planets began to form in the eddies of the nebula, as its rotation had flattened it into a disk of dust and gases, with most of the materials being pulled by gravity into the centre to form the sun and the outer rings rotating about the denser centre.

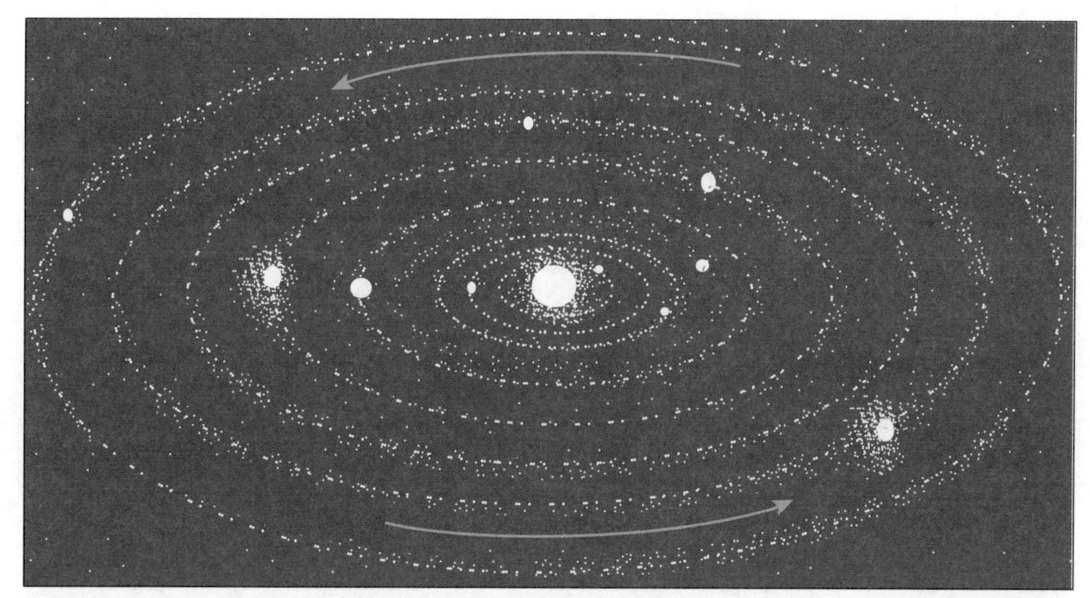

Figure 2.8c The Formation of the Solar System The mass of gases and dust in the centre of the nebula continued to grow in size. In the outer eddies of the rotating disk, planetesimals were growing through the process of accretion. The framework of the future solar system was clearly visible by this stage.

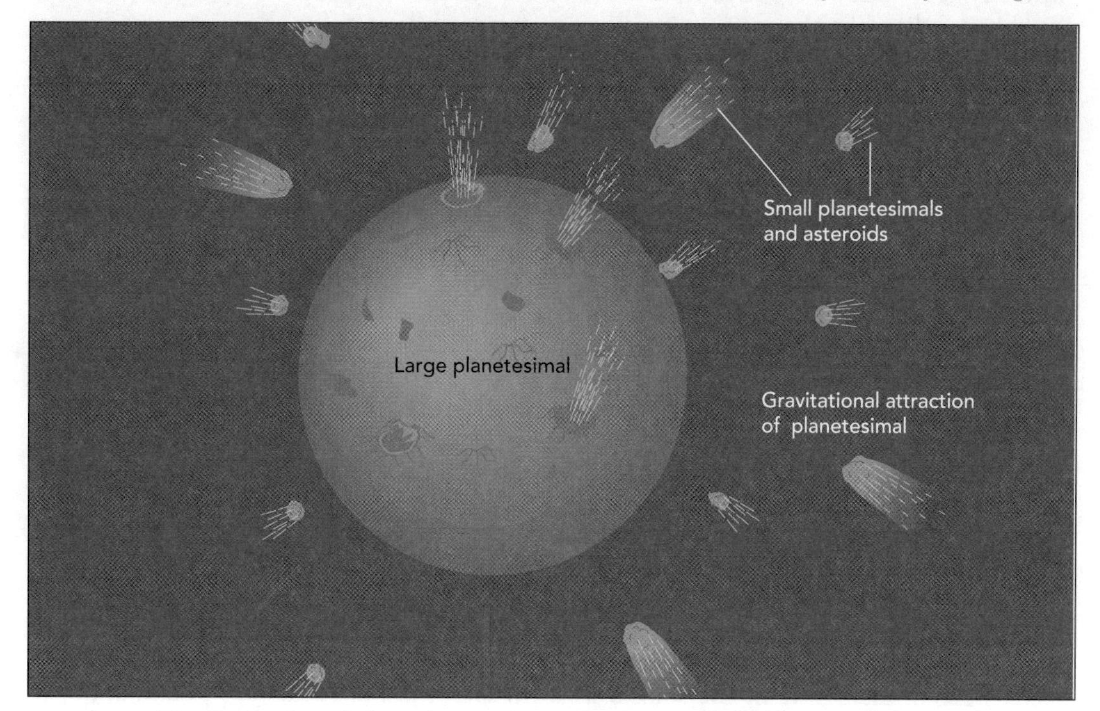

Figure 2.8d The Formation of the Solar System The gravity of the early planetesimals pulled in the smaller asteroids to grow into the planets. This process of accretion occurred quickly so that the planets grew to approximately their present size in ten to thirty million years.

According to the nebular hypothesis, about 5 billion years ago, a giant cloud of dust particles and gases was slowly rotating in an outer arm of the Milky Way galaxy. It was disturbed by some unknown cause and began to contract under its own gravitational pull. A series of eddies or whirlpools formed as the dust and gases moved in toward the centre of the huge cloud. The mass in the centre grew, its gravitational attraction increased, and more particles and gases were drawn inward. As the area occupied by the cloud grew smaller, it began to spin at an increasing rate. This change is best illustrated using the example of a figure skater rotating on the ice. As the skater's arms are pulled in towards the body, the rate of rotation increases; as the skater's arms are extended, the speed of rotation slows. This spinning action flattened the cloud into a disk-like shape.

After about 100 million years, much of the cloud's mass was concentrated in its centre. At the same time, the eddies became centres where lesser amounts of dust and gas collected to form small **planetesimals**. Just as the central part of the cloud had enough mass to pull dust and gas inward, so these smaller planetesimals attracted and held such particles as well. As they grew in size, their increasing gravitational attraction pulled in more materials from their vicinity to eventually form the planets we know today. These planets had a great enough gravitational attraction to capture and hold lighter gases (such as hydrogen and helium), forming primitive atmospheres.

The most explosive event in the history of the solar system was the igniting of the sun's thermonuclear furnace. The huge mass of gas at the centre of the solar system created tremendous pressures that continued to raise the inner temperature of the protosun, the predecessor of the sun. When this temperature reached about 11 million degrees Celsius, it triggered a thermonuclear explosion. The first flash of a nuclear bomb explosion is similar to, but much smaller than, the lighting of the sun's nuclear furnaces.

The sun's explosion sent a powerful solar wind, called the T. Tauri Wind, of charged particles in all directions. The solar wind, travelling at a speed of about 3.3 million km/h and lasting for approximately nine million years, swept away the small gas and dust particles remaining between the planets of the solar system. It also stripped the inner planets (Mercury, Venus, Earth, and Mars) of their first atmospheres. The larger, more distant planets of Jupiter, Saturn, Uranus, and Neptune did not lose their hydrogen and helium atmospheres to this powerful solar wind.

Only the planets and various smaller planetesimals remained within the solar system. Over the remaining 4.5 billion years or so, the planetesimals continued to collide with the planets or their moons, adding to those masses. Many of the craters visible on the moon's surface resulted from the impact of large meteoroids or small planetesimals travelling at speeds of 30 000 to 75 000 km/h. Impacts of these bodies on earth also created craters on the earth's surface, but erosion by water, ice, and wind eventually removed evidence of their existence. The crater shown in Figure 2.9 dramatically illustrates the effect meteorite impacts can have on the earth's surface.

Many of these steps in the nebular hypothesis are very speculative. No one was around to observe the various stages in the origin of the solar system, and it is not yet possible to observe the formation of other stars to see if these steps occur

Figure 2.9 Meteorite Crater A meteor impact created the circular basin occupied by Lake Manicouagan in the Canadian Shield hundreds of millions of years ago.

elsewhere in the universe. The distances are just too great to make such observations possible.

The Sun: The Centre of Our Solar System

The protosun that preceded our present-day sun was composed largely of hydrogen gas, with lesser amounts of helium. Today's sun has a composition of about 72 percent hydrogen, 22 percent helium, and 1 percent of other elements.

The sun's energy comes from nuclear fusion. In this reaction, four hydrogen nuclei, with a total mass of 4.032 atomic mass units, are fused together to form one helium nucleus with a mass of 4.003. This means 0.029 atomic mass units have been converted into energy. This energy powers the sun and supports life on earth.

The surface of the sun, known as the photosphere, is often marked with dark features known as **sunspots**. These are caused by the complex interactions between the sun and its enormous magnetic fields. Sunspots are 1000 to 20 000 km in diameter and are cooler than the photosphere surrounding them. They frequently generate solar flares, "fountains of hot gases" that are the "thunderstorms" of the sun's surface. Sunspots violently blast energy, in the form of X-rays, ultraviolet and visible radiation, and high-speed protons and electrons, into space. The radiation from such flares often disrupts radio communications on earth. Such flares cause the auroras — beautiful coloured arcs of shimmering electrical glow — that appear above the polar regions of the northern and southern hemispheres.

It's a Fact...

Astronomers call the solar wind created when the sun's nuclear furnaces first fired up the "T. Tauri Wind". T. Tauri is the name of a star whose thermonuclear furnace recently ignited to create a massive solar wind made up of a steady stream of hydrogen ions (protons and electrons).

QUESTIONS

10. In a group, discuss reasons why almost every society or culture has an account of the origin of the universe, solar system, and/or the earth.
11. If you could choose to witness only one part of the process of the formation of the solar system, what would it be? Explain your choice.
12. In your own words, write a descriptive narrative of the formation of the solar system from the viewpoint of a distant observer (assuming, of course, that you could live long enough to observe the whole process).
13. If the processes shown in Figure 2.8 are reasonably correct, prepare a defence of the following statement: "It is likely that other solar systems exist in the universe."

2·4 The Origin of the Earth

Explanations about the earth's origin are contained in the stories and legends of many of the world's peoples. The question of how this planet came to be has occupied many great minds over the centuries. However, it has only been during the past century that scientists have begun to fill in the details of earth's 4.6-billion-year history. Recent activity, such as Voyager 2's journey to Jupiter, Saturn, Uranus, and Neptune, continues to expand our knowledge.

Meteorites: Keys to Earth's Age

A not very obvious, but very important, source of information about the earth is meteorites, especially chondrite meteorites, which contain small, round granules of silicate minerals. Analyses of these granules reveal they contain water and other volatile elements that turn into gases at relatively low temperatures. This tells us that these meteorites have not melted since their formation, for if they had, the water and other volatile elements would have evaporated. Measuring rates of radioactive decay indicates they formed up to 4.6 billion years ago, giving scientists further evidence for the date of origin of earth and the solar system.

Earth Gains Mass and Loses an Atmosphere

As shown in Figure 2.8, the young Planet Earth formed from the **accretion** of dust particles and gases. After the solar wind swept away the remaining interplanetary dust and gases, accretion continued as meteorites and planetesimals collided with the earth, adding to its mass. As the earth grew in size, its gravitational attraction increased. This pulled in even more meteorites. The materials that fell to earth were mainly made up of oxides of silicon, iron, magnesium, and metallic iron, with small amounts of radioactive elements also present. It is thought that the earth approached its present size within a few million years of the collapse of the nebula.

Differentiation: Forming an Iron Core

Figures 2.10a-d show the series of events that led to the major changes in the earth after the ignition of the solar furnaces. The melting of the interior occurred as a result of the accumulated heat of gravitational energy (as materials fell onto the earth's surface) and, more importantly, due to heat from the decay of radioactive isotopes of elements such as uranium, thorium, and potassium within the earth's interior. Once the earth's interior

Figure 2.10a The Early Formation of the Earth Meteorite bombardment of the early earth added to its size and the heat of impact kept the earth's surface in a molten condition. This process was much slower after the T. Tauri Wind, created when the sun's nuclear furnace ignited, blew away all but the largest asteroids and meteoroids.

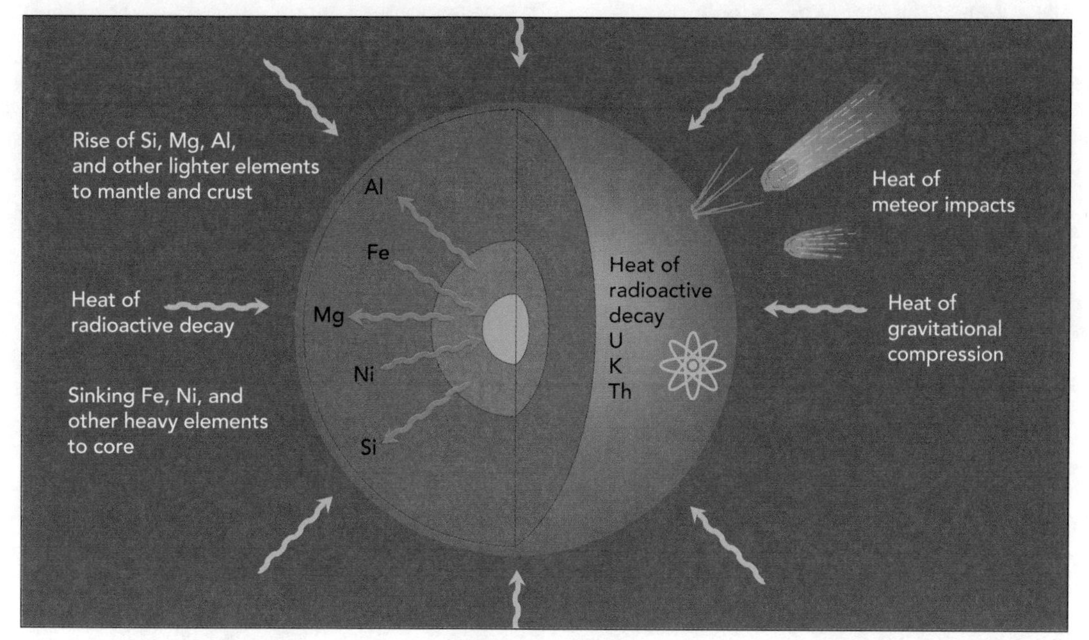

Figure 2.10b The Early Formation of the Earth Heating by meteorite impacts, gravitational compression and, most importantly, radioactive decay of uranium (U), thorium (Th), and potassium (K) melted the interior of the earth. This melting enabled iron (Fe), nickel (Ni), and other heavier elements to sink to the centre to form the core. The lighter elements like silicon (Si), magnesium (Mg), and aluminum (Al) rose to form the mantle and crust; a process known as differentiation.

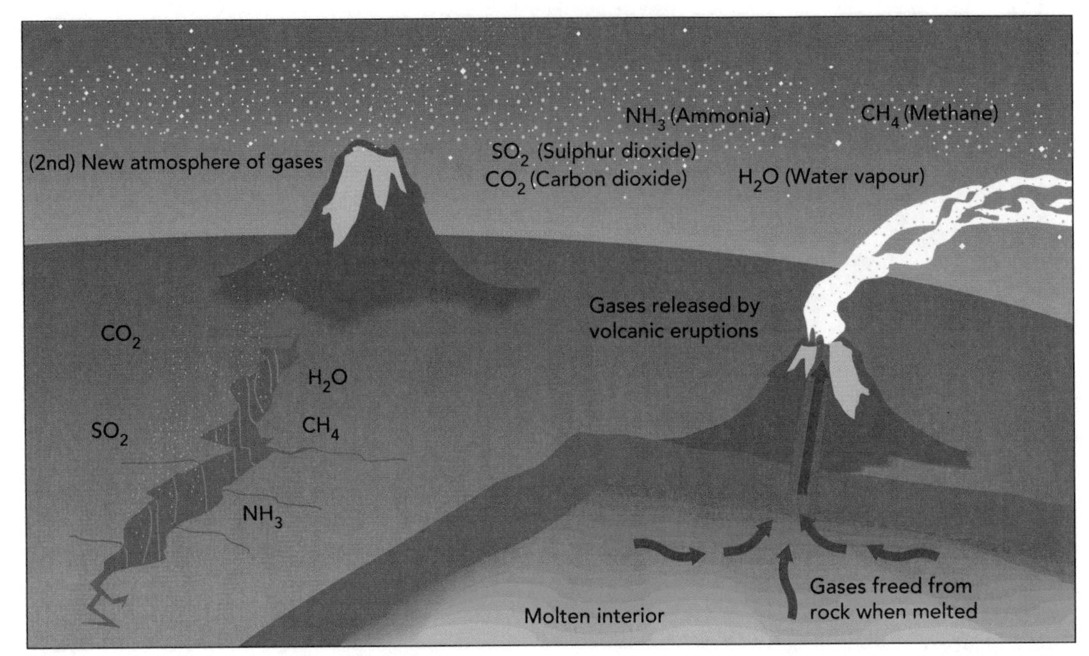

Figure 2.10c The Early Formation of the Earth The melting of the interior also freed many volatile gases such as carbon and sulphur dioxides, hydrogen, ammonia, methane, and water vapour from their rocky traps. These gases worked their way to the surface to form a new atmosphere, replacing the one blown away by the T. Tauri Wind, caused when the sun's nuclear furnaces first ignited.

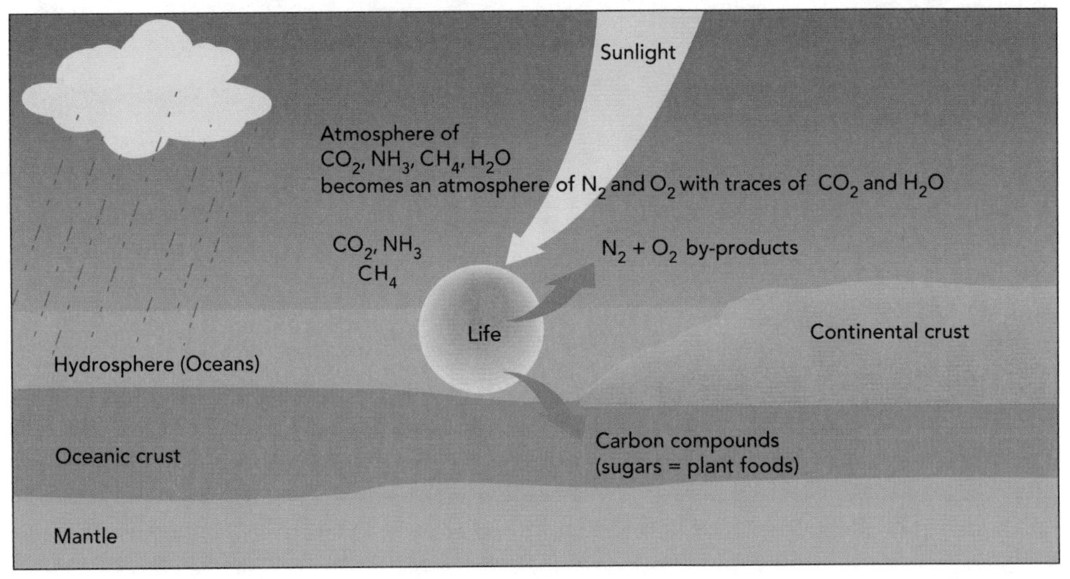

Figure 2.10d The Early Formation of the Earth The arrival of life forms began the transformation of the poisonous anaerobic atmosphere of carbon dioxide, ammonia, methane, and sulphur dioxide and water vapour into an oxygen-rich one through the process of photosynthesis (where sunlight and carbon dioxide are used to produce food, and the by-product is oxygen).

melted, the heavier elements of iron (Fe) and nickel (Ni) were attracted inward to form a very dense core. The lighter elements and compounds such as silicon (Si), magnesium (Mg), and aluminum (Al) moved outward to make up the mantle and crust of the young planet. **Differentiation** of this type, which led to layers within the earth's interior, could only occur in a liquid or semi-liquid body.

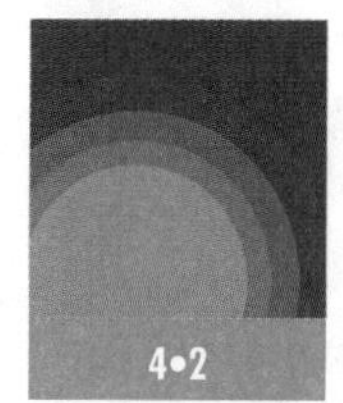
4•2

A Second Atmosphere

The second major event in the earth's history was the formation of a new atmosphere, replacing the early one, which had been removed by the solar wind that accompanied the ignition of the sun's furnace. The gases that made up this new atmosphere came from the melting of the rocks of the earth's mantle and crust. Every volcanic eruption released volatile elements and compounds, including methane, water vapour, ammonia, sulphur dioxide, hydrogen, and carbon dioxide. By this time, the earth's crust had cooled enough to become solid and temperatures were mainly between the freezing and boiling points of water. Under these conditions, the water vapour in the early atmosphere was converted to a liquid and fell as rain. The water collected in the lower-lying basins of the new earth. As a result, one of the most unique features of Planet Earth was created — the oceans.

1•3

The Arrival of Life

The stage was now set for the arrival of life. The earliest signs of life on earth are in rocks 3.5 to 3.8 billion years old. Early life forms had to adapt to a world with little or no **free oxygen** (oxygen not combined with other elements), and the oceans provided the most favourable conditions. According to the evolutionary theory, species appeared that used the process of photosynthesis to create food from sunlight, carbon dioxide, and other minerals. Free oxygen was a by-product gas of this process. These species were responsible for bringing about the third major change in the nature of the earth: an atmosphere with a significant amount of oxygen. While many of the other changes in the earth were the result of physical and chemical processes, this unique change was brought about by life itself.

11•1

QUESTIONS

14. a) Suggest reasons why the record of the early history of the earth is difficult to find in the rocks that make up the present surface of the earth.
    b) How have meteorites come to the rescue of astronomers and earth scientists in finding clues to the early history of our planet?
15. a) Define the terms "accretion" and "differentiation" as they apply to the earth's formation.
    b) What conditions were necessary for these processes to occur?
16. In your opinion, what was the single most important event in the history of the earth? Substantiate your opinion.
17. a) In a group, brainstorm what actions we could/should take if another solar system, with potential for life, were discovered by our scientists. Organize these suggestions into a realistic proposal.
    b) Devise an imaginative way of presenting your group's proposal to the rest of the class.

2·5 Earth's Unique Place in the Solar System

From a great distance, the earth is not a particularly outstanding planet within the solar system. It is very similar in size and composition to the other members of the inner terrestrial planets, such as Mars and Venus. Its orbit around the sun and its 24-hour rotational period are not unusual. The distances between the earth and its inner and outer planetary neighbours follow the general pattern of planet spacing found across the whole solar system.

It is only when you move closer to the planet that its most stunning feature is clearly seen: its blue colour. This is the only planet where water exists in all three states in significant amounts, especially in liquid form. Earth's distance from the sun is such that temperatures fall largely within the range where water can exist in the liquid state. If the planet were any closer to the sun, the water would evaporate and the strength of the sun's rays would break up the water vapour molecules, separating the oxygen and hydrogen atoms. The hydrogen would then likely be lost since it is a light gas that can escape the earth's atmosphere. If the distance to the sun were greater, much more water would be tied up in icefields and glaciers due to lower temperatures.

The vast oceans are a unique feature found nowhere else in the solar system and are partly responsible for the predominantly blue colour of this planet.

As well as distance from the sun, size and rotational period are also linked to the formation of the earth's atmosphere. The planet is just the right size so that gravity can hold onto its gaseous atmosphere. Smaller bodies, such as the earth's moon, or even Mars, have lost parts or all of their atmospheres because their gravity was not powerful enough to hold onto the lighter gases. On the other hand, the giant gas planets of Jupiter, Saturn, Neptune, and Uranus were just the opposite. They retained their original atmospheres of light gases such as hydrogen, helium, and methane because of their size and high gravitational attraction. By holding such gases in abundance, without the presence of free oxygen, these planets have atmospheres poisonous to the life forms found on earth. Earth is the only planet with oxygen in large amounts within its atmosphere.

The twenty-four-hour rotational period of the earth is fast enough that temperatures do not remain below the freezing point for long periods of time during the night, or become extremely high during the day. The regulation of temperatures by the atmosphere and the oceans also reduces the extremes of temperature, allowing most of the earth's water to remain in liquid form.

Earth lies within the narrow **"continuously habitable zone"** or CHZ. The CHZ is the region within the solar system where a planet theoretically could have surface temperatures between -80°C and 100°C over a long enough time for life to evolve, according to the evolutionary theory. It is within this narrow range of temperatures that all known life forms occur. Mars also lies just within the CHZ, but its smaller mass and lower gravitational attraction make it impossible to hold the gases in its atmosphere that are necessary for life. Earth is the only planet in the solar system to have life as we know it.

How small the earth is! Really a tiny planet, whirling around one ordinary star. And no longer the centre of the universe, as people believed for over 40 000 years. But that change in cosmic position does not mean we should value the earth any less. For the earth is our delicate ship, protecting us on our dark passage through space.

From Michael Zeilik, *Astronomy: The Evolving Universe*, 4th Edition (New York: Harper and Row, 1985), p. 120.

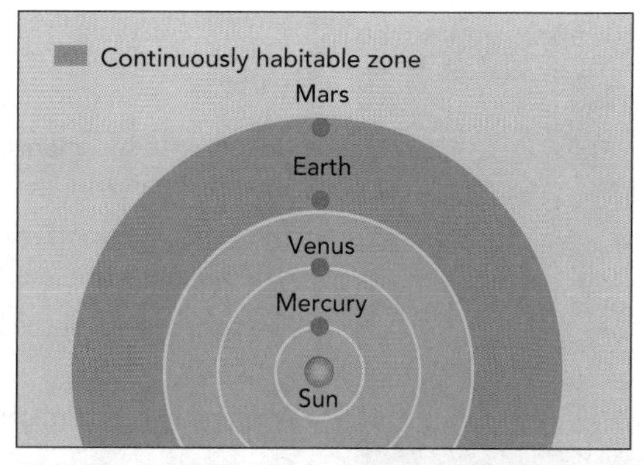

Figure 2.11 The Continuously Habitable Zone of Our Solar System

It's a Fact...

The question of why Mars is too cold for life, Venus far too hot, and Earth just right is known as the "Goldilocks problem of climatology".

QUESTIONS

18. Design a collage or poster to highlight the unique features of the earth. Include in your design a summary of the conditions that permit such features to occur only on earth and not on any other planets in the solar system.
19. a) In a group, brainstorm what changes might occur if the earth's orbit around the sun were altered to place it at a greater or lesser distance from the sun.
    b) Using the ideas generated by the group, write a short story describing the effects such a change would have on life on earth.
20. Consider the following quotation:

    "With chemical and physical activity so abundantly in progress on the surface of the earth, it must be one of the most intensely active regions in our solar system. It is unique, wet, and fermenting, yet all this activity is confined to a very thin zone or skin."

    From David Duneley, *Earth's Voyage Through Time* (England: Chaucer Press, Paladin Books, 1975), p. 47.

    a) How does this quotation point out the fragility of life on earth, within the larger picture of the solar system?
    b) Given that life can survive within a very narrow range of conditions, what activities of our present society threaten our survival?

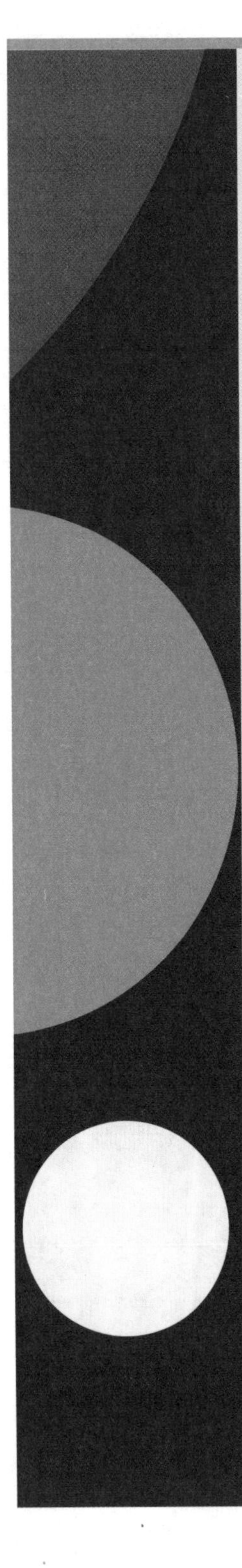

Review

- The universe is composed of an unimaginable number of objects separated by vast distances.
- The Big Bang hypothesis suggests that the universe began as a result of an explosion that sent matter flying at incredible speeds in all directions.
- The solar system developed out of this material, as the nebular hypothesis suggests.
- A thermonuclear explosion ignited the sun's nuclear furnaces.
- Gravity played an important role in drawing together the materials that make up the earth and in differentiating these materials.
- The formation of an atmosphere on this planet created conditions which were suitable for life forms.

Geographic Terms

light year
solar system
asteroid
meteoroid
meteorite
terrestrial planet
gas giant planet
nebular hypothesis
nebula
planetesimal
sunspot
accretion
differentiation
free oxygen
continuously habitable zone

Explorations

1. From what you have learned about the main characteristics of Physical Geography as outlined in Chapter 1, why is this chapter, "Earth: Its Place in the Universe", part of a book on this subject?
2. On the basis of what you have learned in this chapter, discuss the likelihood of life having developed or developing somewhere else in the universe.
3. What do you find most interesting about the field of astronomy and the study of the origins of the universe? Explain.
4. In what ways is the earth a unique planet within the solar system?
5. Why do you think people become astronomers? In what way does the job of an astronomer differ from other jobs?

CHAPTER

3

THE EARTH IN MOTION

OBJECTIVES:

By the end of this chapter, you will be able to:

- describe the major motions within the universe;
- describe the different motions of the earth in space;
- identify the effects of these motions on the physical geography of the planet;
- relate the rotation of the earth to the need for time zones;
- begin to appreciate the interrelationships between the various physical processes and systems operating on the earth.

CHAPTER 1: The Nature of Physical Geography
CHAPTER 2: Earth: Its Place in the Universe
CHAPTER 3: The Earth in Motion
CHAPTER 4: The Earth's Interior
CHAPTER 5: The Earth's Crust
CHAPTER 6: The Lithosphere in Motion: Plate Tectonics
CHAPTER 7: Solar Radiation
CHAPTER 8: Climate
CHAPTER 9: Weather
CHAPTER 10: The Hydrosphere and the Hydrologic Cycle
CHAPTER 11: Natural Vegetation and Soil Systems
CHAPTER 12: Denudation: Weathering and Mass Wasting
CHAPTER 13: Distinctive Landscapes: Humid and Arid Environments
CHAPTER 14: Distinctive Landscapes: Glacial, Periglacial, and Coastal Environments
CHAPTER 15: Natural Hazards: Disrupting Human Systems
CHAPTER 16: The Disruption of Natural Systems
CHAPTER 17: Fragile Environments

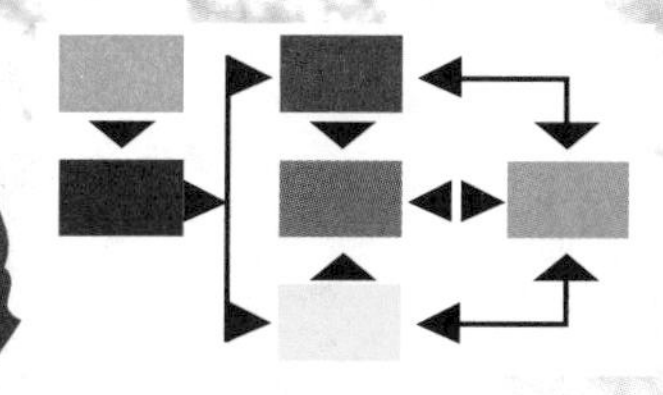

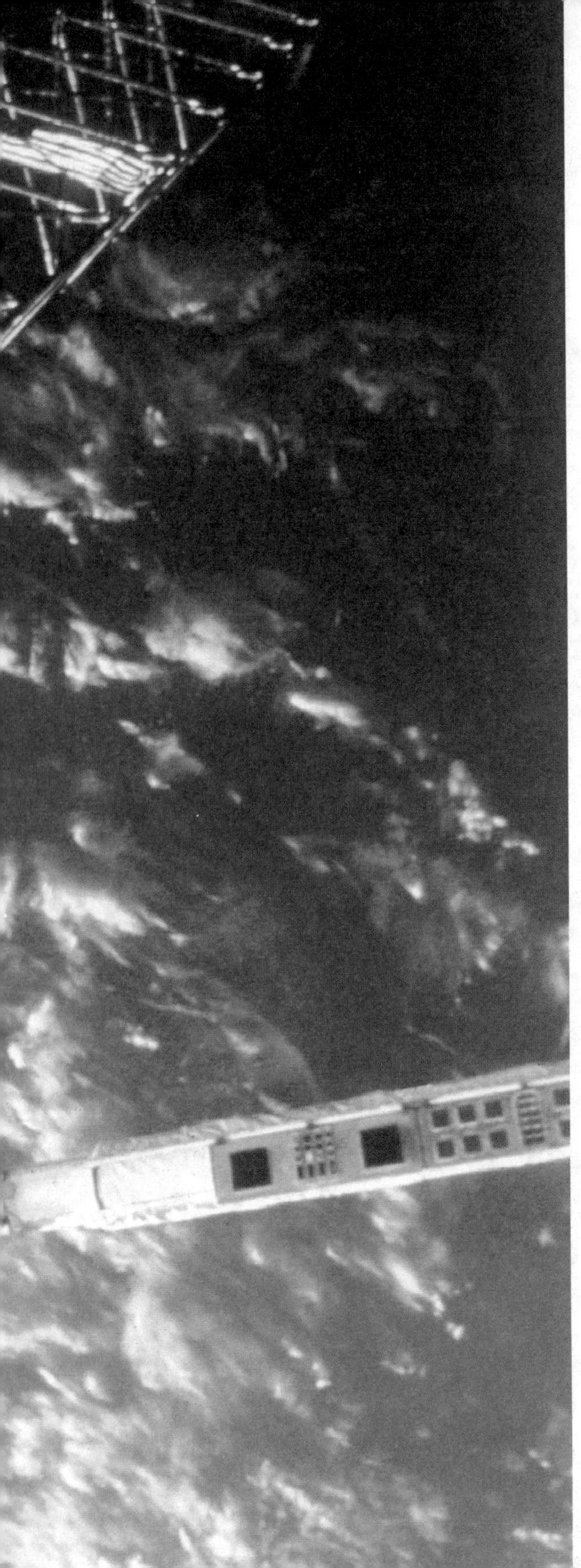

Introduction

Every object or body in the universe is in motion. The earth itself is involved in a number of different motions. These motions are important because they give us day and night and the changing seasons. Scientists have also recently discovered that the motions of the earth in space can bring about changes in the climate and related physical processes on the earth's surface.

3·1 Differing Types of Movement

Just as human beings are constantly in motion, so is Planet Earth. We can study different types of human movements. Our movements within the home, to school each day, on weekend trips to nearby destinations, and extended holiday trips overseas are examples of different types of movement.

Even though we are mostly unaware of it, the earth is going through a number of motions at all times. The biggest movements — those involving the universe and solar system — are generally outside our experiences or senses. We are aware of smaller movements, such as the rotation of the earth on its axis and the orbiting of the earth around the sun, because they give us night and day and the seasons. But even in these instances, we do not have any sense of actual movement beneath our feet.

Moving Galaxies

Most scientists believe that all the known galaxies in the universe are hurtling away from one another at incredible speeds. They theorize that this came about because of a "big bang" that propelled the various galaxies through space, thereby creating an

It's a Fact...

The energy for the revolutions of the planets around the sun came from the spinning motion of the spiralling dust and gas clouds from which the solar system formed.

Moving Objects	Speed of Movement (km/h)
Earth Rotation on Axis (equator)	1 660
Earth Orbit of Sun	108 000
Sun Orbit Around Milky Way	900 000
Movement of Milky Way Galaxy	2 000 000

Figure 3.1
The Movements of the Earth, Sun, and Milky Way

expanding universe. From earth, it appears that all the galaxies are moving away from us. This does not mean our planet lies at the centre of the universe; the same observation would be made at any other point in the universe. It is very difficult to measure the speed at which the galaxies are moving since there are no fixed points to use as reference points.

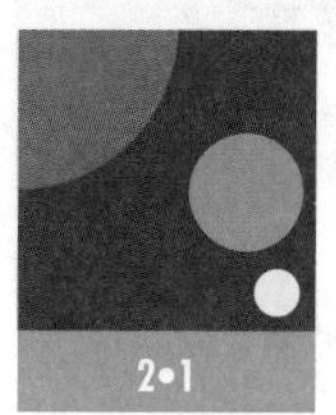

In addition to this general movement, the Milky Way galaxy is spinning in space around a central axis like an enormous pinwheel. Our sun and solar system orbit the centre of this galaxy along with over 100 billion other stars, taking about 175 000 years to complete one **revolution**. It is estimated that the speed of rotation of the Milky Way is about 250 km/s and that the sun's distance from the galactic centre is 24 000 to 26 000 light years.

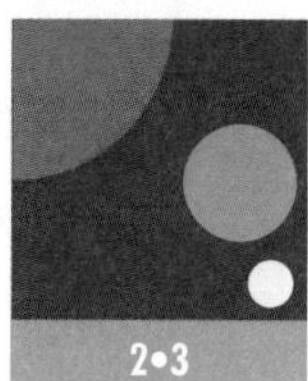

The Moving Earth

As well as the movement of the earth as part of the Milky Way galaxy, our planet goes through other motions. The term "revolution" refers to the earth's orbiting of the sun. The earth revolves around the sun at a speed of approximately 108 000 km/h, travelling 929 million kilometres to complete a single revolution. The orbit of the earth is slightly elliptical; that is, it is almost circular, but not quite. The average distance of the earth from the sun is 150 million kilometres. Presently, the earth is at its greatest distance of 152 million kilometres from the sun on July 4 and at its closest distance of 147 million kilometres on January 3. The term **aphelion** refers to the point in the orbit of the earth where it is farthest from the sun. **Perihelion** refers to the point at which the earth is closest to the sun. This situation changes over an approximately 21 000 – 22 000 year cycle. In 10 500 to 11 000 years, the earth will be closest to the sun in July and farthest

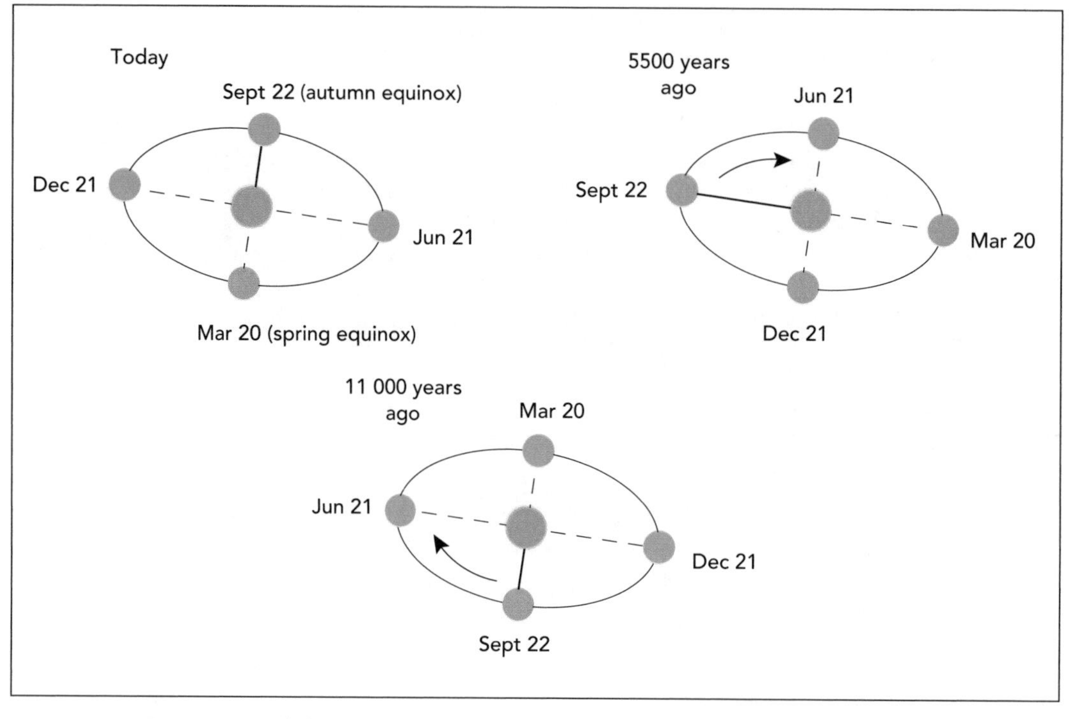

Figure 3.2a **Precession of the Equinoxes (period = 22 000 years)**

away in January. The shift is called the **precession of the equinoxes**. (See Figure 3.2a.) The variations in distances from the sun result in differences in the amount of solar energy received on earth.

The earth experiences a change in the **eccentricity** of its orbit around the sun over a period of approximately 100 000 years. The orbit goes from more nearly circular to much more elliptical and back again during this cycle. (See Figure 3.2b.) When the orbit is nearly circular, annual solar heating is more evenly spread over the earth. When the orbit is elliptical, the earth is closer to the sun at some times than at others, which increases the contrast between seasons.

These changes alone are not great enough to cause the seasons. The seasons result from the tilt of the axis of rotation of the earth.

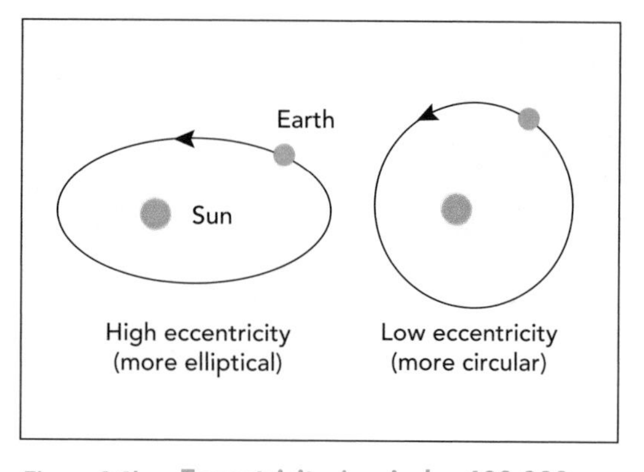

Figure 3.2b **Eccentricity (period = 100 000 years)**

It's a Fact...

The Milky Way galaxy and everything in its vicinity are moving towards the "Great Attractor", a cluster of 75 000 galaxies about 150 million light years from earth, at a speed of 2 000 000 km/h. Don't worry about a collision! The journey will take 100 billion years to complete.

Earth's Tilted Axis

The **axis** of the earth is defined as the line joining the north and south poles, around which the earth rotates. An important characteristic of the axis is that it is tilted at an average angle of 23.5° from the plane of the elliptic, as shown in Figure 3.3. As the earth revolves around the sun, this tilt means the sun shines more fully on one hemisphere than the other. There are only two times in the year (March 21 and September 21) when the sun shines evenly on both hemispheres. At these times — the vernal (spring) and autumnal **equinoxes** — the sun's rays are perpendicular at the equator. The summer **solstice** occurs on approximately June 21 when the sun shines directly over the Tropic of Cancer in the northern hemisphere. This brings higher temperatures since the sun's rays are more concentrated in a smaller area and less reflection occurs in the northern hemisphere. The reverse is true of the southern hemisphere at this time. The winter solstice occurs on or about December 21, at which time the sun's rays are perpendicular to the Tropic of Capricorn.

The tilt of the earth's rotational axis causes the different seasons experienced in the higher latitudes of the northern and southern hemispheres, as illustrated in Figure 3.5. The seasons are based on temperature differences.

Another reason for higher summer temperatures in the northern hemisphere is the fact that the sun shines for much longer during the day in the summer season. The sun is above the horizon for all 24 hours in areas that lie north of the Arctic Circle. In the southern hemisphere, the reverse is true. The days are shorter, the sun's rays strike the hemisphere at a greater angle, and the area south of the Antarctic Circle remains in twilight or almost total darkness for 24 hours a day. Thus, there is an imbalance in solar radiation reaching the higher latitudes.

The earth's angle of tilt from the plane of the elliptic varies from 21.5° to 24.5° over the course of about 40 000 years (See Figure 3.4.) This "wobble" in the axis changes the contrast between seasons. When the tilt is greatest, the earth's winters are coldest and summers hottest. When the tilt is less, there is less difference between summer and winter.

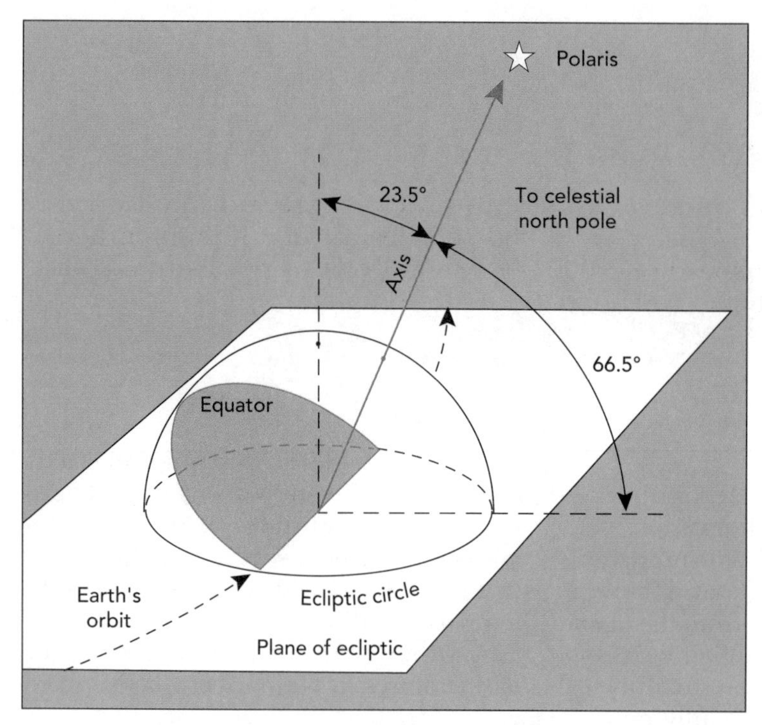

Figure 3.3 The Average Tilt of the Earth's Axis

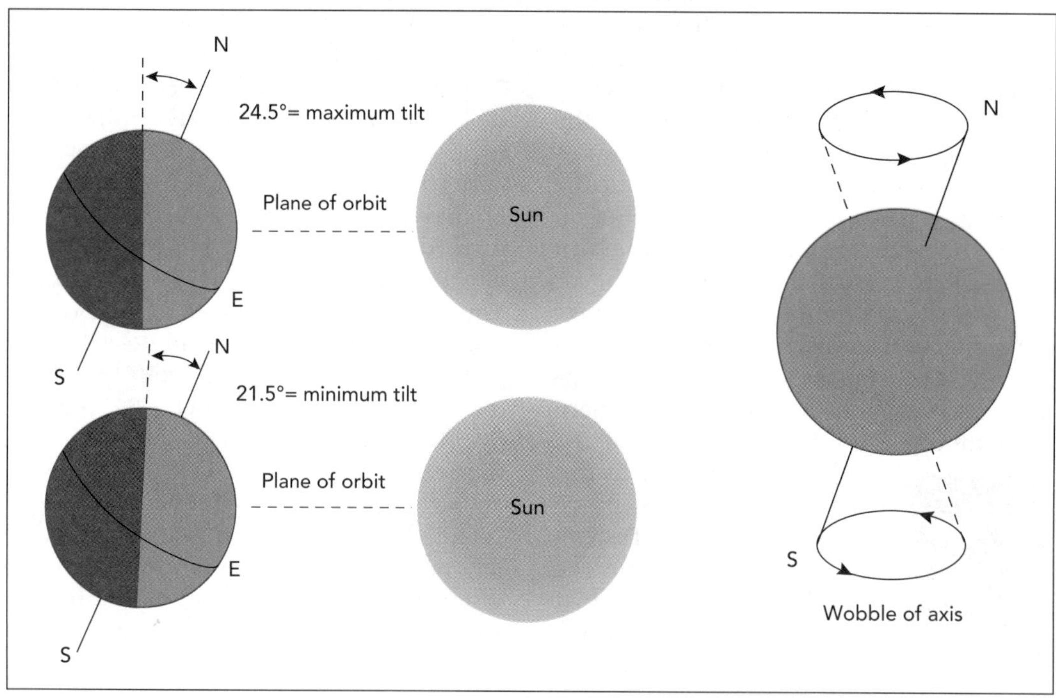

Figure 3.4 The Wobble in the Axis of Earth's Rotation (period = 40 000 years)

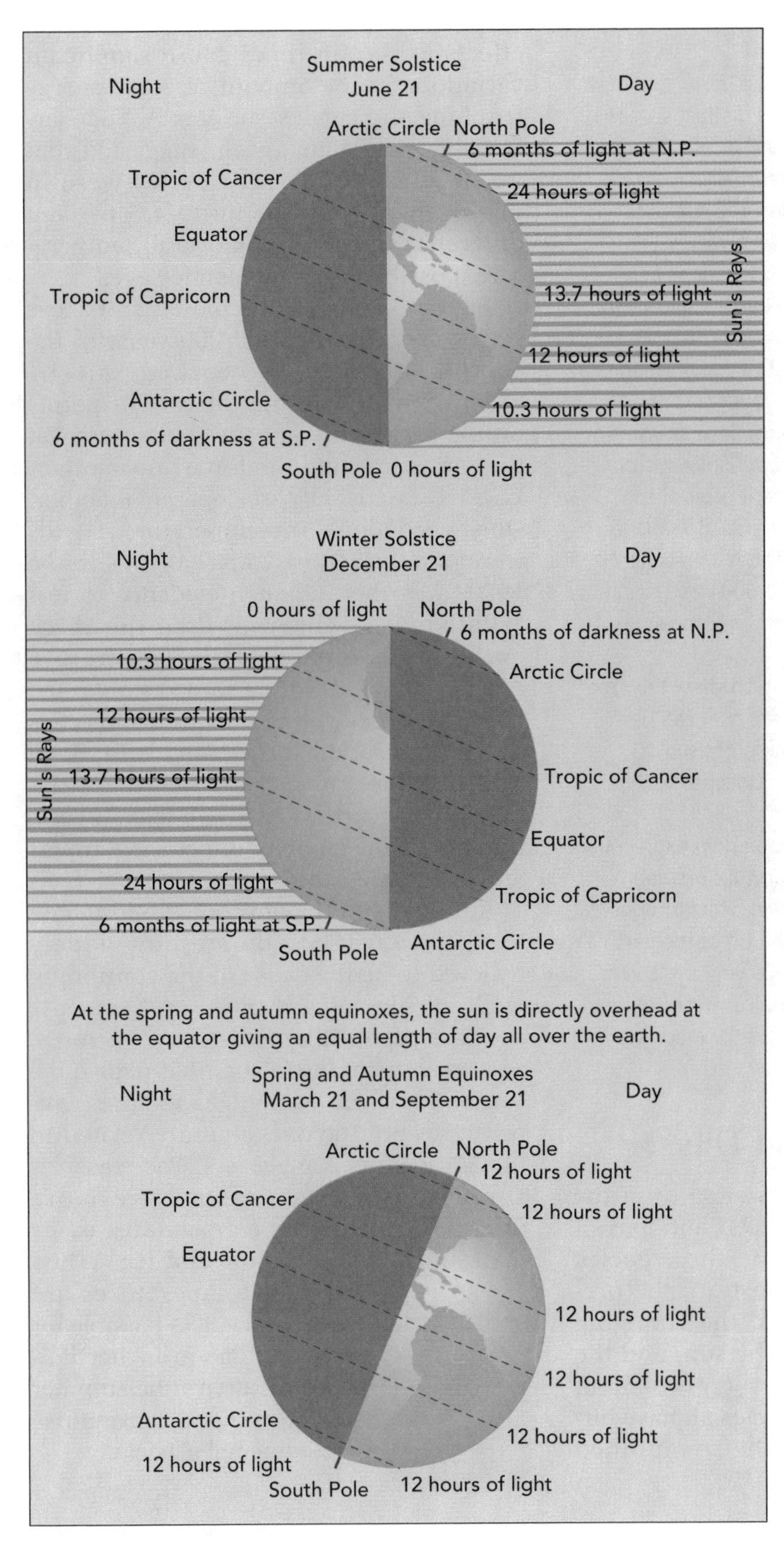

Figure 3.5
The Annual Progression of Solstices and Equinoxes

QUESTIONS

1. Most of Canada experiences distinct seasons. What are the advantages and disadvantages of this situation for you personally?
2. Why is it difficult to establish the speed of movement of objects in the universe?
3. We are used to determining locations and directions of places on the earth's surface in two dimensions using maps and the points of the compass. Develop a list of the problems you would face as a space voyager trying to determine locations and directions as you travel through and beyond our solar system.
4. What effect might the present orbit of the earth around the sun (perihelion in January) have on winters in the northern hemisphere? in Australia? In 10 500 to 11 000 years, what situation will prevail during northern winters? Explain your answers.
5. a) Draw a simple sketch map to show the area of the northern hemisphere that can be called the "land of the midnight sun".
   b) Explain why this area can boast of such a phenomenon.
   c) Why does this phenomenon help explain the great contrasts in temperature between winter and summer at such high latitudes?
6. Explain how variations in the tilt of the earth's axis over the 21 000 – 22 000 year cycle can bring about variations in the contrast between summers and winters. Use sketch diagrams to illustrate your answer.

3·2 Earth Orbits and Tilts

A number of implications arise from the cycles that occur in the orbit and axis of rotation of the earth. The three cycles that are of greatest interest are the precession of the equinoxes, the changing orbit of the earth about the sun, and the variation in the tilt of the axis of rotation. (See Figure 3.4.) These cycles all have different time spans, but their combined effects at certain times cause significant variations in the amount of solar energy reaching the earth's surface. A Yugoslav scientist, M. Milankovich, suggested that when all three of these cycles were in phase, there could be up to a 5 percent difference in the earth's overall temperature, possibly producing an ice age.

Figure 3.6 shows the patterns of these cycles over the past 800 000 years of the earth's history, as worked out by Milankovich. Note that at certain points on the graphs, two of the cycles occur at the same time and reinforce one another. Less frequently, all three operate together. Only when long-term temperature records were obtained from deep sea cores in the 1970s was there enough evidence to test Milankovich's theory. When the three cycles are compared with temperature data based on sediment samples obtained from deep sea drilling operations, there is a striking relationship among them. Such evidence strongly suggests that the variations in the earth's orbit and axis of rotation about the sun can affect the temperatures of the earth's atmosphere.

Although not the only contributing factors, these variations are now widely believed to be the cause of the continuing cycle of glacial advances and retreats over the past million years or so. Strange as it may seem, the factors that reduce the contrast between winter and summer temperatures are the ones that are related to the advances of the glaciers. The reason is simple. When the seasons are more extreme, hot summer temperatures easily melt all the winter snow and ice. When summer temperatures are less extreme, it is possible for snow and ice in higher latitudes to survive the summer season, eventually giving rise to continental glaciers.

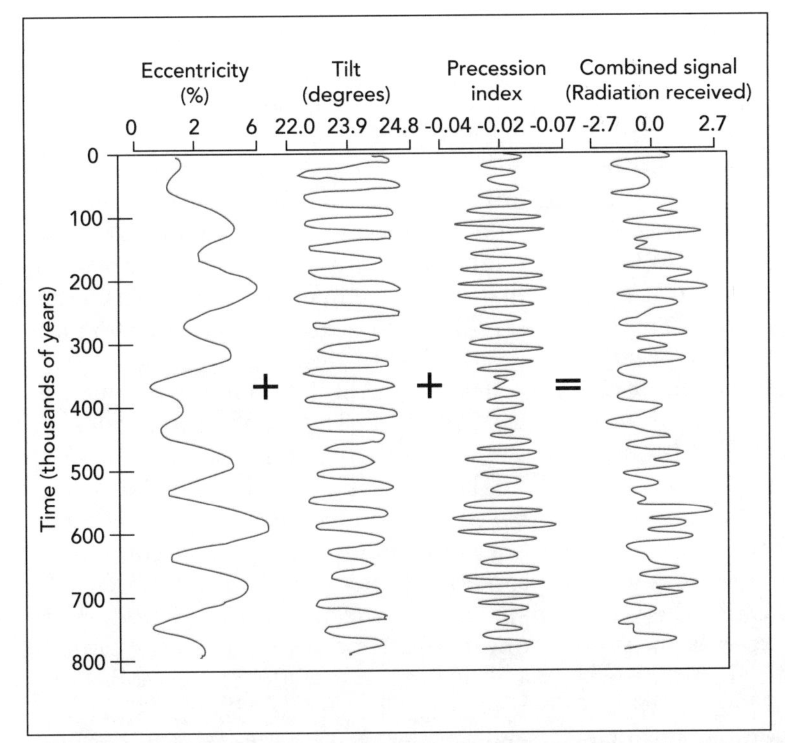

Figure 3.6
Variations in the Earth's Orbits and Tilts
Curves showing variations in orbital eccentricity, tilt, and precession during the last 800 000 years. Summing these factors produces a combined signal that shows the amount of radiation received on the earth at a particular latitude through time.

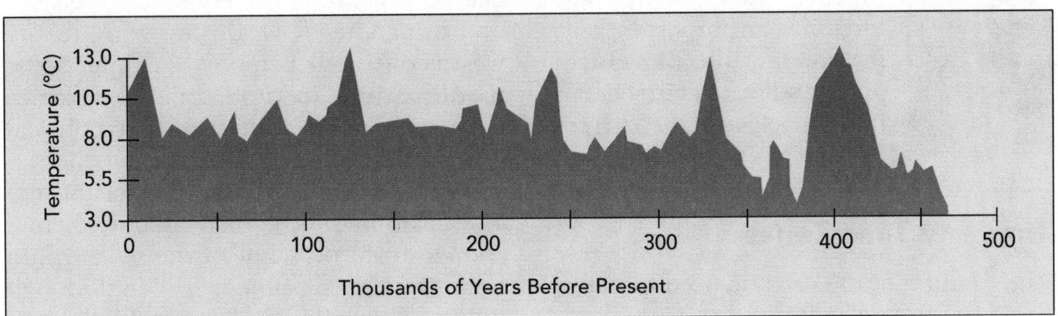

Figure 3.7 **Variations in Temperatures Based on Indian Ocean Deep Sea Cores**

QUESTIONS

7. Describe the conditions of revolution and tilt that would cause the contrast between winter and summer temperatures to be least.
8. Assume the following conditions were true of the earth's orbit and axis of rotation:
   - The orbital path is nearing its most circular path.
   - The tilt of the axis is diminishing from its maximum.
   - Perihelion (earth closest to the sun) now occurs in January.

   a) Write a "climate report" for the earth over the next 50 000 years, given the conditions described above.

   b) Explain the reasons for the conditions predicted in your report.

3·3 Rotation of the Earth

In addition to its revolutions around the sun, the earth spins, or rotates, on its axis. This movement is called **rotation**. Earth rotations have two important effects on our lives. The first is that they give us days and nights. It takes 24 hours for the earth to complete one full rotation, the length of an earth day. The second effect of the rotation is the creation of what is known as the **Coriolis force**. This force causes moving objects to appear to be deflected to the right in the northern hemisphere and to the left in the southern hemisphere. An immediate example of the effects of this force is the spinning motion of water as it goes down the drain in a bathtub or sink. The water spins clockwise as it descends into the drain in the northern hemisphere and counterclockwise in the southern hemisphere. The force has an important role in affecting the movement of air and ocean currents as they travel over the surface of the rotating earth.

Time and Time Zones

The rotation of the earth is used to determine the time of day for any given location. Noon occurs when the sun is directly overhead and midnight arrives when the sun is directly on the other side of the earth. In theory, this means that the place you live in actually has a different time of day from any other place located to the east or west. Depending on the east-west distance, this difference could be a few seconds, minutes, or hours. All places lying along the same line of longitude have the same time.

Before high-speed travel and instant worldwide communications, minor differences in time between places were not too important. In the late nineteenth century, with the advent of the telegraph and telephone, it became increasingly apparent that some system of time zones was necessary. Sir Sanford Fleming, a Canadian, is credited with developing a world time zones scheme to help standardize the time over the earth. The world's **time zones** are shown on the map in Figure 3.8.

The earth is divided in 24 time zones based on the fact that the earth takes 24 hours to complete one full rotation on its axis. Since there are 360° in a circle, the sun moves through 15° of longitude in one hour. Therefore, each time zone is 15° of longitude wide and extends from the north to the south poles. If you choose one zone, the zone to the east is one hour ahead, while the zone to the west is one hour behind your time. Some modifications to time zone boundaries have been made to ensure that political regions remain in the same time zone.

The adoption of time zones meant that a decision had to be made about where a change from one day to the next would occur. Since the prime meridian (0° longitude) is located at Greenwich, England, it was decided that the 180° meridian be used as the International Date Line. By chance, this proved to be a convenient location since it passed down the centre of the Pacific Ocean. As travellers pass over the date line going from North America to Asia, they "lose" a day en route. When they travel back over the Pacific in the opposite direction, they "regain" the day.

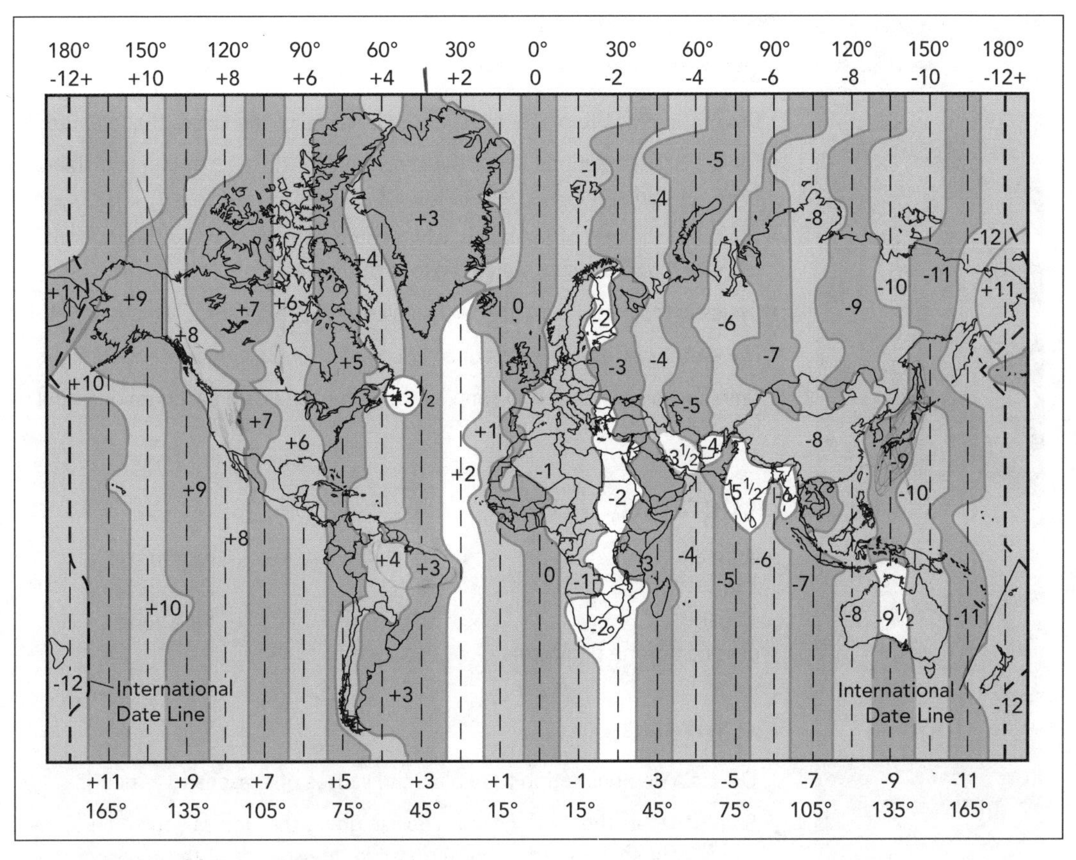

Figure 3.8 **World Time Zones**

QUESTIONS

9. If the earth were to rotate on its axis at twice its current speed, how would you be affected?
10. Using the World Time Zones map in Figure 3.8 and a world atlas, complete the following:
    a) What would people in Paris, France, likely be doing when people in Vancouver are eating breakfast?
    b) In flying from Toronto to Rio de Janeiro, Brazil, what adjustments would you have to make to your watch?
    c) What changes in time would you notice in flying from San Francisco, California, to Tokyo, Japan, on a single 11-hour plane trip? What changes would you notice in making the return trip?
    d) If a hockey game started at 8:00 p.m. in Toronto, at what local time would a fan in Vancouver watch this game on television?
    e) Note the differences between the theoretical and actual time zones as shown in Figure 3.8. Suggest two reasons for the deviations seen in the time zones and in the International Date Line.
11. Why would a Canadian be more likely to develop a world time zone system than a European resident?

Review

- The motions of the planet came about because of the forces that created it.
- The earth's revolutions about the sun vary over time, changing in distance from the sun and in the path of the orbit.
- The tilt of the axis of the earth, when combined with the revolution of the planet around the sun, yields seasons of the year.
- Cycles in the motions of the earth can combine to produce climatic variations.
- The earth rotates on its axis once every 24 hours, giving us day and night.
- Time zones are needed to simplify time; the zones are based on one-hour intervals.

Geographic Terms

revolution
aphelion
perihelion
precession of the equinoxes
eccentricity
axis
equinox
solstice
rotation
Coriolis force
time zone

Explorations

1. Devise a demonstration to explain the causes of seasons to a friend.
2. Summarize all the motions that you are going through as you sit perfectly still in your seat. Include, where possible, the speeds of these motions.
3. Describe the implications of:
   a) a sudden slowing down of the rotation of the earth about its axis
   b) the ending of the earth's rotation about its axis
   c) the ending of the revolution of the earth about the sun
4. Describe the advantages and disadvantages for a Canadian province, or time zone, going on year-round daylight savings time.
5. Research the potential impact of variations in the tilt of the earth's axis over the 21 000 –22 000 years of its cycle on climate. Discuss specific impacts on:
   a) coastal cities
   b) the Far North
   c) desert margins

Related Careers

- astronomer
- meteorologist
- data analyst
- radio technician
- space scientist
- computer programmer
- climatologist
- telecommunications engineer

Energy From Below

UNIT

The earth is a dynamic and changing planet. Each of the layers into which it is differentiated is constantly on the move, slowly, but surely. The energy used to power these movements comes from the decay of radioactive elements in the interior. Even the "solid" crust, the stage upon which life is played out, constantly recycles and transforms itself through powerful tectonic processes that are responsible for the shape and form of the earth's surface.

As you work through this unit, consider these questions:

- Why is the earth composed of distinct layers or spheres and how do they differ from one another?
- How do we uncover the secrets of the interior of the earth?
- What processes create the materials and surface forms of the earth's crust?
- How does the theory of plate tectonics explain the pattern of tectonic processes across the earth?
- What source of energy powers the dynamic processes operating on the surface and in the interior of the earth?

CHAPTER

4

THE EARTH'S INTERIOR

OBJECTIVES:

By the end of this chapter, you will be able to:

- appreciate the difficulty of learning about the interior of the earth;
- recognize that the interior of the earth is divided into layers of different materials and densities;
- identify important sources of information that have been used to create an understanding about the earth's interior;
- relate the nature and composition of the earth to conditions on the surface of the earth.

CHAPTER 1: The Nature of Physical Geography
CHAPTER 2: Earth: Its Place in the Universe
CHAPTER 3: The Earth in Motion
CHAPTER 4: The Earth's Interior
CHAPTER 5: The Earth's Crust
CHAPTER 6: The Lithosphere in Motion: Plate Tectonics
CHAPTER 7: Solar Radiation
CHAPTER 8: Climate
CHAPTER 9: Weather
CHAPTER 10: The Hydrosphere and the Hydrologic Cycle
CHAPTER 11: Natural Vegetation and Soil Systems
CHAPTER 12: Denudation: Weathering and Mass Wasting
CHAPTER 13: Distinctive Landscapes: Humid and Arid Environments
CHAPTER 14: Distinctive Landscapes: Glacial, Periglacial, and Coastal Environments
CHAPTER 15: Natural Hazards: Disrupting Human Systems
CHAPTER 16: The Disruption of Natural Systems
CHAPTER 17: Fragile Environments

Introduction

Ironically, we know more about the planets of our solar system than we do about the interior of our own planet. For much of our history, we have been vastly ignorant of the inside of the sphere on which we live. Only in recent years, with the advent of sophisticated research tools, have we been able to develop an image of the interior of the earth. This chapter explores some of the evidence we have about the earth's interior.

4·1 The Earth's Internal Heat Sources

It is thought that the earth formed about 4.6 billion years ago when a vast cloud of gases and dust was gradually pulled inward by the force of gravity to form the sun and the planets of the solar system. The earth continued to grow because its gravitational attraction acted like a magnet to small planetesimals (early planets) and meteorites. The additional materials must have been solid since the earth's gravity was not strong enough to pull in liquids or gases. This explanation of the origin of the earth is called the cold accretion theory. The term **accretion** refers to a gradual increase in size resulting from the addition of materials from beyond the earth.

Today, it is known that the earth's interior is so hot that it should be in a liquid state. However, the enormous pressures of overlying materials make much of it behave as though it were solid. If the pressures were released, this material would quickly turn into liquid or gaseous states, as volcanic eruptions show.

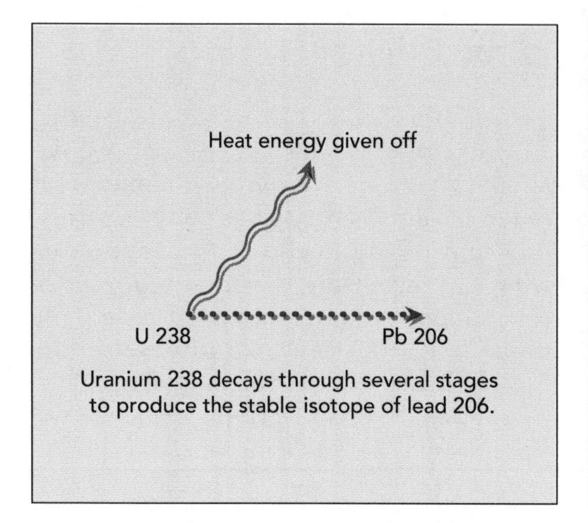

Figure 4.1a Radioactive Decay: A Major Source of Energy From Below

The key long-term radioactive elements:

Uranium (U) 238
Uranium (U) 235
Thorium (Th) 232
Potassium (K) 40

The key shorter-term radioactive elements:

Aluminum (Al) 26
Uranium (U) 236
Samarium (Sm) 146
Plutonium (Pu) 244
Curium (Cm) 247

Figure 4.1b Radioactive Elements in the Earth's Interior

To move from the idea of cold accretion to an understanding of the earth's molten interior, we must address several questions. The first is how the internal temperature of the earth reaches temperatures between 4000°C and 6600°C. A second question is how these temperatures brought about the formation of layers in the interior of the earth. Thirdly, what other effects have these high temperatures had on the earth?

The melting of the materials of the earth would require unthinkable amounts of energy. One source of heat is the kinetic energy of moving bodies striking the earth. A second source is the compression of rock materials inside the earth due to the enormous pressures of the material above. However, the major source of heat was, and continues to be, the decay of unstable, radioactive elements within the rocks of the earth. Figure 4.1b shows the most common radioactive elements that created this internal "heat engine". The short-term elements were present in the rocks of the early earth, but have completely decayed. The longer-term elements are still decaying and heating the rocks of the earth's interior.

It's a Fact...

- Meteorites were scientifically recognized only in the early 1800s. Before that time, people describing stones falling from the sky were thought to be hallucinating or observing fallout from a volcanic eruption.
- It is estimated that about 100 t of dust from burned-up meteorites falls into earth's atmosphere each day.

QUESTIONS

1. List several ways in which your life might be different if the gravitational attraction of the earth were greater, or less, than it is today.
2. a) Use the information in this section to explain why the earth's interior will eventually cool and become solid, like the interior of the moon.
   b) Suggest what general effects a cold, solid interior might have on the surface of the earth.
3. a) Draw a diagram to show how accretion worked to bring about the formation of the earth.
   b) Draw a second diagram to illustrate the three major sources of energy that led to the earth's hot interior.
4. Although much less frequently than in the earth's early period, meteorites still strike the earth. How might this fact be important in your life?

4·2 A Layered Earth

Figure 4.2 lists the proportions of the various elements that make up the earth and their atomic masses. The higher the atomic mass, the greater the density of the element. Studies have shown that the centre of the earth is composed of materials of high density and atomic weight, while the outer layers are made up of less dense materials. How did this "layering" take place?

In a solid body, there is no way elements can separate into layers of different density. But, in a molten and/or semi-molten state, heavier elements begin to differentiate from the surrounding lighter elements and settle towards the centre of the earth under the effects of gravity. This is how the core of the earth was formed in the first billion or so years of the earth's existence (Figure 4.3). Lighter elements

Element	Symbol	Percentage by Mass	Atomic Mass
Iron	Fe	35	55.85
Oxygen	O	30	16.00
Silicon	Si	15	28.09
Magnesium	Mg	13	24.31
Nickel	Ni	2.4	58.69
Sulphur	S	1.9	32.06
Calcium	Ca	1.1	40.08
Aluminum	Al	1.1	26.98
Sodium	Na	0.57	22.99
Chromium	Cr	0.26	52.00
Manganese	Mn	0.22	54.94
Cobalt	Co	0.13	58.93
Phosphorus	P	0.10	30.97
Potassium	K	0.08	39.10
Titanium	Ti	0.05	47.88

Figure 4.2
The Major Elements Making Up Planet Earth

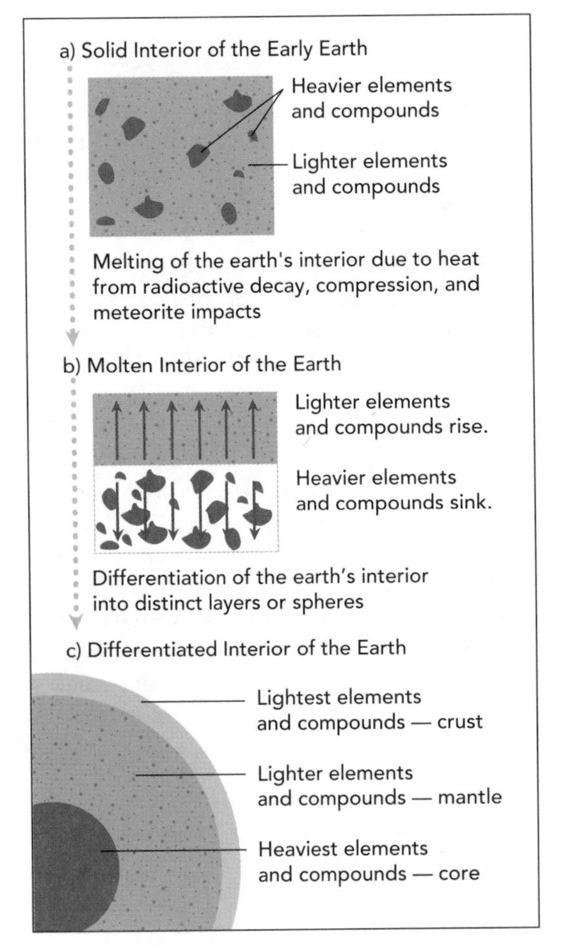

Figure 4.3 **Differentiation of the Earth's Interior Into Layers**

and compounds, on the other hand, rose towards the surface to form the crust and the upper layers of the **mantle**.

Many scientists believe this layering process was a quick one, taking only a few hundred million years. The result was the formation of a dense core of iron and nickel surrounded by a thick mantle of lighter rocks made up largely of various **silicates**. The silicates are minerals that combine the two most abundant elements in the earth's crust, oxygen (O) and silicon (Si). As time passed, the lightest materials of the mantle rose to the surface and began to form the earth's crust.

The process of differentiation first occurred billions of years ago, but the elements and compounds found within the interior of the earth have slowly continued to separate into the various layers shown in Figure 4.4. As this diagram shows, the internal layers of the earth can be determined in two ways: by their chemical composition or by their physical properties. It is important to note the differences in the number, names, and depths of these layers, depending on which criterion is used. For example, the term **"crust"** is used to identify the thin outer layer of rocks that are less dense than those of the mantle. The term **"lithosphere"** refers to the solid outer layer of the earth where the rocks are harder and more rigid than those of the plastic asthenosphere layer below. These layers are now differentiated by only slight differences in density. As a result of this ongoing process, scientists believe that the lighter materials are still being added to the crust of the earth, increasing the size and mass of the continents.

Chemically, the **core** of the earth is believed to be composed of a mixture of iron, nickel, and traces of other heavy metals. The pressures in the core are tremendous. Because of this, and despite extremely high temperatures, physically, the inner core is in a solid state. The outer core, being under less pressure but at a similar high temperature, is in a liquid state.

Chemically, both the upper and lower mantle are composed of silicates of magnesium and iron. These are much lower in density than the materials that make up the core. Physically, the **mesosphere** is largely solid in nature. Even so, it is very hot and will flow slowly under pressure.

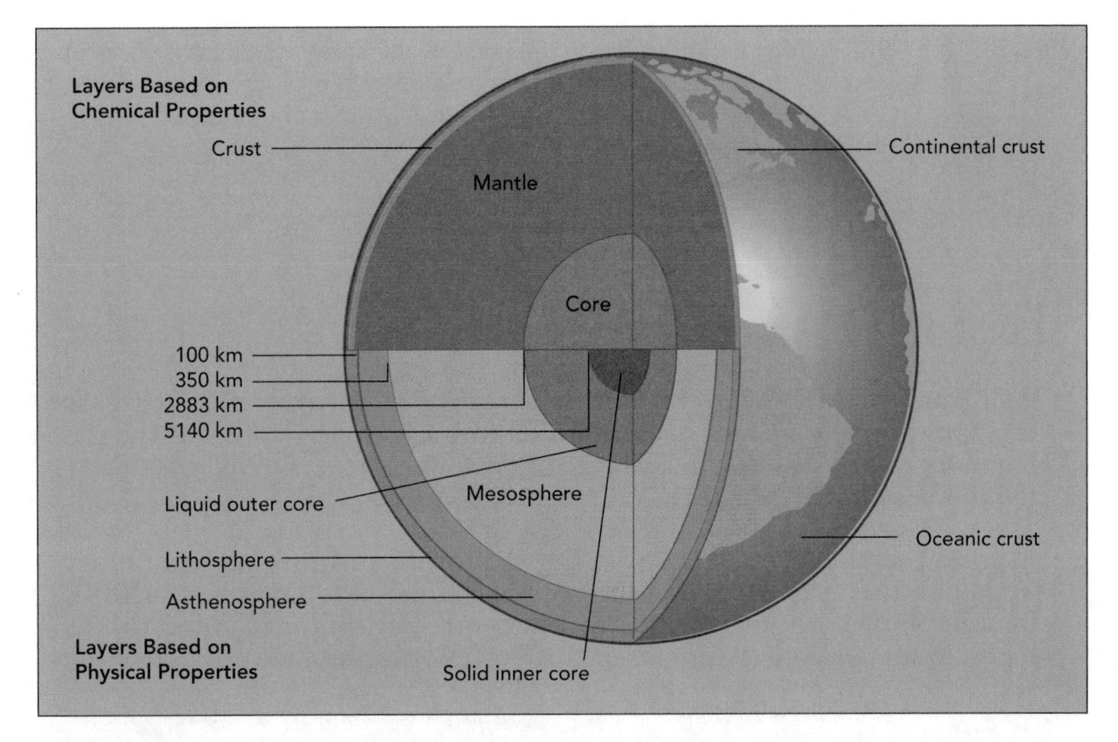

Figure 4.4 The Layered Interior of the Earth

A newly identified physical layer of the earth's interior is the **asthenosphere**, which only began to appear on diagrams of the earth's interior in the 1970s. Its existence was suggested by the need to account for the newly discovered movements of the earth's crust, then called continental drift. Scientists believe the asthenosphere is in a plastic state, part liquid and part solid. If this is not so, it would be difficult to explain the movements of the continents!

The outermost layer of the earth can be defined using either chemical or physical properties. Chemically, it consists of the thin layer (6-70 km) of the lightest materials in the earth and is called the crust. It is proportionally thinner than the skin of an apple or peach. Even this thin layer is differentiated: the deeper ocean floors are mainly composed of basalt, a silicate of iron and magnesium. The continents are mainly composed of somewhat less dense aluminum silicates in the form of granitic rocks. As a result, these lighter rocks "float" higher than the oceanic rocks, elevating the continental blocks. Physically, it is known as the lithosphere and is made up of the hard, rigid outer layer of the earth, approximately 100 km in depth, which consists of a dozen rigid plates that "float" on the underlying, more plastic layer called the asthenosphere. It is important to note that the terms "crust" and "lithosphere" are not interchangeable.

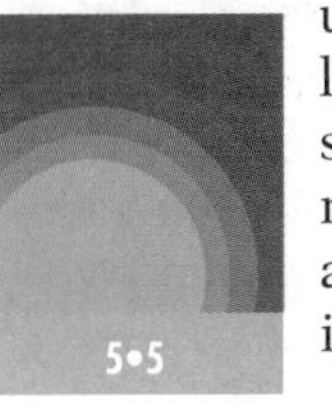

Still lighter is the **hydrosphere** — the oceans. This layer is one of the unique features of earth among the planets of the solar system. Nowhere else in the solar system does water exist in a liquid state in such abundance.

It's a Fact...

- Heat from the earth's interior normally results in a temperature increase of approximately 1°C for each 30 to 60 m increase in depth, for the top 5 km or so. This is known as the geothermal gradient. Below this depth, the increase in temperature is not nearly so rapid.
- In recent studies by scientists at the California Institute of Technology (Caltech), the temperature of the core is estimated to be at least 4500°C for the outer core and 6600°C at the centre of the inner core. This is higher than previous estimates and 1000°C hotter than the surface of the sun!

QUESTIONS

5. What sort of vehicle would you need to travel to the centre of the earth? Explain your ideas.
6. a) Indicate the conditions necessary before the interior of the earth could separate into layers of different densities.
   b) What name is given to this process and why is it an appropriate term to use?
   c) Based on the information given in Figure 4.2, indicate which elements were most likely to contribute to the formation of the core, and which ones would make up the bulk of the mantle and crust.
7. Describe a simple experiment to illustrate the process of differentiation of the earth's interior. If you have the time and equipment, carry out the experiment to see if it actually works.
8. As you move out from the centre of the earth, the layers of materials become increasingly less dense.
   a) Draw a diagram or sketch to prove this statement is correct.
   b) What accounts for this fact?
9. Granite and basalt have a lot to do with the fact that the earth has continents and ocean basins. Explain this statement.

4·3 Uncovering the Earth's Interior

Despite all our advanced technology, we have not been able to drill much more than 16 km into the crust. This is barely enough to puncture the "skin" of the earth. We have more direct evidence about the stars in distant galaxies than we have about the interior of our own planet.

Direct evidence about the upper mantle is found only in scattered places over the surface of the earth where rocks very different from those commonly found at the surface have been thrust up by movements of the ocean floors. One place where unusual upper mantle rocks can be directly seen is in Gros Morne National Park in Newfoundland. They were thrust up above sea level during the formation of the Appalachian Mountain belt, when North America, Europe, and Africa collided in the Paleozoic Era. These rocks contain chemicals that are poisonous to most plants, and thus result in barren hills in a normally forest and tundra landscape. Figure 4.5 shows an aerial view of part of this outcrop and its barren appearance.

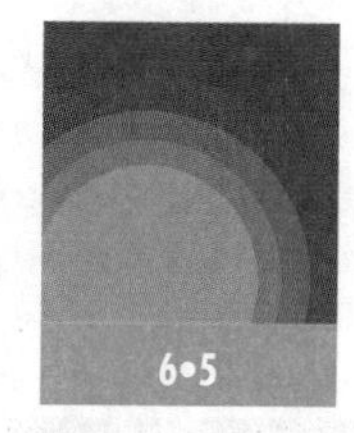

Much of what we know about the earth's interior has been derived from "indirect" evidence. We know, for example, that the average density of the earth is about 4.3 g/cm^3, whereas the average density of rocks that make up the earth's crust ranges from 2.6 to 3.0 g/cm^3. This means that the interior of the earth must differ significantly in density from the rocks at the surface. Differences in density also suggest differences in chemical composition.

Three different sources of indirect information have been used to gain an understanding of the interior of the earth. The first is the study of meteorites; the second comes from the increasingly developed science of seismology; and the third is a very recent, high-tech method known as "seismic tomography".

Meteorites: Evidence About the Interior

Meteorites are a very important source of information about the solar system and its date of origin. In addition, they provide clues about the interior of the earth, as strange as this may seem at first glance. **Meteorites** are the fragments of asteroids and small early planets that broke up on impact with other bodies out in space. It is believed that these asteroids and planetesimals formed in the

Figure 4.5
Upper Mantle Ocean Floor Rocks Exposed, Gros Morne National Park, Newfoundland

same way and from similar materials as the earth. Therefore, by studying meteorites, we can make some inferences about the interior of the earth.

Meteorites are not all the same in composition. In fact, they fit into three main divisions: iron-nickel meteorites (Figure 4.6); stony meteorites (Figure 4.7); and stony-iron meteorites. Scientists believe that the iron-nickel meteorites came from the cores of the original bodies, and the stony meteorites from areas that would correspond to the earth's mantle. The stony-iron meteorites would have originated between the cores and mantles of the asteroids and planetesimals. The densities and characteristics of the meteorites, then, provide clues about the interior of the earth.

Seismology: Key to the Earth's Interior

Seismology is the scientific study of earthquakes, the seismic waves they generate, and the passage of these waves through the earth's interior. It is largely through this science that much of what is presently believed to be true about the earth's interior has been determined.

When an earthquake occurs, the energy it releases causes seismic waves to move outward from its **focus**, or centre, in all directions. The ripples caused by dropping a rock into a still pond are similar to such waves. Seismic waves are detected by seismographs (Figure 4.8), sensitive instruments positioned hundreds and even thousands of kilometres from an earthquake.

Figure 4.6 **An Iron-Nickel Meteorite** Composition: 90% iron, 9% nickel, with traces of other metals

Figure 4.7
A Stony Meteorite
Similar composition to the earth's mantle, with little metal content

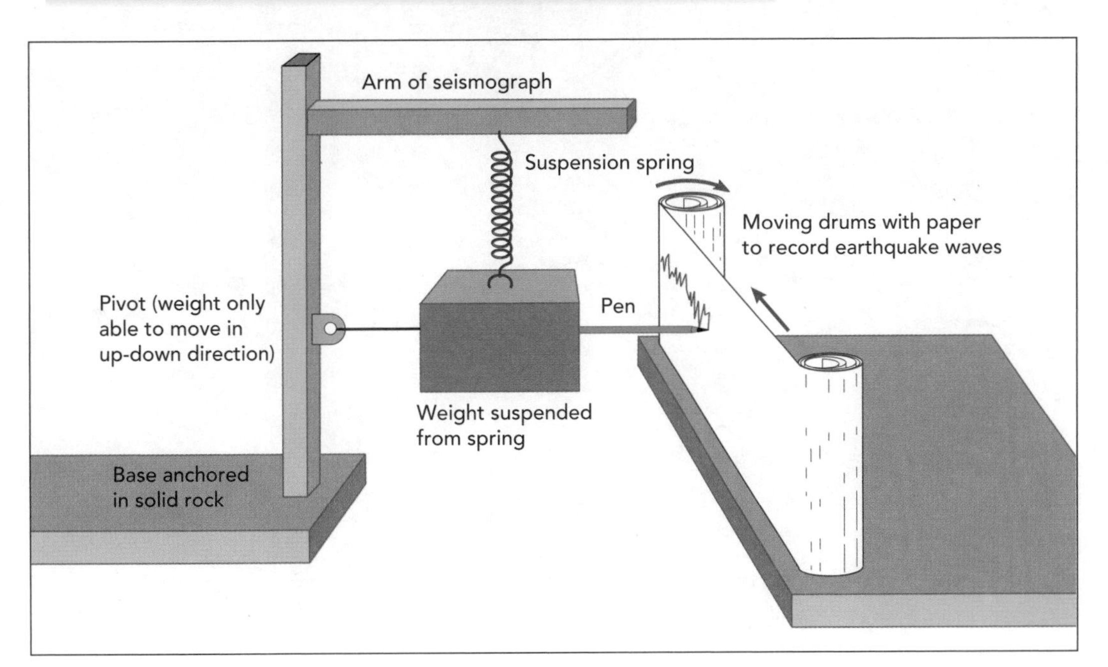

Figure 4.8 **A Simple Seismograph** The weight is isolated from the movements of the earth by the spring, so it tends to remain stationary, while the earth around it moves during a tremor. It is the paper that moves, not the pen.

Three main types of waves are generated by earthquakes: **P** or **primary waves**, **S** or **secondary waves**, and **L** or **long waves** (also known as Love waves). P waves, the fastest of the waves, can penetrate the earth's interior. They are known as compressional waves since they alternately push (compress) and pull (dilate) the rocks through which they pass. S waves are slower, transverse waves that travel through the rock by moving it from side to side (also known as shearing the rock). Because of the type of motion, P waves will travel through both solids and liquids, while S waves can only pass through solids. At some distances from the focus of an earthquake, only P waves are recorded, leading earth scientists to believe that the outer core of the earth is in a liquid state. The areas where S waves are not recorded by seismographs are shown on Figure 4.11. This diagram also shows how seismic waves are bent, or refracted, as they move across the boundaries between the layers with differing densities. This bending creates two shadow zones, where neither P nor S waves are recorded by seismographs. L waves travel on the surface of the earth and shake the rocks sideways as they progress across the surface. Only P and S waves are of great use in finding out about the earth's interior.

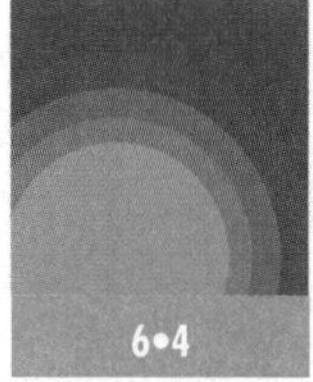

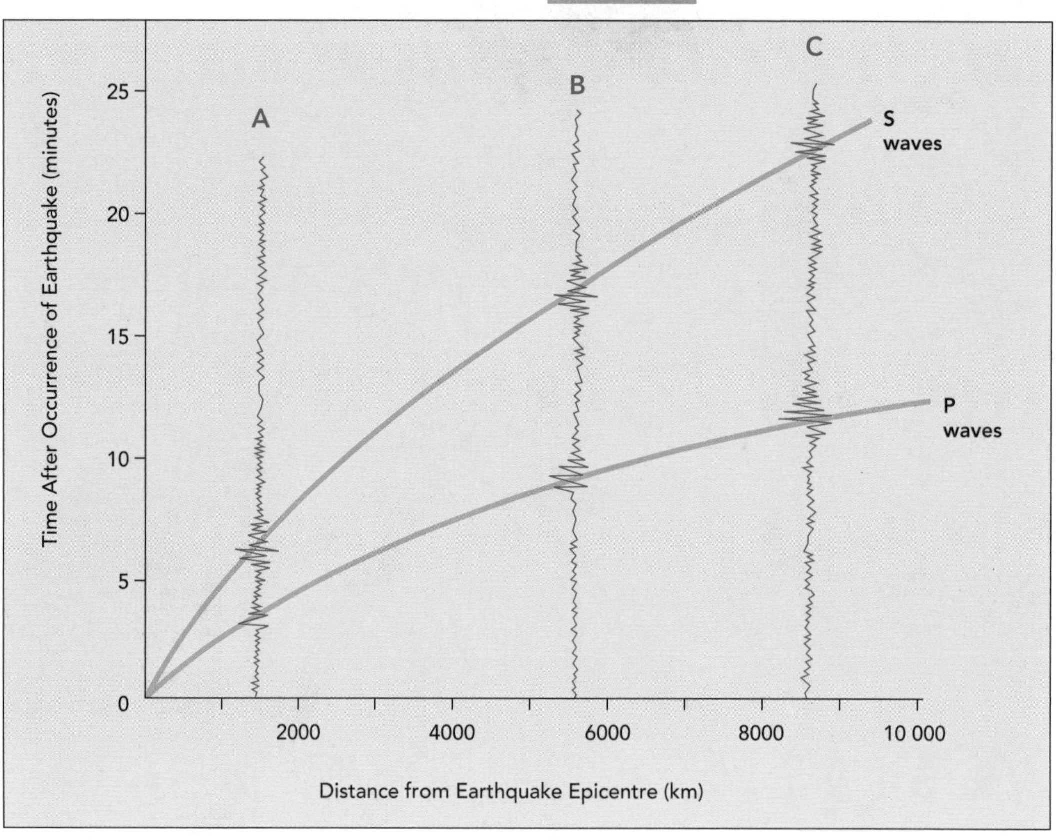

Figure 4.9 Arrival Times of P and S Waves at Various Seismic Stations

Figure 4.9 shows the arrival times of different seismic waves from an earthquake. By measuring the time difference between the arrival of P and S waves, the distance from the seismic station to the earthquake's **epicentre** (the point on the earth's surface directly above the focus) can be determined. The actual location of the earthquake can be plotted on a map using distance calculations from at least three different seismic stations. Earth scientists use the seismic information obtained from earthquakes to determine the nature of the interior of the earth. The speed of the seismic waves gives an indication of density since the denser the material, the faster the wave travels. In addition, as shock waves pass from zones or layers of different densities, they are refracted; that is, the waves change direction, as illustrated in Figure 4.11.

Seismic Tomography: A "Window" Into the Earth's Interior

Seismic tomography is the latest technique used to uncover greater detail about variations in the density and temperature

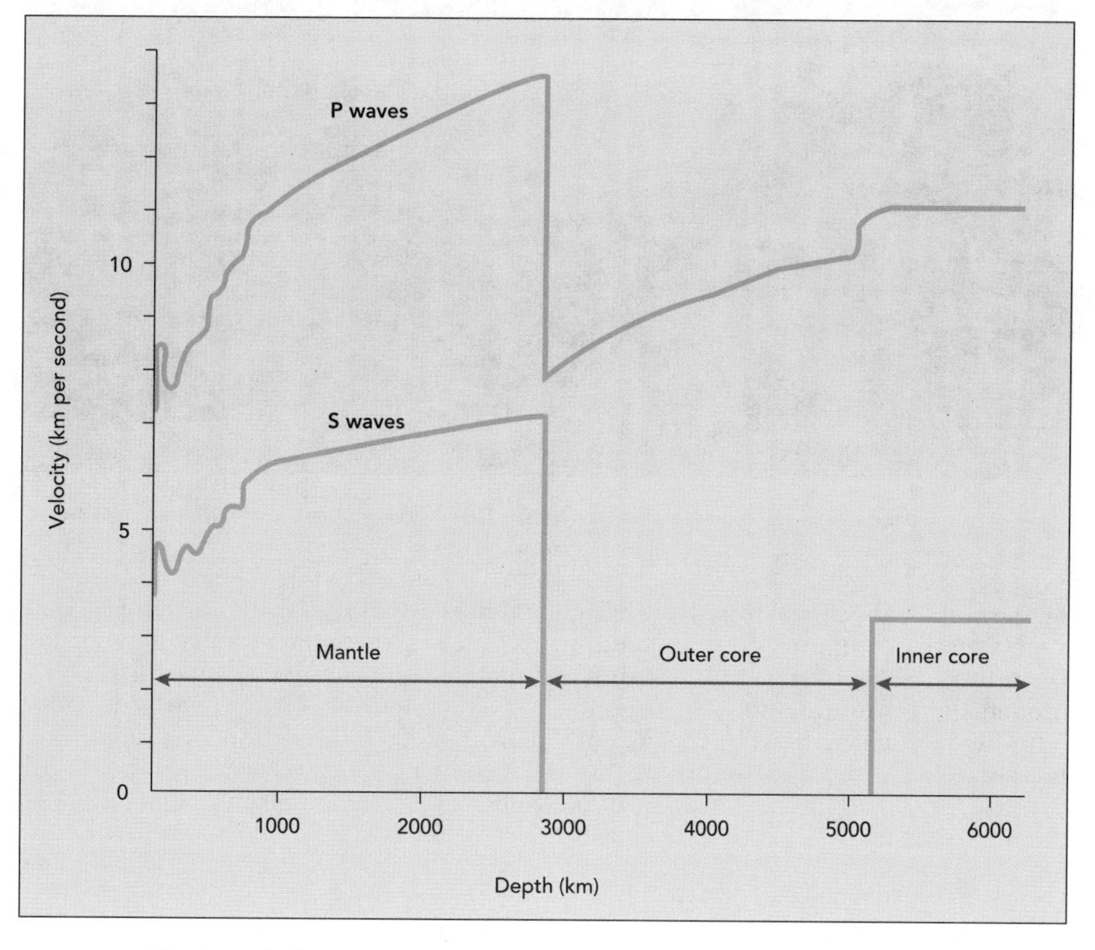

Figure 4.10 The Speed of Seismic Waves in the Earth's Interior

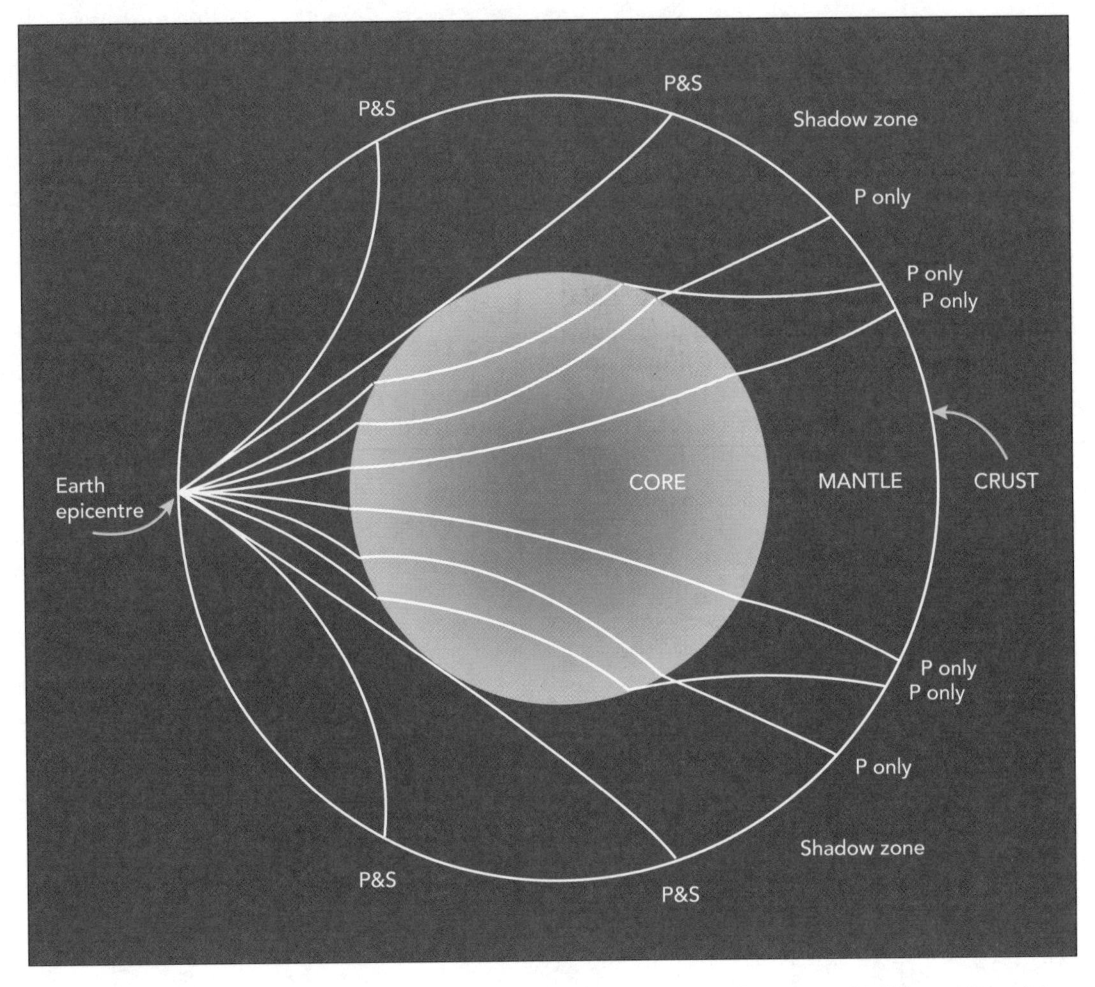

Figure 4.11 **Refraction of Earthquake Waves as They Pass Through Layers of Different Densities**

of the upper and lower mantle. Using sophisticated computer programs, data from hundreds of seismic stations around the world are analysed to point out where these variations in density and temperature occur. They point out, for example, areas of hot, light mantle rocks ("hot spots") as well as cooler, denser mantle. The computers locate such areas by calculating where earthquake or seismic waves either slow down (in hotter, less dense rocks) or speed up (in cooler, denser areas) in the mantle. The end result is a three-dimensional view of the mantle, as illustrated by Figure 4.12.

This method is in the early development stages. A worldwide digital seismic station network is proposed that will allow even more sophisticated analyses of earthquake waves. This, in turn, will lead to an even more complete understanding of the processes at work in the earth's interior.

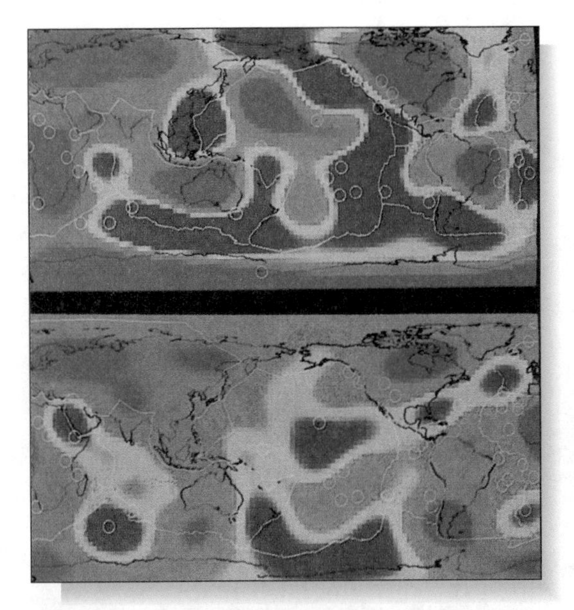

Figure 4.12 **Seismic Tomography** A CAT Scan of the Temperature Variations in the Earth's Interior

The Importance of the Interior Heat Engine

The interior of the earth seems very remote from our daily lives, except, of course, for people who live near volcanoes or hot springs. Otherwise, we are generally unaware of its impact upon us.

Nevertheless, life on earth has been profoundly influenced by the interior and its basic characteristics. Take magnetism, for example. The **magnetic field** of the earth, shown in Figure 4.13, is generated by movements of the molten iron and nickel layer of the outer core. Convection currents in the electrically conducting fluid of the core act as a dynamo, generating and maintaining a magnetic field. Without this dynamo effect, the magnetic field would die out within 10 000 years or so. Physicists are still debating exactly how this magnetic field is actually created, but it is known that the magnetic field has existed over at least the last 3.5 billion years of the earth's history. How would compasses be useful without a magnetic field?

We see the effects of the interior of the earth in other ways, as well. The heat generated by radioactive decay powers very slow moving convection currents in the asthenosphere and mantle. These currents in the asthenosphere cause the movements of the huge, rigid plates that make up the lithosphere, movements that are responsible for the formation of the mountain ranges, deep sea trenches, volcanic belts,

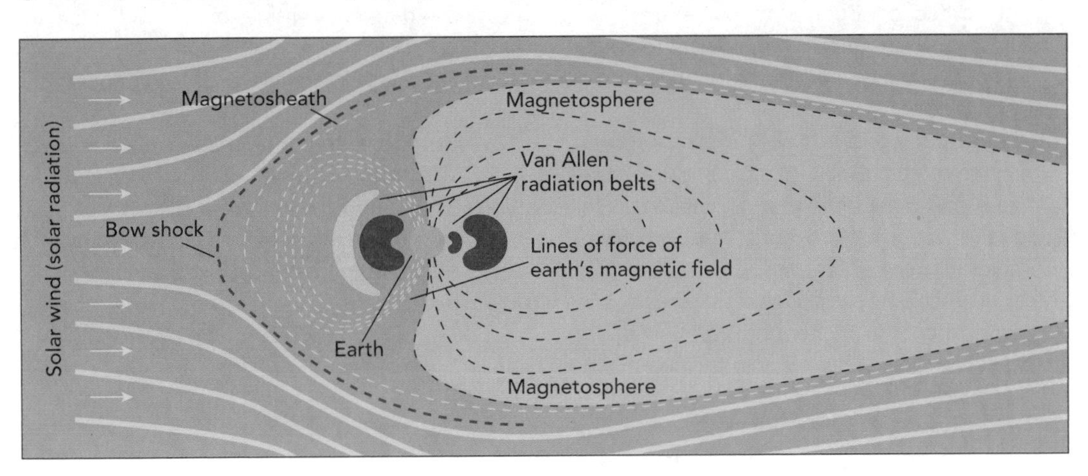

Figure 4.13 The Earth's Magnetosphere and Magnetic Field, Including the Van Allen Radiation Belts

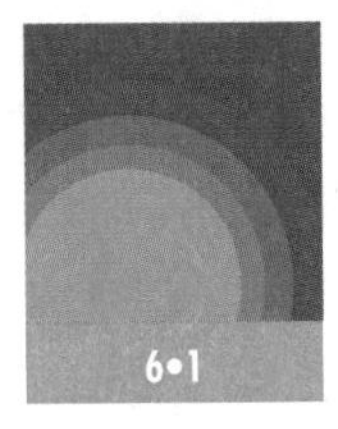

and earthquake zones of the lithosphere. The convection currents might also be associated with long-term climatic changes, as there is a positive correlation between the amount of carbon dioxide in the atmosphere and periods of rapid plate movements.

It's a Fact...

The northern lights, or aurora borealis, that shimmer in polar skies are closely related to the earth's core. The aurora is caused by the interaction of the solar wind, made up of electrically charged particles from the sun, and the earth's magnetic field, which is generated by the core. The solar wind is generated by solar flares on the surface of the sun.

QUESTIONS

10. a) List ways in which you encounter various types of waves in your daily life.
    b) In what ways are such waves captured, recorded, or displayed for your use?
    c) How are earthquake waves similar to these waves and how are they captured and recorded?
11. Draw a series of diagrams or sketches showing the complete life history of a typical iron-nickel meteorite from its origin until it falls to earth in your backyard. Put at least two time indications on the sketches.
12. In a two-column chart, identify sources of evidence about the interior of the earth and what each source reveals about the earth's interior.
13. Using Figure 4.10, give reasons for the great change in the velocities of seismic waves at a depth of approximately 3000 km.
14. Explain how earthquake waves could be used to:
    a) find different densities of rock within the earth's interior;
    b) indicate whether a layer of the interior is in a solid or liquid state.
15. Seismic tomography has been likened to a CAT scan of the human body.
    a) Explain how the two are similar.
    b) How might the information obtained by seismic tomography be used for human benefit?

Review

- The earth grew in size in its period of formation because its gravitational attraction pulled in solid materials from its vicinity.
- Heat from several sources, but most importantly the decay of radioactive elements, caused the planet to develop a hot interior.
- The molten state of the earth's interior allowed layers of different densities and compositions to develop; the core has the greatest density and the crust the least.
- Meteorites and earthquake waves provide indirect evidence about the physical and chemical properties of the earth's interior.
- Seismic activity and magnetism are two ways the earth's interior has an impact on human activities.

Geographic Terms

accretion
mantle
silicate
crust
lithosphere
core
mesosphere
asthenosphere
hydrosphere
meteorite
seismology
earthquake focus
P or primary wave
S or secondary wave
L or long wave
epicentre
seismic tomography
magnetic field

Explorations

1. Draw a labelled sketch or compile a collage of the ways in which the earth's internal heat engine can be tapped for human use.
2. In what ways would energy from the earth's heat engine differ from that derived from fossil fuels such as coal, oil, or natural gas?
3. Write an editorial, or letter to the editor of your local newspaper, indicating the benefits of developing such energy sources over using fossil fuels.
4. Review the chapter for ways in which the interior of the earth has an impact on your life, or might have such an impact in the future.
5. Each of the layers of earth, from the top of the atmosphere to the core, contributes something to the general nature of the planet. Draw up a chart to summarize the contribution(s) of each layer, as far as information in this chapter is concerned. You might wish to refer to Figure 7.4 on page 123 to identify the layers of the atmosphere and their contributions.
6. The differentiation of the heavier and lighter elements into separate layers is an ongoing process on Planet Earth.
   a) What evidence justifies this statement?
   b) If differentiation were complete, explain how your life would be different today.

CHAPTER

5

THE EARTH'S CRUST

OBJECTIVES:

By the end of this chapter, you will be able to:

- recognize that the materials of the crust move slowly through the rock cycle;
- divide the materials of the earth's crust into useful categories;
- understand the different characteristics of rocks in the ocean basins and on the continents;
- identify the characteristics and conditions of origin of igneous, sedimentary, and metamorphic rocks;
- understand the basic relationship between mineral deposits and geological structure;
- know the basic categories for grouping minerals of economic importance.

CHAPTER 1: The Nature of Physical Geography
CHAPTER 2: Earth: Its Place in the Universe
CHAPTER 3: The Earth in Motion
CHAPTER 4: The Earth's Interior
CHAPTER 5: The Earth's Crust
CHAPTER 6: The Lithosphere in Motion: Plate Tectonics
CHAPTER 7: Solar Radiation
CHAPTER 8: Climate
CHAPTER 9: Weather
CHAPTER 10: The Hydrosphere and the Hydrologic Cycle
CHAPTER 11: Natural Vegetation and Soil Systems
CHAPTER 12: Denudation: Weathering and Mass Wasting
CHAPTER 13: Distinctive Landscapes: Humid and Arid Environments
CHAPTER 14: Distinctive Landscapes: Glacial, Periglacial, and Coastal Environments
CHAPTER 15: Natural Hazards: Disrupting Human Systems
CHAPTER 16: The Disruption of Natural Systems
CHAPTER 17: Fragile Environments

Introduction

The earth's surface provides the ingredients from which life is formed and is the stage upon which the drama of life is played out. It is composed of solid rock materials that are lighter and cooler than the materials of the earth's interior. Its surface is divided into lower-lying ocean basins and the more elevated blocks of the continents. Over 70 percent of its surface is covered by the hydrosphere, which completely covers the ocean basins and washes over the lowest-lying margins of the continents. Its rocks hold the clues that tell the story of its multi-billion year past and reveal the processes that were responsible for their formation. Its ever-changing surface is pushed and pulled by forces powered by energy from the interior and moulded and reshaped by erosive agents powered by the radiant energy of the sun. This chapter focuses on the origin and makeup of the chemically distinct, solid surface layer known as the earth's crust.

5·1 The Minerals and Rocks of the Earth's Crust

If you had a giant chemistry set and attempted to duplicate the minerals that make up the crust of the earth, you would not need a great number of ingredients. Figure 5.1 lists those ingredients that account for over 99 percent of the rocks of the crust. By far the most important elements are oxygen and silicon. **Elements** are materials that cannot be subdivided by ordinary chemical methods into simpler materials. There are 91 naturally occurring elements in the rocks that make up the earth. When two or more elements combine in a crystalline structure, they form **minerals**. The 91 elements combine to form at least 2000 different minerals.

Element	Percentage by Mass
Oxygen	45.2
Silicon	27.2
Aluminum	8.0
Iron	5.8
Calcium	5.1
Magnesium	2.8
Sodium	2.3
Potassium	1.7
Titanium	0.9
Hydrogen	0.14

Figure 5.1 The Average Composition of the Earth's Crust

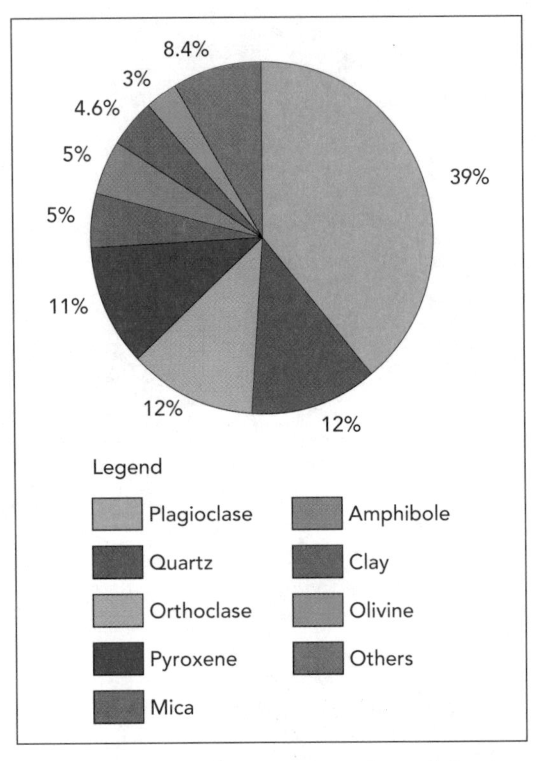

Figure 5.2 The Common Minerals: Building Blocks of the Crust

A **rock** is any consolidated mixture of one or more minerals. Oxygen (O) and silicon (Si) combine to form the silica tetrahedron (SiO_4), the basic building block of the earth's crust. Figure 5.2 shows the most common minerals. Oxygen and silicon are common to almost all of these minerals. Even clay and the minerals in the "others" category are largely constructed from the SiO_4 tetrahedron. The variety of rocks is only limited by the various combinations of minerals that occur in the earth.

The composition of crustal rocks gives us a good deal of information about both the crust and interior of the earth. The composition of the rocks is "locked" in as they form. By analysing the crystalline structure and physical properties of rocks, we can determine something about the processes that formed them as well as their age.

As scientists gained more information about the ages of rocks on the earth's surface, a puzzle emerged. The rocks were simply not as old as the earth. Some rocks were very old, even more than 3 billion years, while others were very recent. No rocks were close to the estimated 4.6 billion year age of the earth.

Where did all the old rocks of the earth's surface go? The answer comes from the nature of the earth's interior. In Chapter 4, we saw that the decay of radioactive elements in the interior of the earth drives convection currents in the outer core and mantle. These currents put pressure on the crust, which causes the rocks of the ocean floors to move. In other places, portions of the crust disappear into the mantle at deep trenches on the ocean floor. This constant recycling of the ocean crust means there are almost no ocean-floor rocks older than 150 to 200 million years. Since the continental parts of the crust are made of less dense materials than the ocean floors, the continents are not recycled into the mantle. The rocks of the continents are, therefore, older.

A second part of the answer lies in the unique feature of the earth: its hydrosphere. No other body in our solar system has so much as a lake-sized deposit of water on its surface. The movements of water and ice play a major role in the eroding of older rocks and the deposition of materials to form newer rocks. The constant erosion and deposition helps account for the youthful age of many of the rocks of the earth's crust.

The process of replacing older rocks with new ones is known as the **rock cycle**, which is summarized in Figure 5.3. Three rock classes, igneous, sedimentary, and metamorphic, are identified in the cycle. The percentage of the earth's crust made up of these rock classes is shown in Figure 5.4.

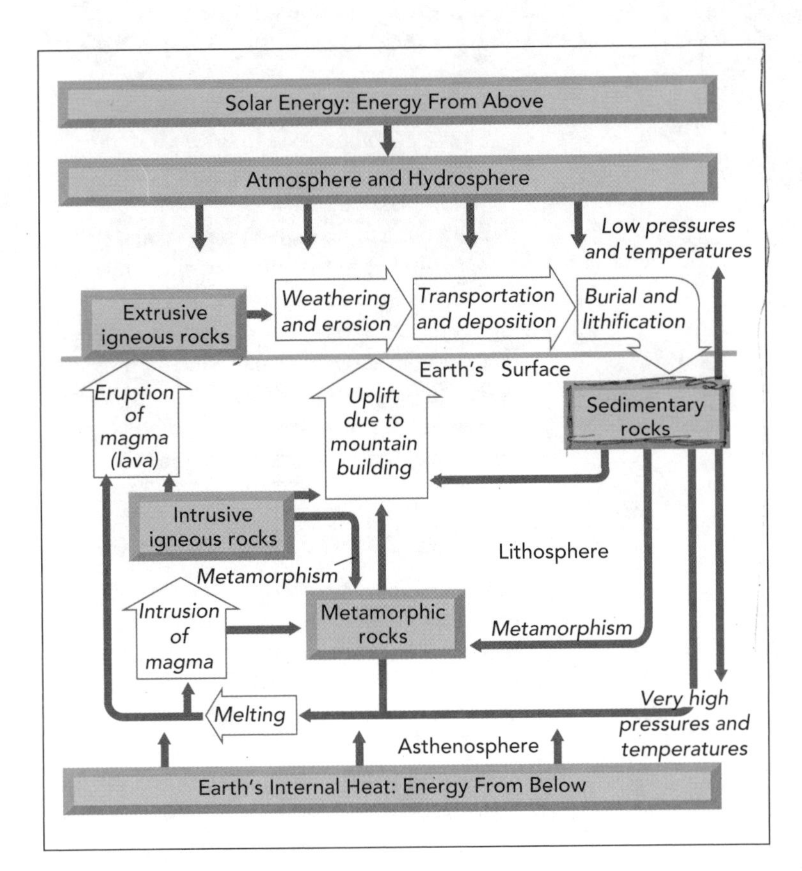

Figure 5.3
The Rock Cycle: Driven by Energy From Above and Below

It's a Fact...

The earth and the moon were formed at approximately the same time. However, earth scientists have not discovered any rocks dating back to their origin 4.6 billion years ago. The oldest known minerals, from Australia, are 4.3 billion years old. The oldest rocks in the world are 3.96 billion years old; they were discovered in the Northwest Territories of Canada. On the other hand, the moon has few young rocks. The rocks brought back from the moon by the Apollo astronauts date back 4.5 billion years. There are few moon rocks less than 3 billion years old.

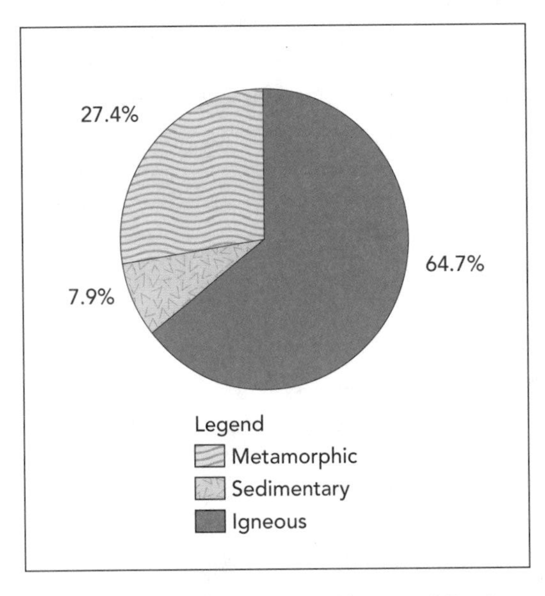

Figure 5.4 The Three Major Classes of Rocks Making Up the Earth's Crust

QUESTIONS

1. Make a sketch or chart to show the differences between elements, minerals, and rocks.
2. a) Use the data in Figure 5.1 to draw a pie graph showing the average composition of the earth's crust.
   b) Use the data for the top eight elements in Figure 4.2 on page 53 to draw a second pie graph to show the average composition of the whole earth.
   c) Point out and explain the significant differences between the composition of the crust and the earth as a whole.
3. In what ways is the rock cycle powered by energy from above (the sun) and energy from below (the interior of the earth)?
4. Give examples of events that could occur on a day-to-day basis that are part of the rock cycle.

5·2 Born of Fire: Igneous Rocks

Igneous rocks are formed by the cooling and solidification of molten materials, or **magma**, found beneath the earth's surface. These rocks make up the largest percentage of the rocks of the crust (Figure 5.4). The term **igneous** comes from the Latin word meaning "fire", and refers to the origin of such rocks in volcanic eruptions common in southern Italy. The energy that is involved in the formation of such rocks comes almost entirely from the earth's interior.

Physical geographers most frequently classify igneous rocks according to the location of the magma as it solidified into rock. The two major divisions within this classification system are **extrusive**, or **volcanic**, rocks and **intrusive**, or **plutonic**, rocks. Extrusive igneous rocks cool on the surface, while intrusive igneous rocks cool below the surface of the earth.

Extrusive Igneous Rocks

Rapid cooling occurs when molten magma is erupted onto the earth's surface and exposed to the atmosphere. Magma which has reached, or been erupted onto, the earth's surface is known as **lava**. Heat is given off quickly and the lava solidifies into hard rock. At times, smooth volcanic glass, or **obsidian**, is formed (Figure 5.6). Sometimes the lava cools so quickly that even gases do not escape before the lava solidifies, forming a rock with many holes (Figure 5.7). Pumice, an extrusive igneous rock, is filled with so many holes that it floats!

Intrusive Igneous Rocks

Where magmas cool slowly below the earth's surface, the igneous rocks are much different in appearance. Slow cooling allows the various elements to gather together to form crystals of pure minerals that are large enough to be visible to the eye. Thus, any igneous rock with large crystals is probably a member of the intrusive category. Granite is the most common intrusive igneous rock (Figure 5.8).

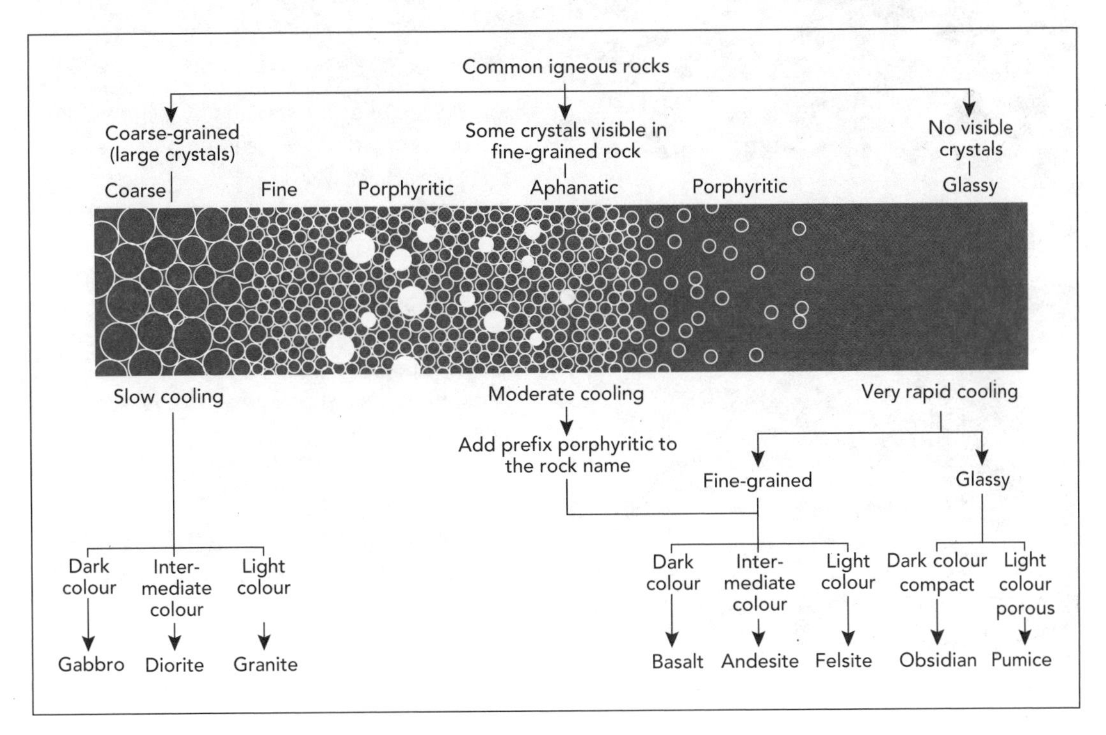

Figure 5.5 Flow Diagram for Identifying Igneous Rocks

Figure 5.6 **Obsidian** Rapid cooling formed volcanic glass.

Figure 5.7 **Gas Holes in Extrusive Igneous Rock** Fast cooling trapped the gases.

Figure 5.8 **Granite Rock: An Igneous Intrusive Rock**

Metallic Mineral Deposits

Metallic mineral deposits are commonly associated with intrusive igneous rock formations. Many of the metals that are of most use to us are relatively rare within the earth's crust. They are usually found in trace amounts in igneous rocks. However, they do occur in concentrations where magma has cooled slowly. In such situations, the metals, dissolved in superheated brines or liquids, are deposited as veins that are sometimes large or rich enough in metallic minerals to be profitably mined, as illustrated in Figure 5.9. For example, many of the copper, lead, zinc, silver, gold, iron, and nickel deposits of the Canadian Shield and Western Cordillera of North America are closely associated with intrusive rock areas.

QUESTIONS

5. a) Why are people fascinated by pictures and stories describing volcanic eruptions, such as those in Hawaii?
   b) Describe your own feelings on seeing a volcanic eruption.
6. Why is the term "born of fire" a suitable title for this section on igneous rocks?
7. Prepare a simple chart to summarize the different characteristics or features that could be used to distinguish extrusive from intrusive igneous rocks.

5·3 Born of Erosion and Deposition: Sedimentary Rocks

Unlike igneous rocks, the origins of sedimentary rocks are usually rather quiet and not very spectacular. The energy for the creation of sedimentary rock comes largely from the sun, rather than from

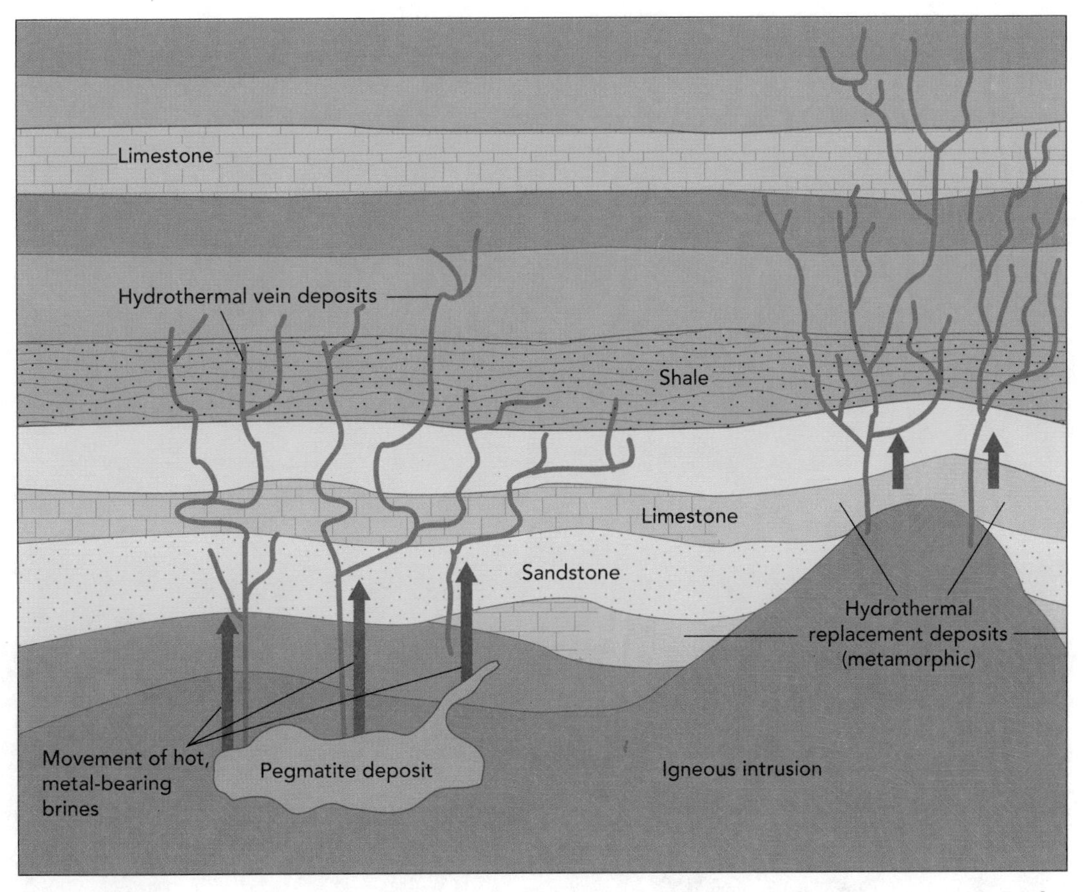

Figure 5.9 The Formation of Metallic Mineral Deposits in Association With Igneous Intrusive Rocks

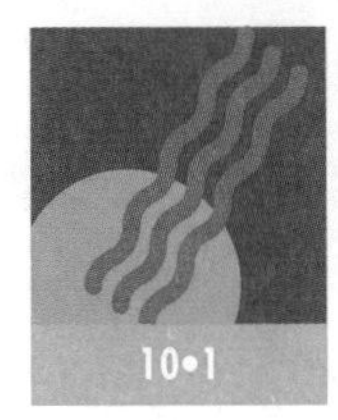

the earth's interior, as with igneous rocks. The creation of sedimentary rocks is closely related to the movements of water and the hydrological cycle.

Sedimentary rocks are found over much of the surface areas of the continents and continental shelves. However, many areas of level sedimentary rocks are only skin deep. They overlie much thicker layers of igneous and metamorphic rocks.

As their name indicates, **sedimentary rocks** are formed from **sediments**, particles of rock materials that have been transported to new locations. These sediments have various origins and sizes and are classified into two types. Inorganic sediments, such as gravels, sands, silts, and clays, are carried by the world's rivers into lakes, seas, and oceans. These sediments are produced by the breaking down of rocks, and are known as **clastic** materials or sediments. Clastic sediments range in size from silt to boulders. **Nonclastic** chemical or organic sediments are precipitated as solids from ocean waters or are made up of animal or plant remains. Figure 5.11 shows the materials from which three common sedimentary rocks are made.

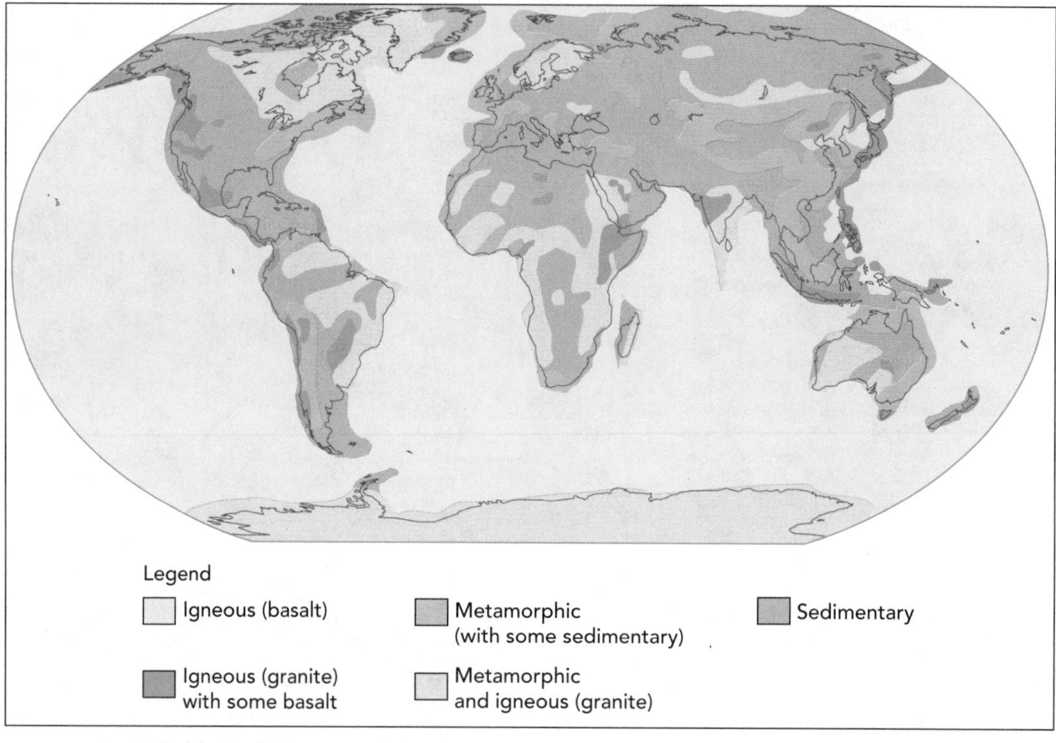

Figure 5.10 **World: Rock Types**

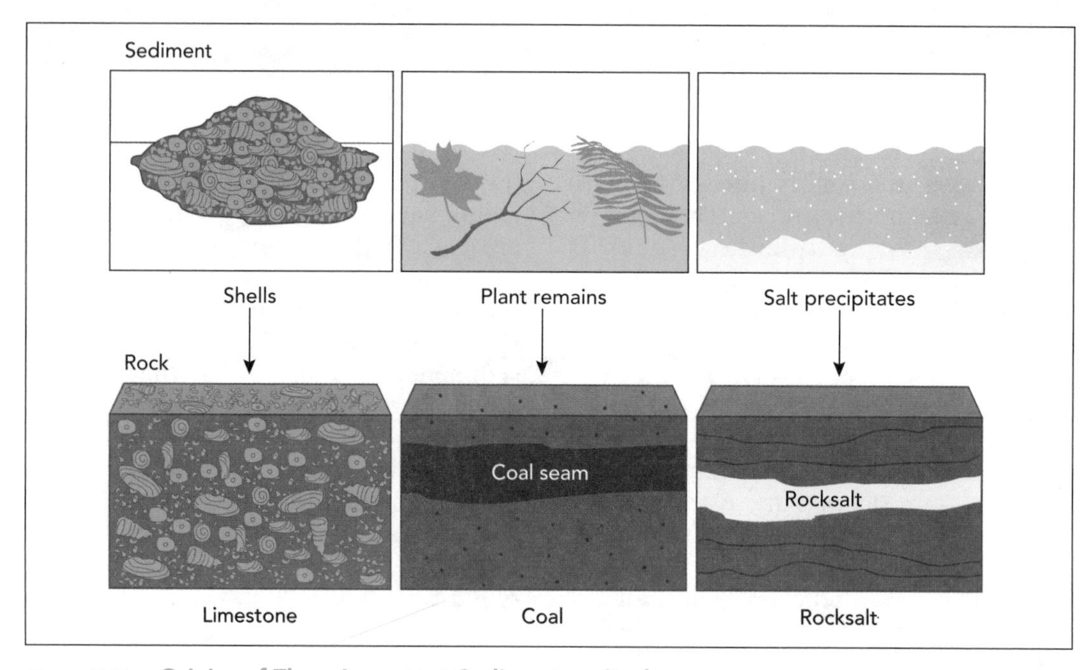

Figure 5.11 **Origins of Three Important Sedimentary Rocks**

Clastic Sedimentary Rocks

As clastic sediments settle out of relatively calm waters, they build up horizontal layers on the floors of these lakes or oceans. These sediments are eventually turned into clastic sedimentary rocks by two main processes: the first is pressure from more recent, overlying layers of sediments, and the second, and more important, process is the cementing together of the sediments by various chemicals, such as calcium carbonate, that seep into the layers after they are deposited on the sea floor. The process of compacting and cementing of sediments is called **lithification**.

Today, clastic sedimentary rocks are slowly forming in many parts of the world. However, the most productive rock-forming areas occur where major rivers deposit sediments into the ocean, forming huge deltas. The mouths of rivers such as the Mississippi, Nile, Rhine, Ganges, Yangtze, Huanghe, Amazon, Niger, Zaire, and Mackenzie are particularly productive environments.

Sedimentary rocks are also formed as streams deposit sediments along their courses, such as at the bases of mountain ranges or in areas of interior drainage. Under certain desert conditions, winds deposit large amounts of very fine wind-blown sediments that can eventually be turned into sedimentary rocks.

Non-Clastic Sedimentary Rocks

Non-clastic sediments originate from the remains of plants and animals (usually the skeletons or shells) that accumulate on the floors of oceans and seas, or from the chemical precipitation of minerals from seawater. Although they take much longer to build up, these sediments are the main ingredients in the formation of such common non-clastic sedimentary rocks as limestone and dolomite. The Great Barrier Reef of Australia and the reefs of the Caribbean Sea and Pacific Ocean are areas where limestone rocks are presently being formed from the remains of sea creatures and coral reefs.

a) Limestone

b) Shale

c) Sandstone

d) Coal

Figure 5.12 **Typical Sedimentary Rock Types** It is estimated that shale accounts for 53 percent of all sedimentary rocks of the earth's crust, while limestone (24 percent), sandstone (22 percent), and others (1 percent) make up the rest. The layers within sedimentary rocks are called strata.

Fossil Fuel Deposits

Deposits of **fossil fuels** are found in areas of sedimentary rock formation. Various forms of plants and animals lived and died in the swamps and poorly drained areas of coastal and deltaic regions of the past. Since there was a shortage of oxygen in the muds where the remains of these plants and animals were deposited, they did not decay. Rather, they accumulated in the mud as layers of thick plant remains or as small globules of animal remains (organic fluids). Later deposits of sands, silts, and clays buried these organic remains. As the sediments were consolidated into rock, the organic materials were transformed into valuable fossil fuel deposits, specifically, coal (plant remains), oil (organic fluids), and gas (organic gases). The coal layers, known as seams, were turned from organic material into coal with little movement of the mineral taking place.

Natural gas and oil deposits formed in a different way. Drops of oil and bubbles of gas were widely scattered through the source rocks. Only when squeezed out of the **source rocks** and concentrated in **traps** or **reservoirs** do these two minerals become economically useful as fossil fuels. Some common types of oil and gas traps are shown in Figure 5.13. Each of these traps has a common set of characteristics: a source rock from which the oil and/or gas are squeezed, a porous, permeable **reservoir rock** (usually a sandstone or coral reef formation), and an impermeable **cap rock** that is able to hold in the gas and oil. Groundwater forces the lighter gas and oil to rise to the highest point in the trap. When such traps are punctured by drilling equipment, the built-up pressure within the trap causes a "gusher" of oil or gas to pour out of the well.

Most large oil and gas deposits are found in areas where originally flat-lying sedimentary rocks were gently folded or

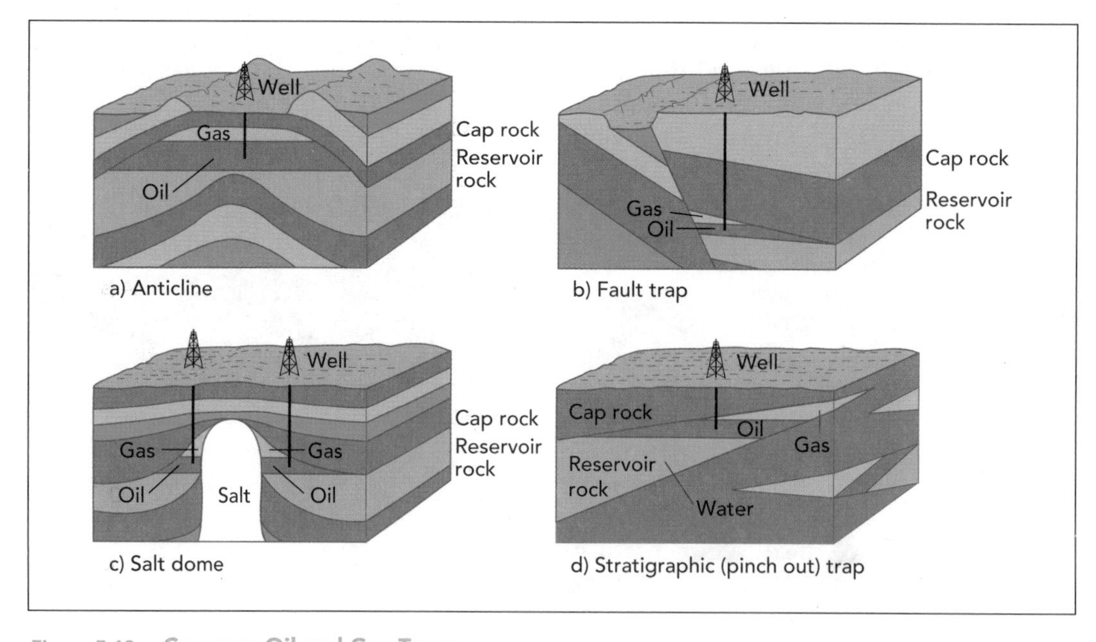

Figure 5.13 Common Oil and Gas Traps

faulted to create geological traps. Examples of such areas are the foothills of the Rocky Mountains of Canada and the Persian Gulf. Traps are also found where **salt domes** occur. Such traps are commonly located along the Gulf Coast of the United States and in Iran. The third type of trap is in fossilized coral reef formations that contain many pores created by the coral polyps. The oil and gas can easily seep into and concentrate in such rocks. Oil bearing reefs occur in Alberta and Ontario.

Non-Metallic Industrial Mineral Deposits

Because of their creation through deposition, minerals in sedimentary rocks are non-metallic. They include a wide variety of salts that were deposited as shallow seas evaporated from basins within hot desert areas. The basins often had small inlets from the sea so water flowed in and was continuously evaporated, leaving thick beds of salts behind. Although they are not as valuable as fossil fuels, these minerals are vital. Fortunately, they are very abundant in the sedimentary rocks of the world, far more so than the rapidly depleting supplies of oil and gas. Of the major non-metallic industrial minerals, only a few, such as asbestos, are not found within sedimentary rocks.

Rock Type	Minerals
Sedimentary Rocks	Gypsum Rock Salt Calcium Chloride Sodium Chloride Phosphate Potash
Igneous and Metamorphic Rocks	Asbestos Talc

Figure 5.14 **Common Non-Metallic, Industrial Minerals**

It's a Fact...

The salt on your dinner table is produced from a soluble sedimentary rock, known as halite. Chemically, it is called sodium chloride (NaCl). Rock salt is deposited when shallow seas are evaporated, usually under desert conditions.

QUESTIONS

8. a) List the ways in which minerals derived from sedimentary rocks play a part in your life.
   b) Describe two or three major effects you would notice if these minerals were unavailable for your use for the rest of the day.
9. a) Using the figures given in the caption to Figure 5.12, draw a pie graph to show the percentages of the major sedimentary rocks of the earth's crust.
   b) Suggest the names of sedimentary rocks that would make up the 1 percent in the "other" category.
10. a) The deltas of the following rivers are locations where major deposits of clastic sedimentary rocks are being formed:
    - Mississippi
    - Colorado
    - Mackenzie
    - Volga
    - Amazon
    - Po
    - Rhone
    - Rhine
    - Nile
    - Niger
    - Danube
    - Indus
    - Mekong
    - Zaire
    - Huanghe
    - Yangtze
    - Fraser
    - Irriwaddy
    - Tigris-Euphrates
    - Lena
    - Ganges-Brahmaputra
    - Syr-Darya/Amu Darya
    - Sacramento-San Joaquin

    Mark eight of these deltas and the ocean or sea areas adjacent to them on a world map.
    b) One method of classifying rocks is to place them in the geological period in which they form. Using the geologic time scale shown in Figure 1.6 on page 11, give the name of the period into which currently forming rocks would be placed.
11. If you were a geologist attempting to discover places where oil and gas deposits are found in the world, what key conditions would you use to select possible sites for your drilling crews?
12. Explain why sedimentary rocks usually do not contain metallic minerals.
13. While non-metallic industrial minerals are widely distributed throughout the world, fossil fuels are not. Identify three political, economic, and environmental consequences of this fact.

5·4 Born of Great Heat and Pressure: Metamorphic Rocks

The name of the **metamorphic rock** class comes from the word "metamorphosis", meaning change. Rocks that have been greatly altered from their original forms through tremendous heat and pressure fit into this category. Metamorphosis takes place while the rocks are still in the solid state. If the rocks melt, and then resolidify, they are classified as igneous rocks.

We cannot directly observe metamorphic rocks forming, as with some igneous and sedimentary rocks. In most cases, the tremendous heat and pressures that change igneous or sedimentary rocks into metamorphic rocks are found deep below the earth's surface where mountain ranges are forming. It is only when the overlying materials are eroded away that the metamorphic rocks formed millions of years ago become visible. It was just this situation that occurred in the Canadian Shield, where the roots of ancient mountain ranges are now exposed at the earth's surface.

Metamorphic rocks form where the earth's crustal plates collide, pushing up new mountain ranges, such as the Himalayas, Andes, Rocky Mountains, and Alps. They also are formed where massive amounts of hot magma move up and through rocks of the crust, along converging plate boundaries or above hot spots in the earth's mantle. Such areas are found around the Pacific Rim of Fire, in Iceland, and in Yellowstone Park.

Some common metamorphic rocks, and the original rocks that were transformed by heat and pressure to make them, are shown in Figure 5.16. Under different amounts of heat and pressure, different metamorphic rocks are produced from the same original rocks. The term **"foliated"** is used to describe the banded structure caused by the gathering together of different minerals into parallel bands. This is a common characteristic of many metamorphic rocks. As a rule, metamorphic rocks are extremely compact and very hard to break up, a reflection of their origins in the roots of mountains or next to intrusions of hot magma deep below the earth's surface.

The banding seen in foliated metamorphic rocks should not be confused with the layering found in sedimentary rocks. In the case of the metamorphic rocks, this banding is a result of the rearrangement of specific crystals or minerals, while the rock was under heat and pressure. The rock does not readily break along such bands. Sedimentary rocks do, however, tend to break along their layers.

a) Gneiss

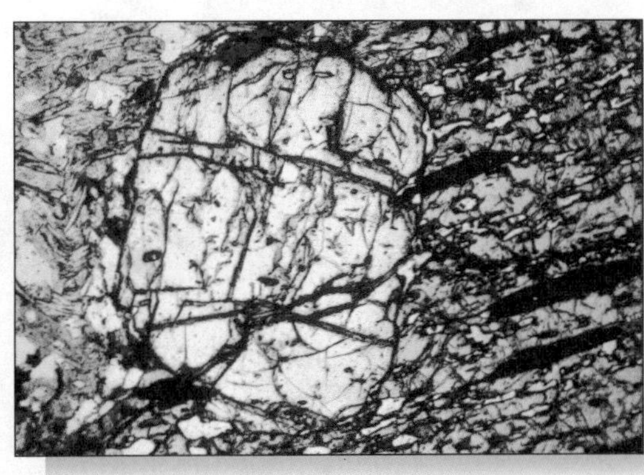

b) Schist

c) Slate

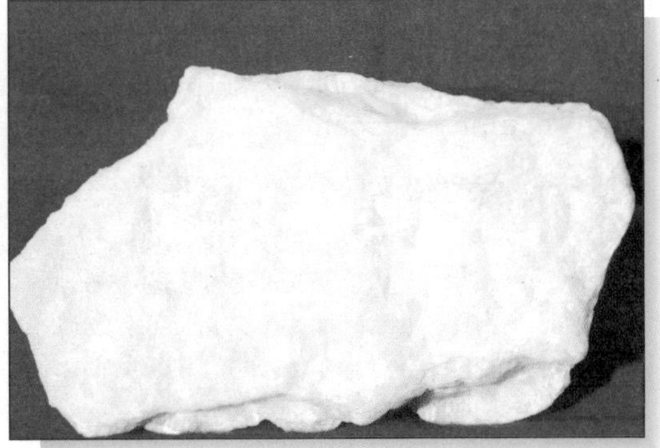

d) Marble

Figure 5.15 **Common Metamorphic Rock Samples** Gneiss (pronounced "nice") accounts for 77 percent of all metamorphic rocks, while the remainder is made up of schist (18 percent), marble (3 percent), and others (2 percent).

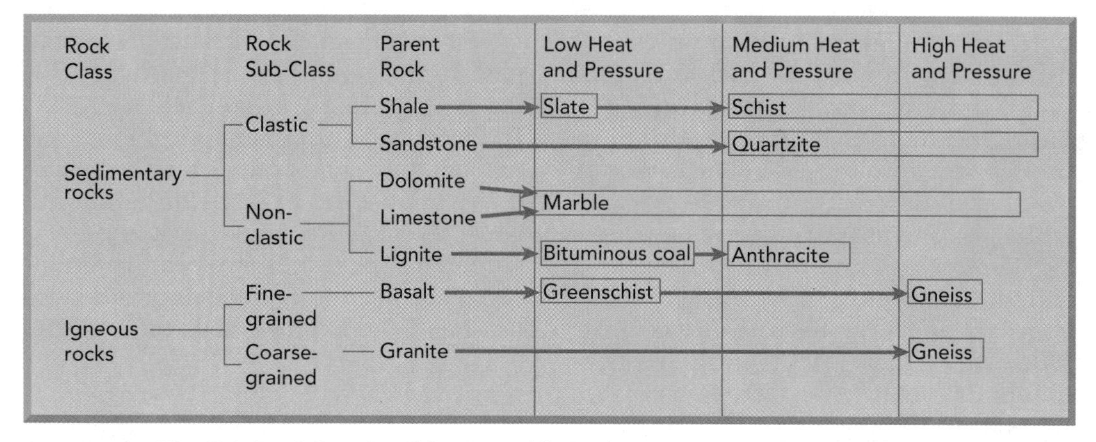

Figure 5.16 **The Origin of Common Metamorphic Rocks**

Uses of Metamorphic Rocks

Slate, marble, and gneiss are metamorphic rocks useful for building purposes, as durable exterior surfaces, for monuments, or for attractive effects inside buildings. Metamorphic rocks seldom contain oil, natural gas, or coal deposits because such organic materials would be burned up or squeezed out of the rocks as they formed. On the other hand, metals are found in metamorphic rocks, formed from igneous rocks that originally contained such minerals. This situation is common in the Canadian Shield, where both igneous and metamorphic rocks are found together in many complex formations.

QUESTIONS

14. a) Give examples of where heat and/or pressure change things in your daily life. One example is the changes that result when an egg is fried on the stove.
    b) How would such examples compare with the heat and pressure needed to change a sedimentary or igneous rock into a metamorphic rock?
    c) Devise a model to illustrate how heat and pressure metamorphose rocks.
15. a) Using the percentages given in the caption to Figure 5.15, draw a pie graph to show the major types of metamorphic rocks found within the earth's crust.
    b) Name at least three metamorphic rocks that would be included in the "other" category.
16. Explain why most of the metamorphic rocks being formed on the earth today lie below mountain ranges such as the Alps, the Himalayas, and the Andes.
17. Point out how it is possible to tell the difference between sedimentary and metamorphic rocks, even though both appear to be "layered".

5·5 Oceans and Continents

The division of the earth's materials into the three major rock classes (igneous, sedimentary, and metamorphic) is only one way to consider them. A second approach is to divide rocks into only two divisions: those found on the continents and those on the ocean floors, as shown in Figure 5.17. These two groups of rocks have different densities and generally different colours.

The term **"sial"** is used to identify the granitic rocks of the continents since they are largely composed of the elements **si**licon and **al**uminum. These rocks are less dense than the basaltic rocks that underlie the oceans because they contain large amounts of light elements such as aluminum. The average density of continental rocks is 2.8 g/cm^3. The aluminum content also makes the minerals light coloured.

On the other hand, the term **"sima"** is used for the denser basaltic rocks of the ocean basins that are made up largely of **si**licon and **ma**gnesium. Iron is also present in greater amounts in these rocks. They have an average density of 3.0 g/cm^3. These basaltic rocks are generally darker in colour than the granitic continental rocks.

Theory of Isostasy: Balance Between Crust and Mantle

The difference in density between granitic and basaltic rocks is the key to the development of continents and ocean basins. It is believed that the various blocks of the crust float on the softer materials of the asthenosphere. The principle of displacement developed by Archimedes states that an object floating in a liquid displaces a weight of the liquid equal to its own weight. The denser ocean floor rocks must sink more deeply into the plastic asthenosphere to displace their own weight than the lighter rocks of the continents. This is called the theory of **isostasy**.

The theory can be visualized using a common example. Two logs are floating in a lake. If one log is a dense tropical hardwood, it will float more deeply in the water than a lighter softwood of pine or spruce. In the same way, the denser rocks of the oceans sink more deeply, creating the ocean basins. Using Archimedes' principle, the denser ocean floor rocks

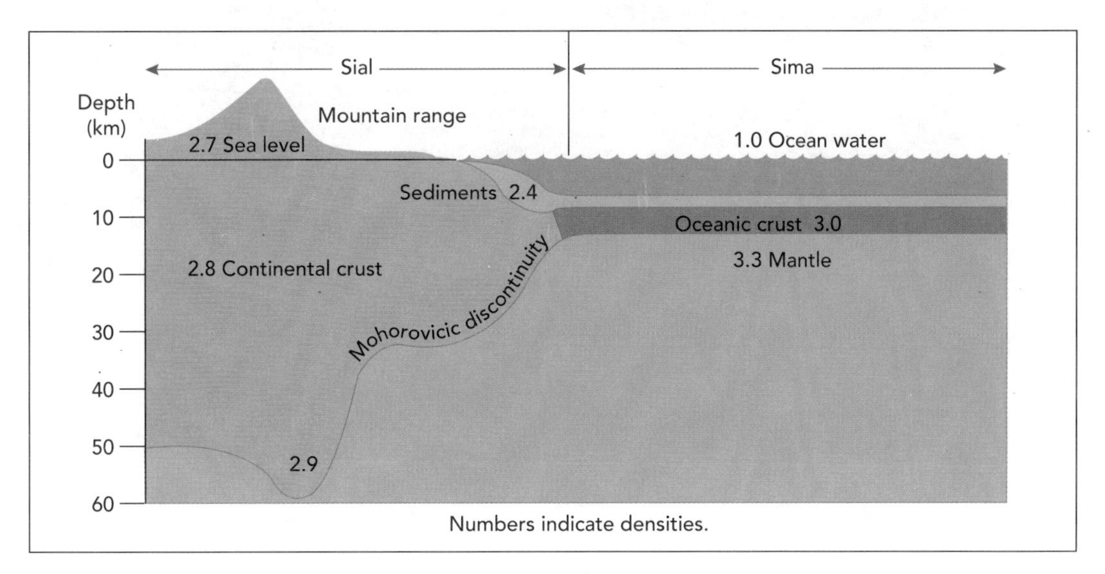

Figure 5.17 Sial and Sima: Continents and Ocean Floors

sink more deeply into the plastic asthenosphere to displace their own weight than the lighter rocks of the continents, as illustrated in Figure 5.18.

The present coastlines between the oceans and continents do not mark the boundary line between continental and ocean floor rocks. This boundary lies along the edges of the continental shelves where they slope steeply into the ocean basins. The shelves, shown in Figure 5.19, are areas of thinner, lower-lying granitic rocks that have been inundated by oceanic waters.

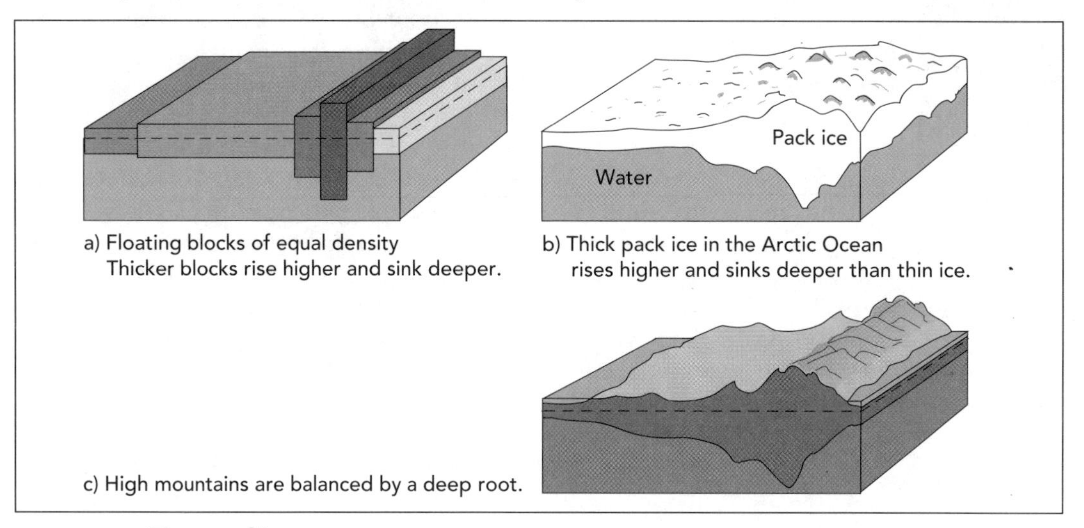

Figure 5.18 **Theory of Isostasy**

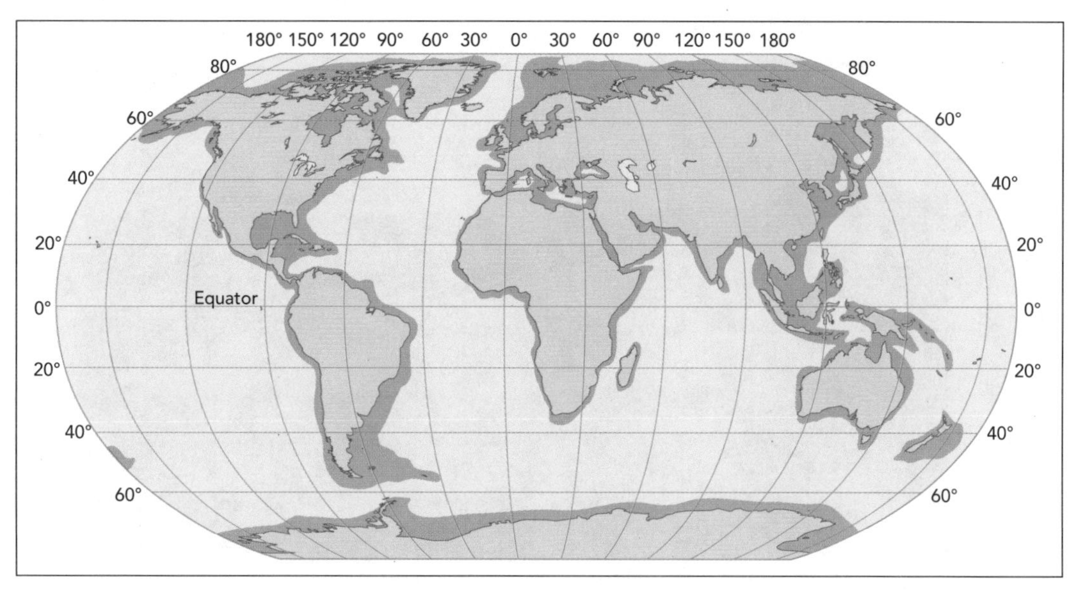

Figure 5.19 **The Continental Shelves of the World** The continental shelves and slopes (shown in purple) together form about 25 percent of the continents.

Differences Between Ocean Basins and Continents

A fundamental difference between the oceans and continents is in the ages of the rocks that make them up. The ocean basins are made up of rocks that are less than 200 million years old. On the other hand, the oldest rocks known on the continents date back as far as 3.96 billion years. This is a result of the processes of sea-floor spreading and plate convergence considered in more detail on pages 92-96 and 103-110.

Another difference between the rocks of the ocean basins and the continents relates to their origins. The rocks of the ocean basins are almost entirely made up of basalt, an extrusive igneous rock. Only near the edges of the continents are these extrusive rocks covered with a thin veneer of sediments. The continents, on the other hand, are made up of complicated mixtures of intrusive and extrusive igneous, sedimentary, and metamorphic rocks of various ages and types. This complexity is a result of the variety of processes at work on the surface, particularly those related to mountain building and plate tectonics.

QUESTIONS

18. a) Describe what would happen to a small motor boat as you and a large group of friends got on board, one at a time.
    b) What would happen when you get out of the boat, one at a time?
    c) Relate this experience to the theory of isostasy described above.
19. Point out how the validity of isostasy as applied to the crust is proof that the upper mantle is in a plastic state that allows it to move slowly.
20. Describe the earth's surface if only one type of rock were to exist in the crust, either light granitic or denser basaltic rocks. Put your description in the form of an account written by an alien from a distant solar system.

Review

- Oxygen and silicon combine to create the silica tetrahedron, the basic building block of the earth's crust.
- Rocks are combinations of naturally occurring minerals.
- The rock cycle summarizes the processes that create igneous, sedimentary, and metamorphic rocks.
- Igneous rocks are formed from magma and may be either intrusive or extrusive based on where, and how quickly, they cool.
- Metallic mineral deposits are frequently associated with igneous rocks.
- Sedimentary rocks are classified as either clastic or non-clastic.
- Fossil fuels are almost exclusively found in sedimentary rocks.
- Metamorphic rocks are created by the great heat and pressure associated with mountain building and volcanic activity.
- The rocks of the ocean floors are denser, younger, and darker in colour than continental rocks.
- The different densities of oceanic and continental rocks explain why the continents "float" higher on the soft materials of the mantle than oceanic rocks.

Geographic Terms

element
mineral
rock
rock cycle
magma
igneous
extrusive
volcanic rock
intrusive
plutonic rock
lava
obsidian
sedimentary rock
sediment
clastic
non-clastic
lithification
fossil fuel
source rock
trap or reservoir
reservoir rock
cap rock
salt dome
metamorphic rock
foliated
sial
sima
isostasy

Explorations

1. Create a collage to illustrate the importance of each of the three major rock classes.
2. a) Based on the three major rock classes, design a chart to point out the visible differences between rocks which could be used in placing individual rock samples into one of the rock classes. Divide the igneous rock class into its two subclasses.
   b) Use the rock classification chart developed above to identify a set of rock samples collected from the field or available in your school.
3. In a group, design a model to demonstrate to a class of Grade 7 students the reasons why we have continents and ocean basins on the face of the earth. If possible, visit a class and give your demonstration. Evaluate its effectiveness in a follow-up report.

CHAPTER

6

THE LITHOSPHERE IN MOTION: PLATE TECTONICS

OBJECTIVES:

By the end of this chapter, you will be able to:

- appreciate the power and scope of tectonic processes and their effects;
- appreciate the slowness of tectonic processes based on a human time scale;
- understand that the lithosphere is an ever-changing part of a dynamic planet;
- understand the general pattern of tectonic activity over geologic time and explain the location pattern of tectonic activity over the earth's surface;
- explain the tectonic processes that shape the earth's surface, including folding, faulting, and volcanic activity;
- describe and explain the pattern of major surface features created by tectonic processes;
- predict the nature and general patterns of occurrence of tectonic activities and processes, especially earthquakes and volcanic eruptions;
- describe the positive and negative aspects of tectonic activities.

CHAPTER 1: The Nature of Physical Geography
CHAPTER 2: Earth: Its Place in the Universe
CHAPTER 3: The Earth in Motion
CHAPTER 4: The Earth's Interior
CHAPTER 5: The Earth's Crust
CHAPTER 6: The Lithosphere in Motion: Plate Tectonics
CHAPTER 7: Solar Radiation
CHAPTER 8: Climate
CHAPTER 9: Weather
CHAPTER 10: The Hydrosphere and the Hydrologic Cycle
CHAPTER 11: Natural Vegetation and Soil Systems
CHAPTER 12: Denudation: Weathering and Mass Wasting
CHAPTER 13: Distinctive Landscapes: Humid and Arid Environments
CHAPTER 14: Distinctive Landscapes: Glacial, Periglacial, and Coastal Environments
CHAPTER 15: Natural Hazards: Disrupting Human Systems
CHAPTER 16: The Disruption of Natural Systems
CHAPTER 17: Fragile Environments

Introduction

Moving continents? "Impossible!" said earth scientists in the early twentieth century. "How could they move?" But many new discoveries about the earth in the 1960s, and especially about the ocean floors, led to the idea that the lithosphere was made up of large slabs of solid rock that were being created, moved, and destroyed. The theory was proposed, tested, and accepted by most earth scientists, literally revolutionizing the way we look at the surface of the earth. It explained why mountain belts, volcanoes, earthquakes, deep ocean trenches, and mid-ocean ridges existed, why they were located where they are, and how they formed. This chapter will look in detail at the theory of plate tectonics and what it has to tell us about our dynamic planet.

6•1 A Stable Earth?

Many people, when alighting from an airplane or disembarking after a long sea journey, have been overheard to say, "Thank goodness I'm back on solid earth." Or, you have probably heard the expression: "She's got her feet solidly on the ground." Both reveal a common feeling that the earth's lithosphere is a solid and firm foundation upon which humans carry out their activities. The reason for the "solid earth" feeling is very understandable.

However, earth scientists believe the rocks of the lithosphere are moving slowly but steadily across the earth's surface. These motions occur on a geological time scale of millions of years rather than on a human time scale of days, months, years, and decades.

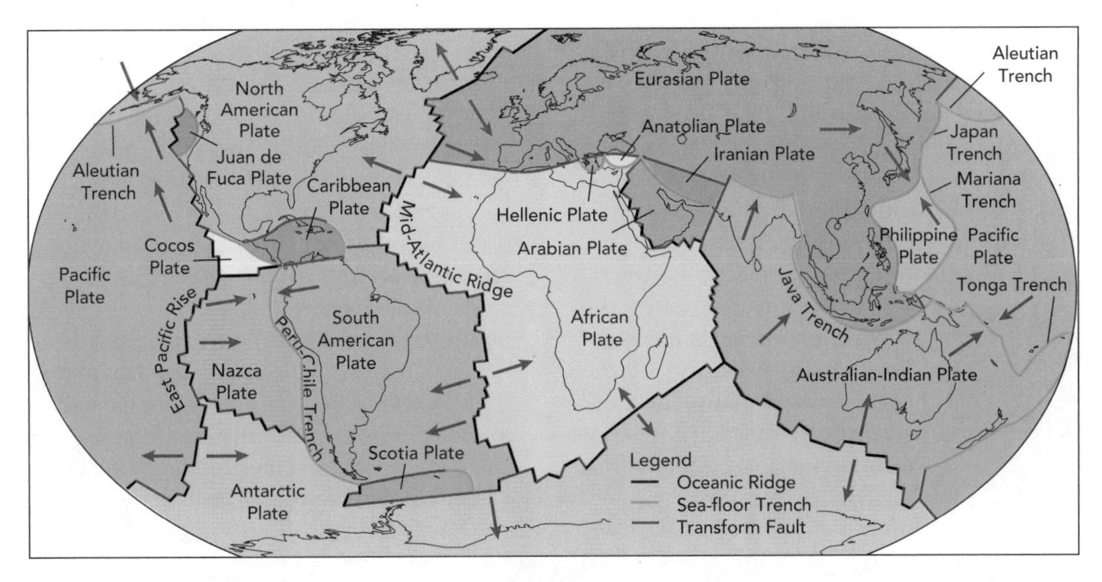

Figure 6.1 World: The Major Plates of the Lithosphere

The lithosphere does not move as a single unit, but is broken up into approximately twenty large plates (Figure 6.1) that float on the denser rocks of the asthenosphere. A **plate** is a rigid slab of solid lithosphere rock that has clearly defined boundaries or edges. Each plate is approximately 100 km thick and consists of a thinner, upper layer of crustal rock and a thicker, underlying layer of upper mantle rock. In places, continents of lighter granitic rocks ride on the surface of the plates, increasing their thickness. The plates move independently of one another, as shown by the arrows in Figure 6.1.

The **asthenosphere** is the layer of the upper mantle that lies directly below the lithosphere. Although considered a solid, it is so hot that it is made up of a small (10 percent) portion of melted materials that give it the properties of a plastic, meaning it can flow slowly when put under constant pressure.

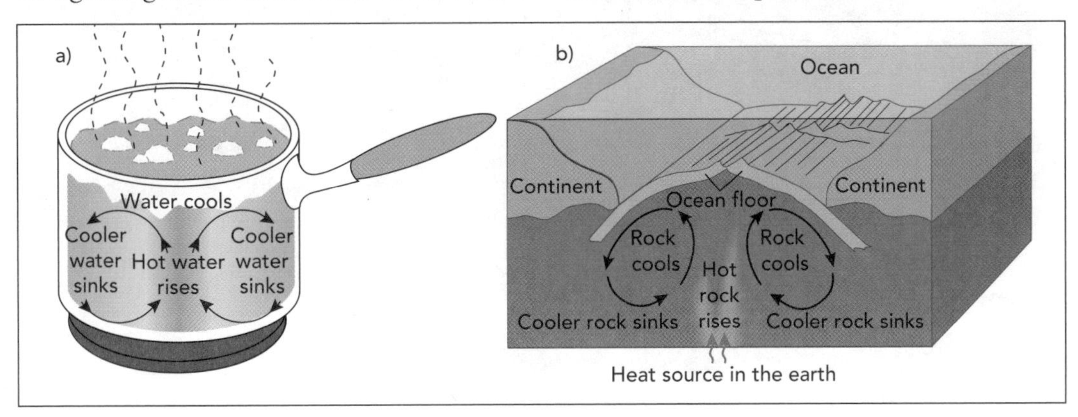

Figure 6.2 Mantle Convection Currents

The plates are set in motion by **convection currents** in the asthenosphere, as illustrated in Figure 6.2. The convection currents are similar to those set in motion when a pot of soup is heated. The hottest parts of the soup heat up, expand, and begin to rise to the surface, while the cooler soup sinks, setting up a convection current.

The energy that creates the convection currents to power the constant motion of the plates comes from the heat generated by the decay of radioactive elements within the earth's interior.

The study of the movement of plates and the effects they have on the surface features of the lithosphere is known as **plate tectonics**. **"Tectonics"** refers to the processes that change and deform the earth's lithosphere and to the rock structures and surface features that are produced by these processes.

QUESTIONS

1. Why is it important to understand the differences between a human time scale and the geological time scale when studying topics such as plate tectonics?
2. Describe an experiment that you could conduct to show what convection currents are and how they are created.
3. a) Why is the term "plate" an appropriate name to apply to the large, rigid slabs of rock that make up the earth's lithosphere?
   b) Explain why the plastic nature of the asthenosphere and the presence of convection currents are key elements in the theory of plate tectonics.

6•2 Lithosphere Plates: A Simple But Elegant Idea

The idea that the lithosphere is divided into a number of rigid plates was first proposed in 1967. The plate boundaries were based on newly available world

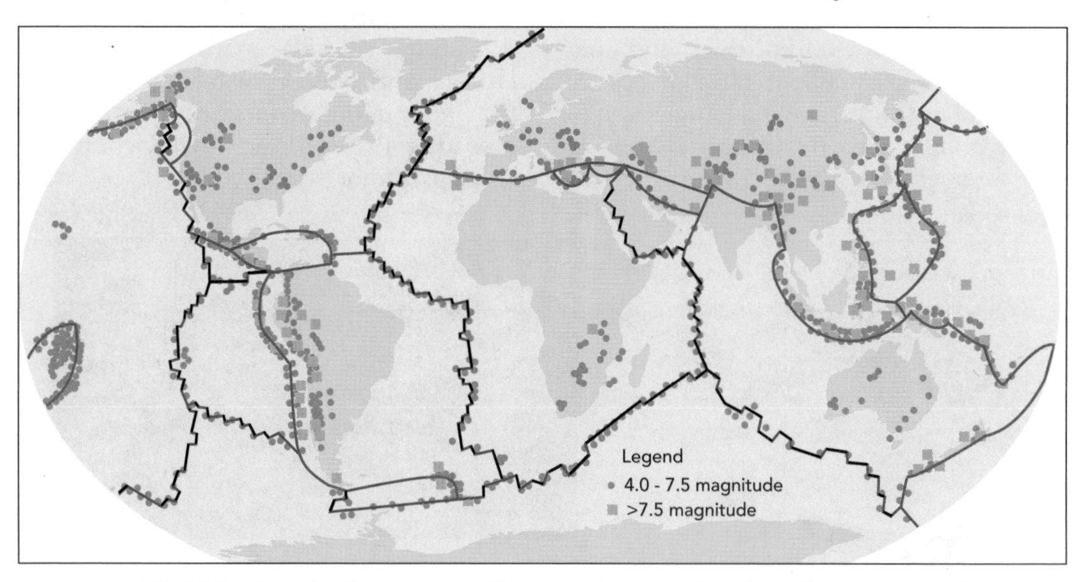

Figure 6.3 **World Earthquake Epicentres** This map shows earthquakes of magnitude 4.0 or greater from 1960 to 1989.

maps showing earthquake locations (Figure 6.3). These maps revealed that earthquakes commonly occurred in narrow, clearly defined belts that overlapped the major active mountain chains and mid-ocean mountain ridges of the world.

Earth scientists soon realized that these earthquake belts marked the boundaries of large, moving plates that were colliding, separating, and slipping against one another. The interiors of the plates were areas where quakes were infrequent. The theory of plate tectonics gave geologists powerful explanations for many facts about the earth's surface that had long defied understanding.

The Major Tectonic Processes

Plate movements produce three tectonic processes — folding, faulting, and vulcanism — that warp, buckle, and break the rocks of the lithosphere. **Folding** is the process that bends and twists rocks, usually due to compression or squeezing. It is commonly found where plates move together. **Faulting**, the process where rocks move past each other along a fracture or crack, occurs where plates are separating, sliding past one another, or moving together. **Vulcanism** is the term used to describe the movement of molten rock, or magma, beneath or above the earth's surface. Volcanic activity is commonly found where plates are separating or coming together. All three tectonic processes combine to create major mountain chains.

Movements of the crust release energy, triggering shock waves, known as earthquakes. Where plates are moving together, large amounts of energy are released, resulting in major earthquakes. Where plates are spreading apart, the energy released is relatively small and only minor quakes are recorded.

Plate Boundaries: Zones of Tectonic Activity

Before the theory of plate tectonics was proposed, geologists knew how, where, and when mountains were formed, but couldn't explain their patterns of location or why they formed. The plate theory suggests that mountain-building activities are located mainly along plate boundaries, where the plates bump into and jostle one another, causing earthquakes, volcanic eruptions, and the folding and faulting of rocks. The theory also explains the origin of many older, inactive mountain belts, such as the Appalachian Mountains of eastern North America and the Ural Mountains of Russia. They were formed hundreds of millions of years ago, when two continental plates were "welded" together as they collided to form a larger plate. These areas are no longer active plate boundaries and, thus, the mountain-building processes have stopped.

The only mountain-building process not always related to plate boundaries is the volcanic activity found within the interiors of plates in places such as Hawaii, Cameroon in Africa, and Yellowstone Park in the United States. These unusual volcanic areas mark the points where **hot spots**, or strong upward convection currents of hot magma in the upper mantle, are pushing up below the interior areas of plates, as shown in Figure 6.4. The heat and pressure of these currents of magma "burn a hole" through the plate, resulting in the creation of volcanic features.

Plate Boundaries

The three types of plate boundaries shown in Figure 6.5 occur because the

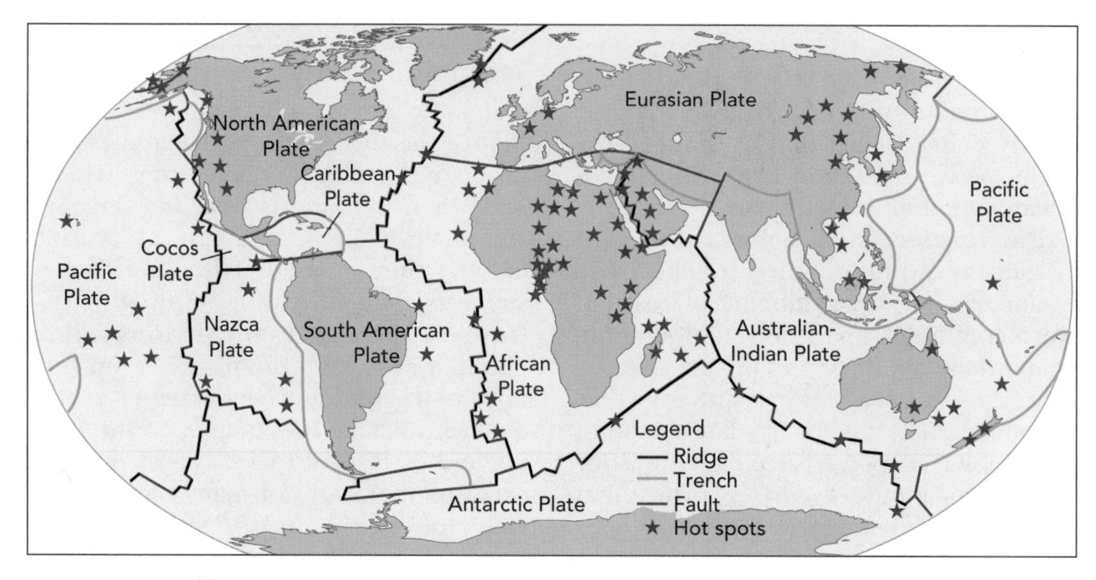

Figure 6.4 World Hot Spots

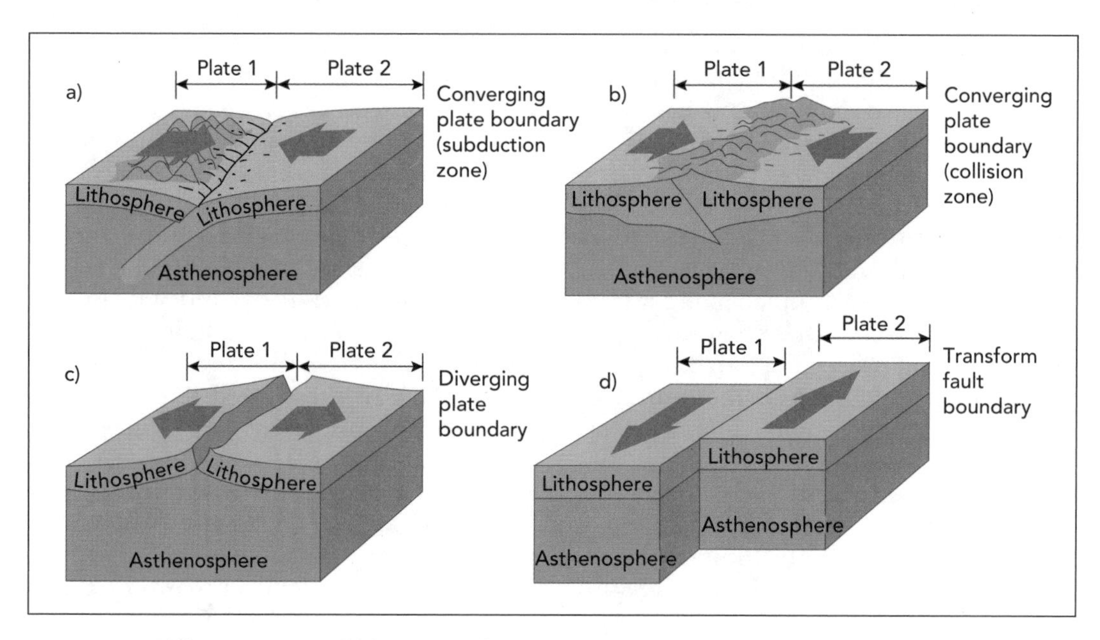

Figure 6.5 Different Types of Plate Boundaries

plates are moving in different directions and at different speeds. The boundaries include:

- **converging plate boundaries** where two plates are moving towards one another;
- **diverging plate boundaries** that develop where plates are moving apart or separating from one another;
- **transform fault boundaries** where plates are slipping and sliding past one another.

Type of Plate Boundary	Plate Actions	Major Tectonic Processes	Associated Features	Example/ Locations
Subduction (Converging) ⟶	oceanic plate descends beneath continental plate	folding faulting vulcanism major quakes	•deep oceanic trench •island or continental arc volcanoes •young, active mountain ranges	Aleutians, Japan (island arcs) Andes, Cascades (continental arcs)
Collision (Converging) ⟶	continental rocks are deformed to create complex mountain ranges	folding faulting major quakes	•young, active fold mountain ranges	Himalayas (south Asia)
Diverging	plates are moving apart at oceanic ridge, creating or widening ocean basin	vulcanism faulting minor quakes	•oceanic ridge •hot, deep ocean vents	Iceland Mid-Atlantic Ocean East Pacific Ocean
Transform Fault	plates slide past one another along transform faults	faulting major quakes	fault lines	San Andreas Fault (California)

Figure 6.6 **The Types of Plate Boundaries: A Comparison Chart**

Converging plate boundaries are of two types: **collision zones**, where two plates containing continents are meeting, and **subduction zones**, where an oceanic plate is sinking below a plate containing continents. The western and eastern edges of the Pacific plate are largely made up of converging plate boundaries. The important features associated with this type of plate boundary are summarized in Figure 6.6.

Diverging plate boundaries are located beneath the oceans of the world. Only in a few places, such as Iceland, can they be directly observed above sea level.

Transform fault boundaries occur largely beneath the oceans. Where this type of boundary does cut across continental areas, such as the San Andreas fault of California, earthquake activity is frequent and severe.

It's a Fact . . .

During the late 1950s, the United States established a network of sensitive seismic stations to detect and precisely locate the source of seismic waves. Its purpose was to detect seismic waves generated by underground atomic tests carried out by communist countries. Two side effects of this network were the ability to map the worldwide distribution of earthquakes and the discovery of the existence of lithospheric plates and the theory of plate tectonics.

QUESTIONS

4. a) Using Figure 6.1, name the plate on which your community is located.
   b) Based on Figure 6.3, is your community located in a zone of high risk earthquake and/or volcanic zone? Explain.
   c) Identify two countries that lie almost entirely within zones of:
      i) high risk earthquake and/or volcanic activity
      ii) low risk earthquake and/or volcanic activity
5. Suggest why the close relationship between earthquake locations and the mid-ocean ridges was not discovered until the 1960s.
6. Not all volcanic activity occurs along plate boundaries. Explain why.
7. a) Draw a diagram to illustrate each of the three major tectonic processes discussed on page 89.
   b) Below each diagram, indicate the name of the process and the type, or types, of plate boundary where each commonly occurs.
   c) What is the source of energy that powers these processes?
8. a) Name and briefly describe the differences between the types of plate boundaries shown in Figure 6.5.
   b) Which type of plate boundary is:
      i) found at mid-ocean ridges?
      ii) most often associated with newly forming mountain belts on the continents?
      iii) associated with the San Andreas fault of California?
      iv) related to deep ocean trenches?
      v) found closest to your community?
9. Assuming the direction of plate movements shown in Figure 6.1 continues into the future, draw a map of the world as it might appear millions of years from now.

6·3 Mid-Ocean Ridges: Diverging Plate Boundaries

A 84 000 km long mountain range is very hard to hide. Yet, the world's oceans hid its full extent until the late 1940s and 1950s. The first full explorations of this mountain range were carried out where it runs down the centre of the Atlantic Ocean. The feature was labelled a **mid-ocean ridge**. Later explorations of the Indian, Pacific, and Antarctic oceans revealed the mountain range's total extent. Studies revealed that the ridge rises one to three kilometres above the ocean floors, is 1500 to 2500 km wide, and has a steep-sided valley, up to a kilometre or more deep, marking its centre.

Initially, earth scientists could not explain why mid-ocean ridges exist, since the evidence to provide such an explanation was hidden deep below the oceans. However, new technologies, including radar imaging, remote sensing from satellites and instruments towed behind surface ships, and new deepwater research submarines or submersibles soon revealed the answers.

Sea-Floor Spreading

As the new technologies uncovered the characteristics of the deep ocean floors, a new hypothesis, known as **sea-floor spreading**, emerged. It suggested that the sea floors are splitting and moving apart at the mid-ocean ridges, powered by convection currents in the asthenosphere. As the spreading occurs, magma from the upper mantle fills the opening cracks, cools, and hardens, adding to the ocean floor on each side of the ridge (Figure 6.7d). The ridge stands above the general level of the ocean floor due to the upwelling of the convection current from

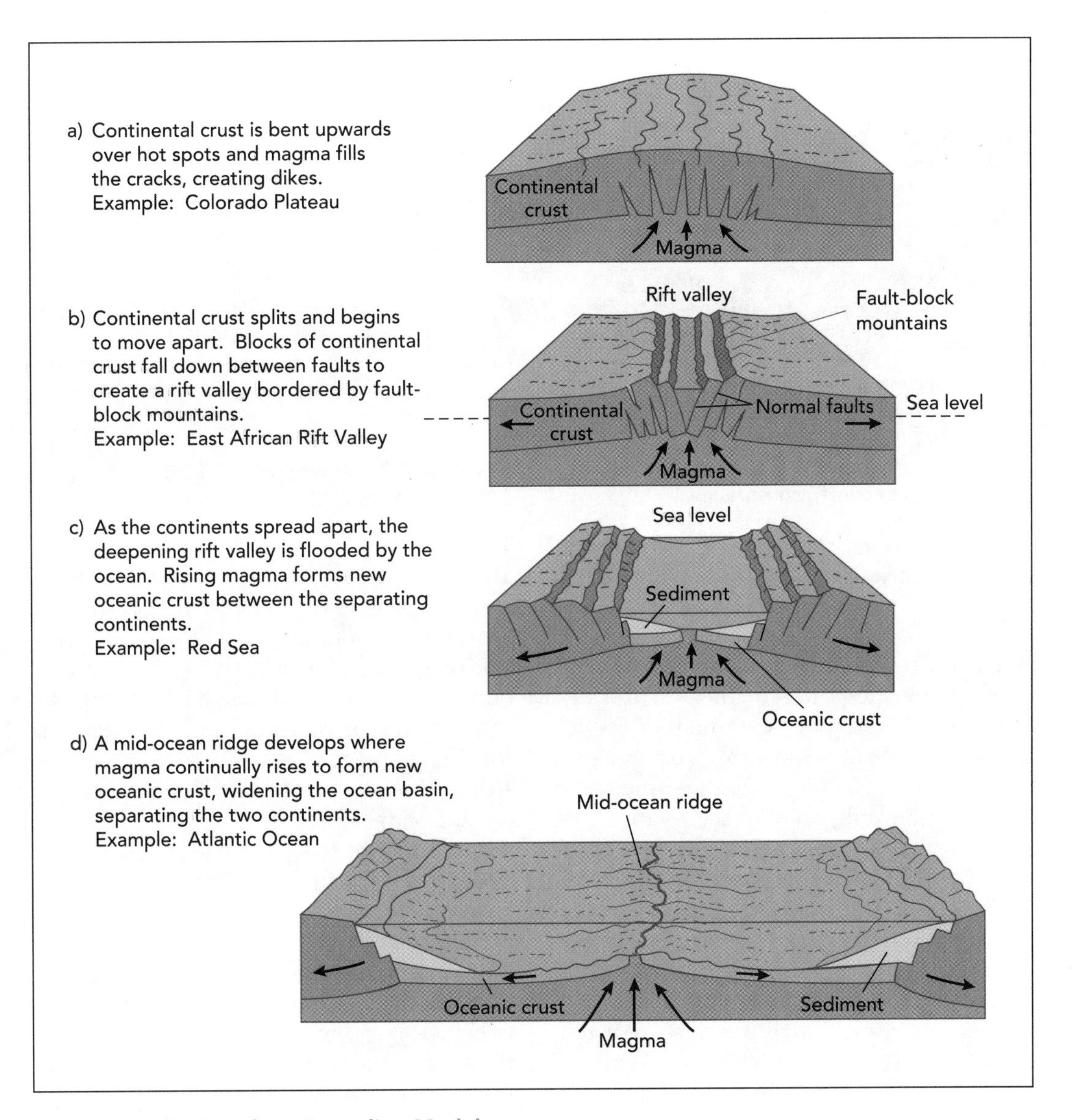

Figure 6.7 **The Sea-Floor Spreading Model**

below. Evidence to support this hypothesis was not long in coming.

The first evidence of sea-floor spreading was the pattern of magnetic variations on the sea floor on either side of the mid-ocean ridges in both the Atlantic and Pacific oceans. These variations revealed a striped pattern that was parallel to the mid-ocean ridge. The pattern also showed a mirror image on both sides of the mid-ocean ridge (Figure 6.8). When first discovered in the 1950s, these magnetic variations could not be explained.

The second important discovery dealt with the **magnetic reversals** of the polarity of the earth's magnetic field. In "normal"

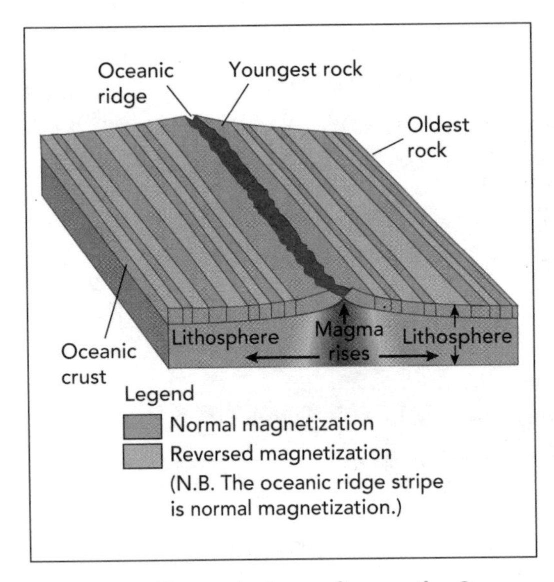

Figure 6.8 **Magnetic Anomalies on the Sea Floor Near the Mid-Atlantic Ridge**

magnetic periods, such as the present, a compass needle points to the North Magnetic Pole, located in the Canadian Arctic. During a reversal, which happens approximately every 500 000 years, the same compass needle would point to the South Magnetic Pole in Antarctica. As basaltic magma solidifies into rock at the mid-ocean ridge, the magnetite grains in the cooling magma become tiny, permanent magnets with the same alignment and polarity as the earth's magnetic field when the magma cools. Earth scientists realized that magnetic variations on each side of the ridge were caused by the normal and reversed magnetic polarity of the rocks of the ocean floors. As the plates on either side of the ridge moved apart and new magma rose to fill the gap between them, the newly forming rocks recorded the earth's changing magnetic polarity going back millions of years from the present. This process is illustrated in Figure 6.9.

If the sea-floor spreading hypothesis was valid, the age of rocks should increase as one moves farther away from the mid-ocean ridge. Drilling ships sent out to sample ocean floor rocks soon found that the youngest rocks were at the mid-ocean ridges and that rocks became increasingly older away from the ridge. It was concluded that new oceanic crust was being created at the mid-ocean ridges, or diverging plate boundaries. On-going research using satellites has put the rate of sea-floor spreading at between three and seven centimetres per year.

Tectonic Processes at the Mid-Ocean Ridge

The formation of new ocean floor is created by a great deal of tectonic activity, especially volcanic eruptions, faulting, and earthquake activity. However, the tectonic processes at diverging plate boundaries are much less violent than at other types of boundaries. The reasons for this include:

- the lithosphere is thinnest near the mid-ocean ridges;
- the plates are splitting apart, allowing hot magma from the upper mantle to move easily toward the earth's surface; and
- the low **viscosity** (resistance to flow) of the magma itself.

The viscosity of magma is influenced by two factors: its composition and its temperature. Since the basaltic magma at mid-ocean ridges contains very little silica, it flows more freely than other magmas. Since it travels a short distance from its source to the earth's surface, the erupting magma is very hot, reducing its viscosity even further. Volcanic eruptions along mid-ocean ridges are gentle compared to the explosive eruptions at converging plate boundaries.

The eruption of magma at the crests of mid-ocean ridges is the most frequent type

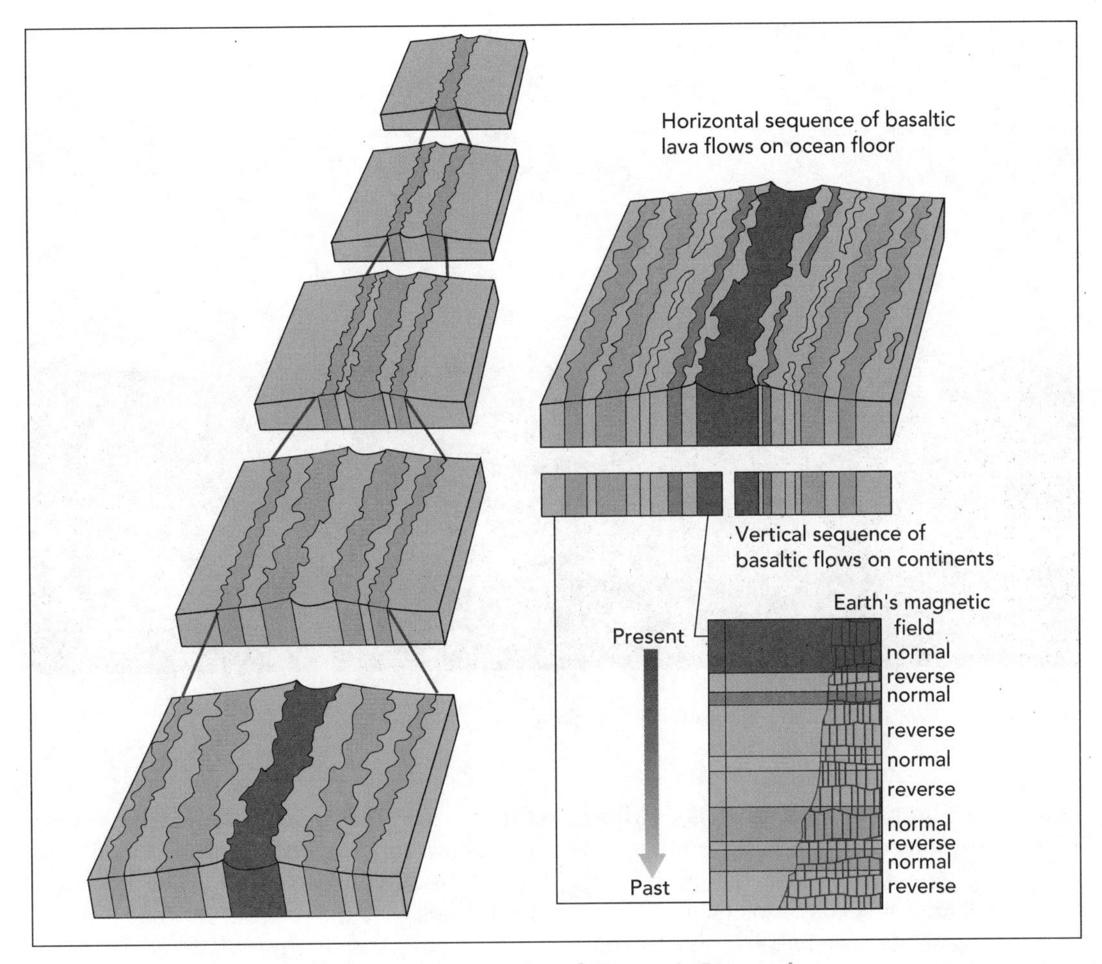

Figure 6.9 **Sea-Floor Spreading: Tape Recorder of Magnetic Reversals**

of volcanic activity in the world. Such eruptions create **pillow lavas**, features formed by the rapid surface cooling of magma in direct contact with cold ocean water. Pillow lavas have distinct, pillow-like shapes, and have been identified along the crest of all the oceanic ridges.

Where hot spots rise beneath mid-ocean ridges, the added magma forms large volcanoes. Since these volcanoes are built up by flows of very fluid basaltic lava, they have gently rising, smooth slopes that flatten near the top, giving them a dome-shaped cross-section, as

Figure 6.10 **Pillow Lavas on the Summit of MOK Seamount, a Deep Sea Volcano West of the East Pacific Rise**

Figure 6.11 Profile View of a Shield Volcano (Mauna Loa, Hawaii)

shown in Figure 6.11. Known as **shield volcanoes**, they are found above hot spots in Iceland, the Azores, Ascension, Tristan da Cunha, and the Galapagos Islands.

Faulting is responsible for the formation of a rift valley, or graben, along the crest of the mid-ocean ridge. A **rift valley** is formed when a block of the earth's crust falls down between two parallel fault lines. The rift valley found along the crest of most mid-ocean ridges, shown in Figure 6.7, marks the zone where two plates are moving apart and blocks of the ocean floor have dropped down between fault lines.

Earthquakes at the ridges occur near the surface and are generally low in magnitude since there is little buildup of stress along the plate boundaries. Earthquakes are produced by movements along faults, or when magma moves upward into cracks caused by separating plates.

It's a Fact...

Research has shown the rates of sea-floor spreading can be divided into three groups: slow (less than 3 cm/year), medium (between 3 and 7 cm/year), and fast (over 7 cm/year).

QUESTIONS

10. a) Using an atlas, measure the longest east-west extent of Canada. How many times longer than this measurement is the mid-ocean ridge?
    b) Explain how such a major mountain system as the mid-ocean ridge could have remained undiscovered until after 1945.
11. Create a demonstration to show how the process of sea-floor spreading works.
12. In your own words, explain how the pattern of magnetic variations on the ocean floor and magnetic reversals were able to prove that the hypothesis of sea-floor spreading was correct.
13. a) List reasons why volcanic eruptions at mid-ocean ridges are less explosive than those at other plate boundaries.
    b) Draw a diagram to illustrate these reasons in a graphic way.
14. a) Why might mid-ocean ridges be major sources of geothermal power in the future?
    b) What obstacles would have to be overcome to tap such energy resources?

6·4 Transform Fault Boundaries: Plates Sliding Past One Another

A second type of plate boundary occurs where plates are sliding past one another along **transform faults**. The movement is horizontal and extends deep into the lithosphere.

Most transform faults are associated with mid-ocean ridges and are rather short. The mid-ocean ridge is broken up into segments that are offset from one another, as Figure 6.12b illustrates. The

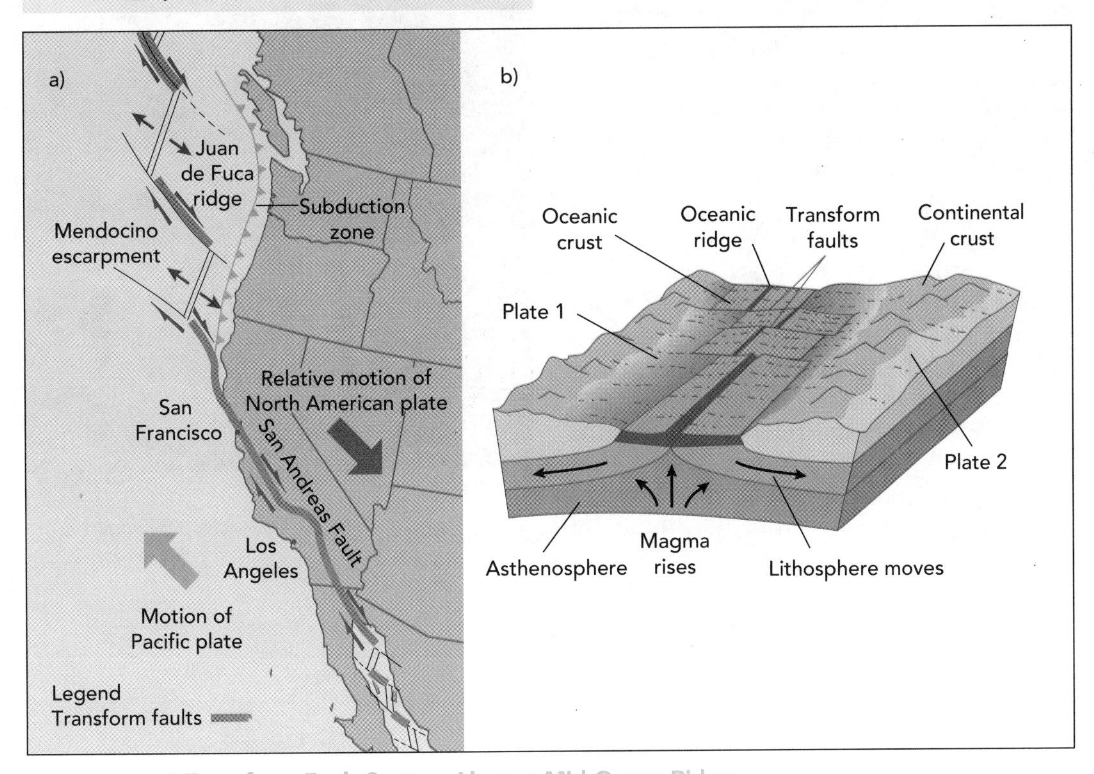

Figure 6.12 A Transform Fault System Along a Mid-Ocean Ridge

offset sections are sliding past one another in opposite directions. This is a characteristic of ridges in all of the world's ocean basins. Longer transform faults also link mid-ocean ridges with convergent plate boundaries. Examples include the transform faults along the north and south edges of the Scotia and Caribbean plates, and much of the boundary between the Eurasian plate and the African and Iranian plates, as shown in Figure 6.1 on page 87.

Tectonic Processes Along Transform Fault Boundaries

Faulting, with its associated earthquake activity, is the dominant tectonic process occuring along transform plate boundaries. As the plates slide past one another, the enormous pressures between them shatter the rocks along the fault line. On land, these shattered rocks are easily eroded to create a narrow valley along the fault line, as illustrated by Figure 6.13. Since movement along transform faults is largely horizontal, there is little uplift on either side of the fault line.

Earthquakes along transform faults are shallow and include a range of magnitudes. When short movements along a fault occur frequently as a result of microearthquakes, the fault is said to "creep". The gradual release of pressure along a fault line means that earthquakes are lower in magnitude and less destructive. Many may not even be detectable by human beings under normal conditions.

Where rock surfaces along the fault are rough, the plates may become "locked", and earthquakes do not occur with any frequency. Even though such faults may appear to be inactive, pressure or strain is constantly building up, eventually to be released as an earthquake. The longer a fault remains locked, the more pressure builds up and the stronger and more destructive the earthquakes will be when they do occur.

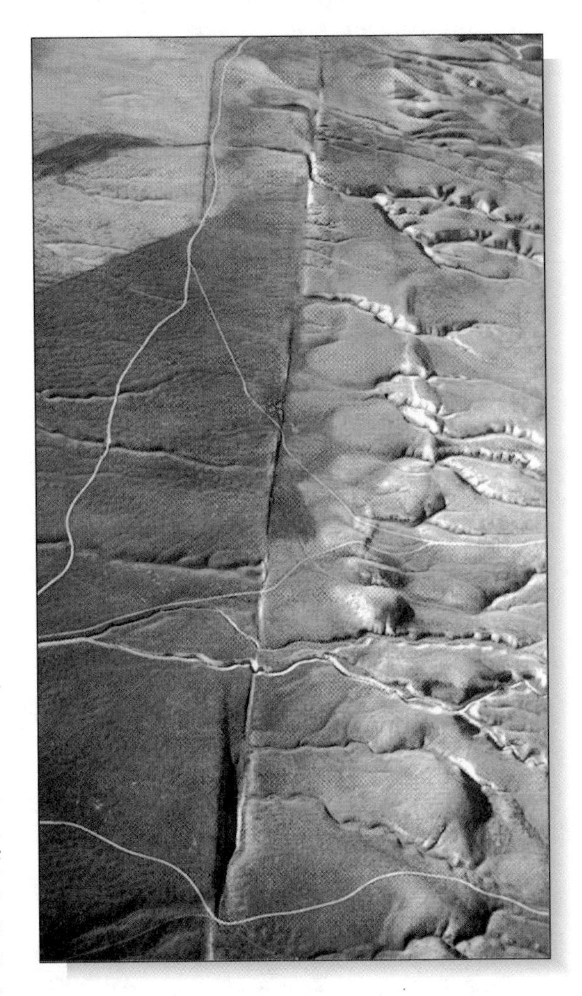

Figure 6.13 **The Transform San Andreas Fault From the Air**

Fault Landforms

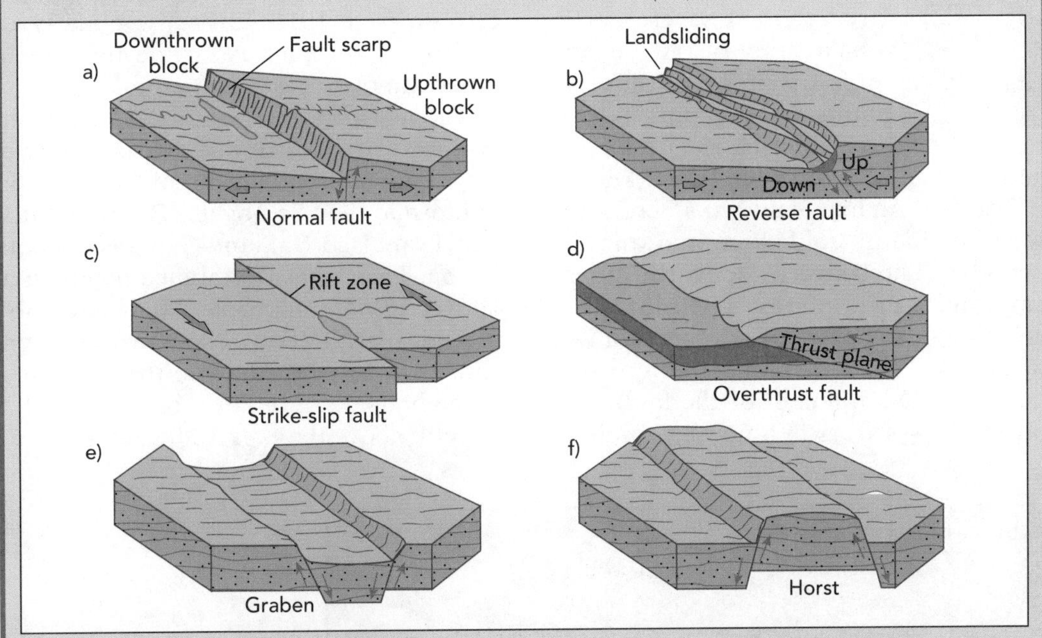

Figure 6.14 Types of Fault Features

The most common types of faults and associated landforms are illustrated in Figure 6.14. **Normal faults** result from the upward movement on one side of a fault line and/or the downward movement on the other. Normal faults often occur where blocks of the crust are moving apart. **Reverse faults** occur when one block moves upward against another. This type of fault occurs where the blocks are being pushed together. **Fault scarps** are created by both normal or reverse faulting and usually run in a straight and continuous line across a landscape for considerable distances.

Strike-slip faults develop where two sections of rock move horizontally past each other. They occur frequently along transform fault boundaries where plates are moving horizontally. Fault scarps are seldom formed along such faults.

Grabens or **rift valleys** are created where sections of the crust move apart and a block of land drops down between two parallel fault lines. The Rhine Valley south from Frankfurt, West Germany, to Basel, Switzerland, Death Valley in California, and the Great Rift Valley of East Africa are three well-known rift valleys.

Horst or **block mountains** form where parts of the crust move together forcing blocks upward between two parallel faults. When a block moves up at an angle, a **tilted block mountain** is created. The Sierra Nevada Mountains of California and the Grand Teton Mountains of Wyoming are two excellent examples of tilted block mountains.

The San Andreas Fault

Transform fault boundaries are zones of severe earthquake activity. Two examples of such destructive faults are the San Andreas in California (Figures 6.12a and 6.15) and the Anatolian in Turkey.

The San Andreas fault cuts across the western side of California, marking the boundary line between the North American and Pacific plates. The plates are grinding slowly past one another at a rate of about 2 cm/year along the approximately 1000 km length of the fault. The offset rows of trees in an orange orchard in southern California is clear evidence of the movement of these two plates along the fault line. In the photograph in Figure 6.16 on page 101, the Pacific plate is at the top of the picture, while the North American plate is at the bottom.

The notoriety of the San Andreas fault is due to its location close to the two major population clusters of the state of California: the Los Angeles Basin and the San Francisco-Oakland-San José urban region. They have a combined population of over 15 million people. Many new subdivisions in the San José-San Francisco area are built on or next to the fault line. Some houses are built on unstable bluffs directly above the San Andreas fault line,

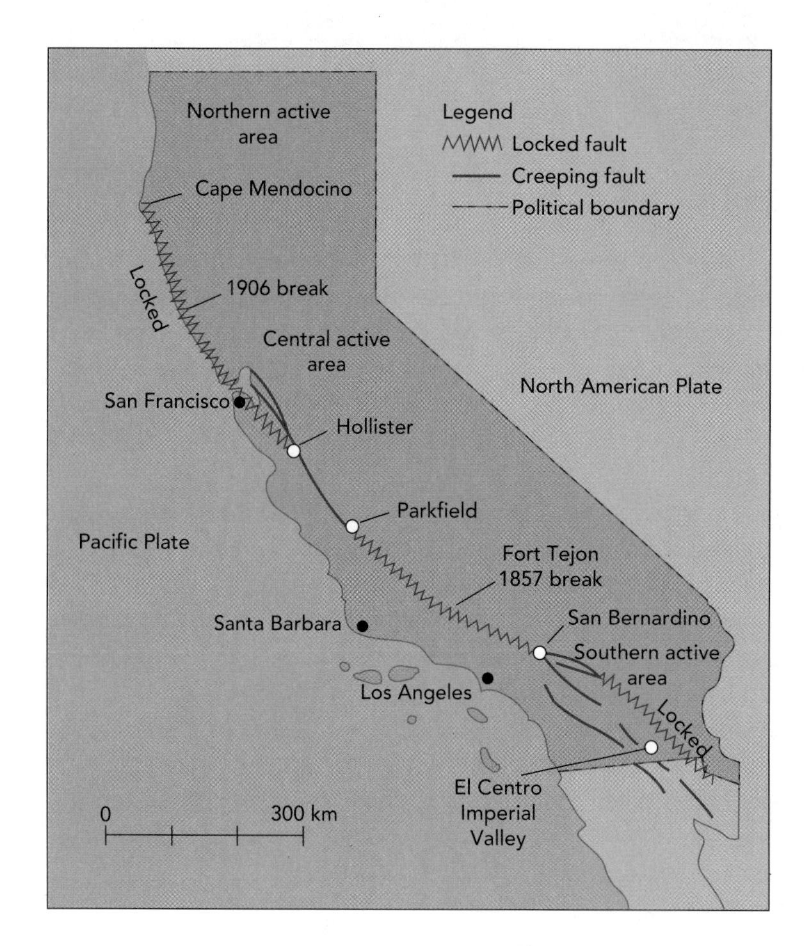

Figure 6.15
San Andreas Fault in California

Figure 6.16
Orchard Offset Along San Andreas Fault, California

overlooking the Pacific Ocean. When a quake does occur, many of these homes may end up in the ocean.

The portion of the San Andreas fault near San Francisco has been locked since the destructive earthquake of 1906. The strain has been building up since that time and a similar-sized quake has been forecast. The longer it takes to occur, the larger and more destructive the quake will be.

Forecasting Earthquakes

An understanding of plate tectonics has given earth scientists some ability to forecast earthquakes. New instruments, such as laser sensors, sensitive seismographs, and strain gauges, can detect shifts and strains along a fault. From this, researchers can decide if a fault line is locked or creeping steadily. Through a study of historical accounts, as well as the records in the rocks and sediments along known faults, they can look for patterns of earthquake activities over long periods of time.

Using information from many sources, seismologists can develop forecasts about the long-term chances of earthquakes in a given region. The map in Figure 6.17 on page 102 shows the probabilities of major earthquakes along various sectors of the San Andreas fault for a thirty-year period, as estimated by the U.S. Geological Survey. It is interesting to note that the map, published in 1985, indicated a 60 percent probability of a major earthquake in the area where the Loma Prieta (World Series) earthquake struck in 1989.

While seismologists can make general forecasts, it is difficult to be more precise about the actual year, date, hour, or minute when and where quakes will occur. Such information would be very useful; for example, in hazardous locations, buildings could be evacuated and many lives saved.

Predicting earthquakes is a risky business. What would happen if scientists predict an earthquake at a particular time and it does not occur? Millions of lives would have been needlessly disrupted and businesses affected. On the other hand, what happens if scientists expect a quake but fail to issue a warning? This is a real dilemma for those involved in this branch of science.

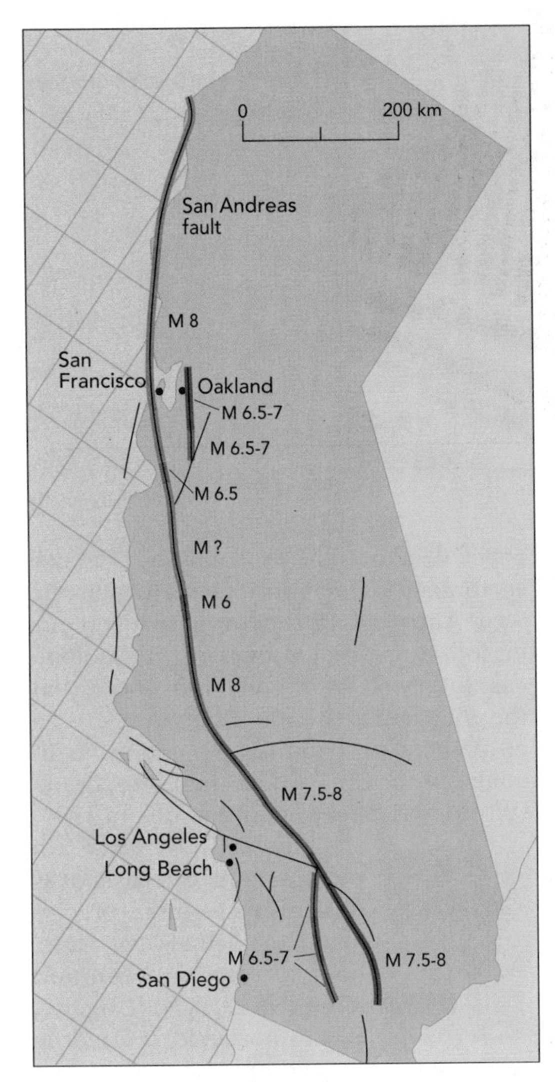

Figure 6.17
San Andreas Fault Earthquake Probability Map This map shows the probability of earthquakes along segments of the fault over the next 30 years, as estimated by the U.S. Geological Survey. The likely magnitudes (M) of quakes along each segment are indicated, using the Richter scale.

QUESTIONS

15. Suggest reasons why volcanic activity is relatively rare along transform fault boundaries.
16. Is it better to live near a fault line where small earthquakes occur frequently or where the fault appears to be inactive and less threatening? Explain your answer.
17. a) Describe the dilemma of earth scientists and their responsibilities in warning the general public of potential earthquakes, or other natural events, given the uncertainty of such predictions.
    b) Outline a possible procedure that might allow them to warn the public of the danger of such events but still not make them responsible if their predictions are inaccurate.

It's a Fact...

One problem in predicting earthquakes in California is the large number of different faults found within the state. In the past few decades, many earthquakes have occurred on fault lines that were not known to geologists until they "announced themselves" by generating their own earthquakes.

6•5 Collision and Subduction Zones: Converging Plate Boundaries

After being created at the mid-ocean ridges and sliding past one another at transform faults, plates come together and are destroyed at converging plate boundaries. The great pressures exerted as plates collide create spectacular features on the ocean floors and along the margins of continents. Converging plate boundaries are divided into two types:

- subduction zones, where an oceanic plate is slipping below a continental plate; and
- collision zones, where two continental plates are meeting.

The world distribution of these boundaries is shown in Figure 6.1.

Subduction zones develop where relatively thin oceanic plates, composed of heavier basaltic rocks, descend into the plastic asthenosphere beneath lighter, thicker granitic rocks of continental plates, as shown in Figure 6.18. Since descending plates are cooler and denser than the surrounding asthenosphere into which they are moving, they sink under their own weight once they begin their downward path. This process is known as **subduction**, and it completes the recycling of oceanic plate materials back into the asthenosphere, a process which began at the mid-ocean ridges millions of years before. The total lifespan of an oceanic plate from mid-ocean ridge to subduction zone ranges from 150 to 200 million years.

Where oceanic plates begin their descent, the sea floor is often marked by deep oceanic trenches, as shown in Figure 6.18. The descending oceanic plates pull down the edges of the continental plates with them, creating deep oceanic trenches. As the sinking plates intrude into the hot asthenosphere, their upper surfaces begin to melt. The molten rock rises to the surface through cracks in the crust to form volcanoes. Earthquakes occur along the contact zone between the plates.

Collision zones, the second type of converging plate boundaries, occur where two continental plates collide. Since the rocks of the continental plates are lighter and less dense than those of the asthenosphere, they do not sink downward. As a result, they collide to form massive mountain ranges of twisted rocks (Figure 6.19). The Himalayas, the best example of this type of mountain, formed when the subcontinent of India collided with the Eurasian plate about

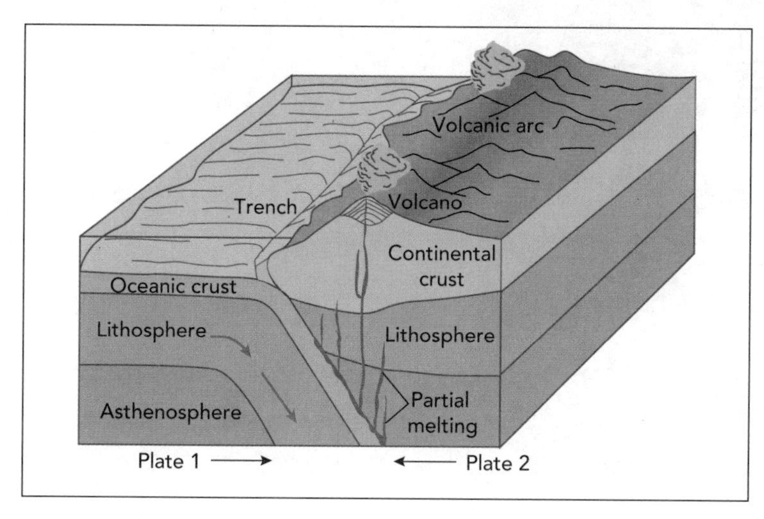

Figure 6.18
A Subduction Zone Between Oceanic and Continental Plates

45 million years ago. In fact, the whole series of mountains from New Guinea in the east to Pakistan in the west are a result of the collision between these two plates.

The Appalachian Mountains were created over 200 million years ago when the continents of North America, Africa, and Europe collided. In the same way, the Ural mountains of Russia were formed when two continent-bearing plates (Europe and Asia) collided and were "welded" together. Over the next few million years, new mountain ranges will likely be created as the African plate collides with the Eurasian plate in southern Europe. The Alps have been formed in the early stages of this collision.

Tectonic Processes at Converging Plate Boundaries

All three major tectonic processes — volcanic activity, folding, and faulting — are at work along converging plate boundaries. The amount of energy released through volcanic eruptions is considerable, as magma produced by the melting of descending plates reaches the earth's surface in often violent eruptions. In addition, the energy released as the massive plates move into one another creates more mountain systems of folded and faulted rocks than anywhere else on the earth's surface.

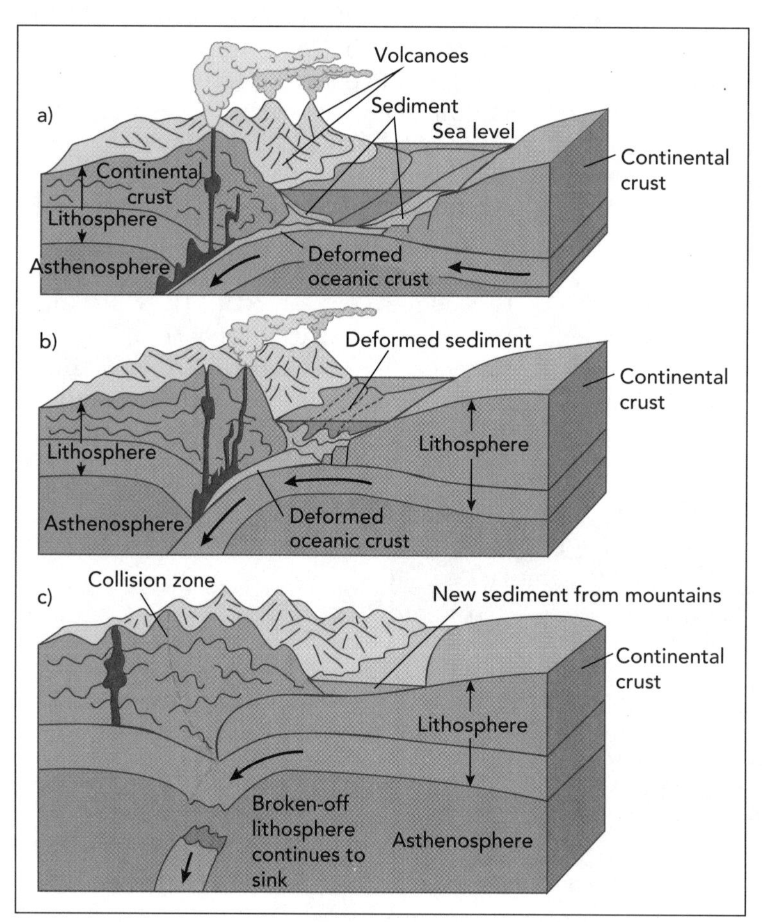

Figure 6.19
A Collision Zone Between Two Continental Plates

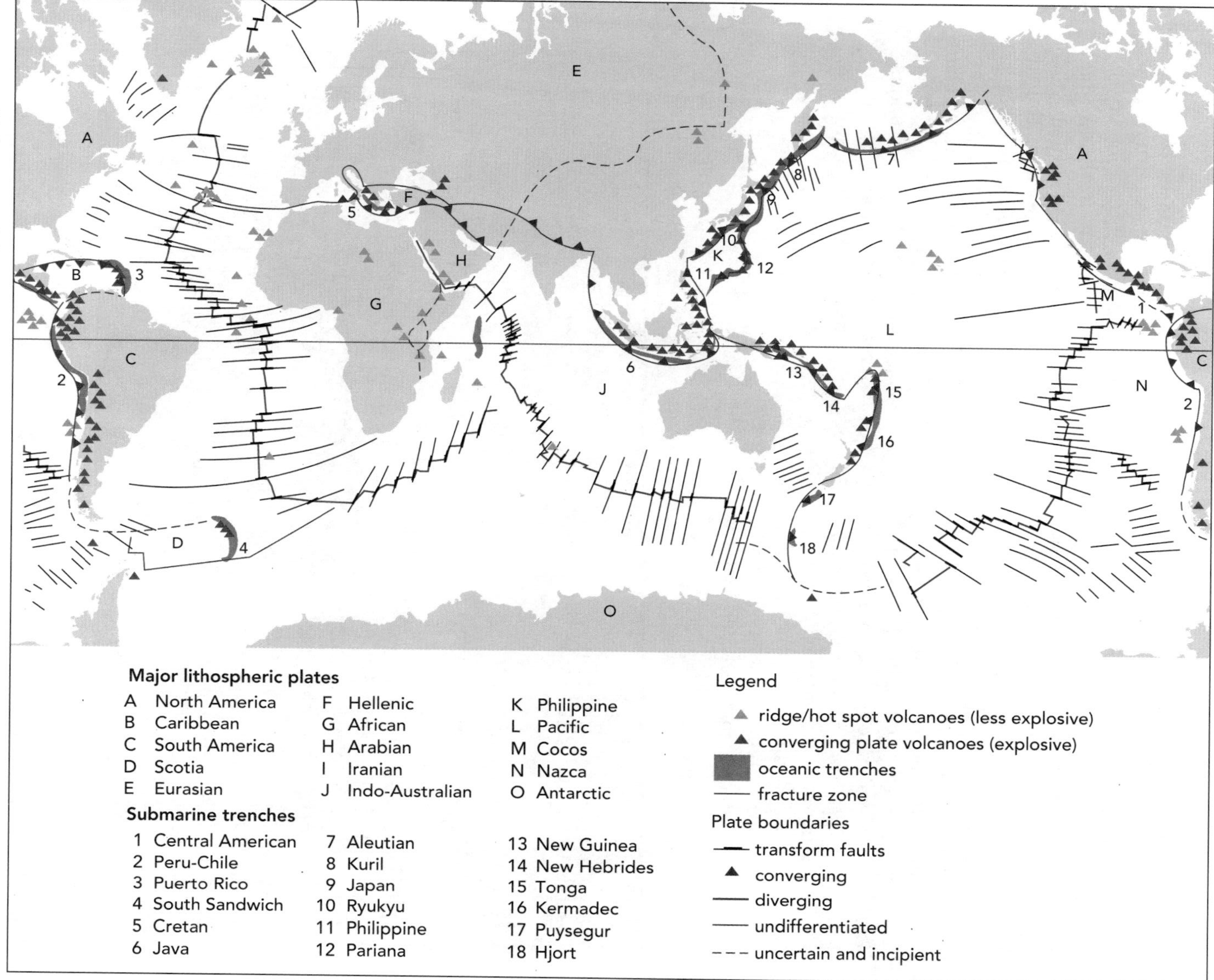

Figure 6.20 Volcanoes of the World A large majority of the world's active, destructive volcanoes lie along converging plate boundaries.

Vulcanism

Violent volcanic activity is a spectacular feature of converging plate boundaries, particularly subduction zones. Produced by the melting of subducting oceanic plates in the asthenosphere, the **andesitic magmas** erupted at these boundaries have more silica and dissolved gases and lower temperatures than the basaltic magmas of ridges and hot spots. As a result, andesitic magmas have a higher viscosity and are more likely to clog up the vents of volcanoes, leading to the buildup of pressure from below. When the pressure becomes great enough, a major explosion hurls magma onto the earth's surface and into the atmosphere in the form of ash, volcanic rocks, and lava.

The volcanic peaks built up on continents form arc-like chains known as **continental volcanic arcs**. The string of volcanoes stretching down the Andes Mountains from Colombia to Peru and the Cascade volcanoes of the western United States are typical continental volcanic arcs. When the volcanoes build up on the ocean floor, they form **island arcs**. The Aleutian Islands, Japan, the Caribbean islands north from Trinidad, and the southern islands of Indonesia are examples of island arcs.

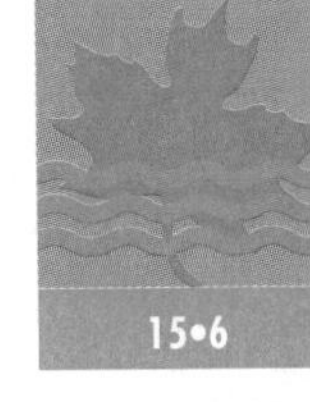

Volcanic Landforms

The wide variety of landforms associated with volcanic activity at or near subduction zones is shown in Figure 6.21. Some are formed when magma reaches the surface and cools quickly (igneous extrusive rocks), while others cool slowly beneath the earth's surface to form crystalline intrusive igneous rocks. At the same time, the heat from volcanic activity and pressure from colliding plates create metamorphic rocks.

The most spectacular volcanic landforms are the cones of **composite volcanoes** (Figure 6.22).

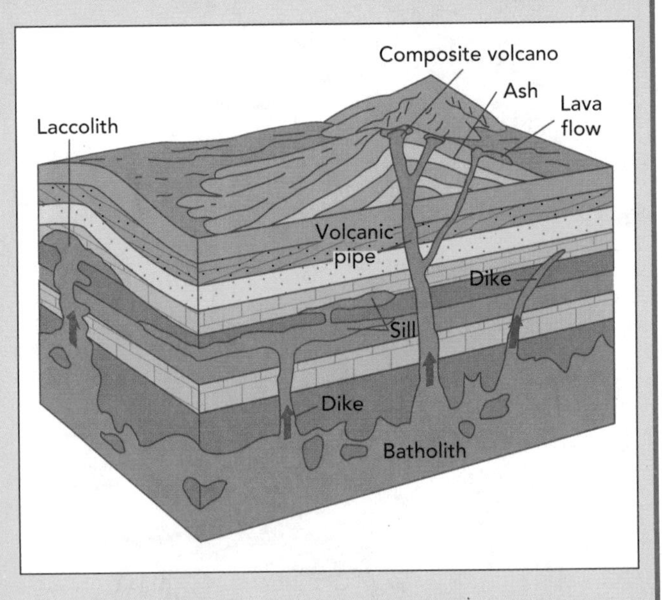

Figure 6.21 Volcanic Landforms and Subsurface Forms

Because they are made of layers of alternating ash and lava, they have long, rising slopes with a summit crater. Where such mountains have been eroded over millions of years, the harder pipe rocks are left standing above the landscape as **volcanic necks**. An example is Shiprock, New Mexico, a remnant of a long extinct composite volcanic peak (Figure 6.23). Other volcanic hills, usually lower in elevation than composite cones and composed primarily of volcanic ash and rock spewed out of the vent during explosive eruptions, are known as **cinder cones**.

Figure 6.22 Composite Volcanic Cone, Mount Fuji, Japan

Many volcanic formations are created when magma, unable to make its way to the surface, cools and hardens into intrusive igneous rock. Dykes and sills are hardened sheets of magma that made their way between the beds of the local rock. **Dykes** cut across the rock beds, whereas **sills** follow the layers of the original rocks. Where magma created a chamber by forcing apart the local rocks, laccoliths and batholiths are formed. **Laccoliths** are generally smaller than the often bottomless **batholiths**.

Figure 6.23 A Volcanic Neck, Shiprock, New Mexico

In some places, volcanic intrusions have been exposed by erosive activity. The Sierra Nevada and Coast Range mountains of western North America are two examples of exposed batholiths.

Folding

Folding is caused when rocks are compressed into mountain ranges by converging plates. Such mountains, largely composed of sedimentary rocks, extend along the western sides of North and South America, and across the Eurasian and north African landmasses from Morocco to Indonesia. The Alps and Himalayas are examples of such mountain ranges. Collision of two continental plates usually welds together the plates along the zone of mountain-building activity.

Where one of the converging plates contains only a small amount of continental rocks, such as a series of island arcs or microcontinents, a slightly different process occurs. In this case, the continental rocks are scraped off the surface of the descending ocean floor plate, and are severely folded as they are plastered onto the edge of the advancing continental plate.

Fold Landforms

The folding process can be demonstrated by placing two hands on a tablecloth and then bringing them together. The tablecloth is bent up and folds over as your hands move together. When rock layers are compressed under great stress, they too are deformed into folds (Figure 6.24). **Anticlines** are rocks that have been bent upward in an arch, creating long, rounded hills. Where rocks are bent downward, into a basin shape, they form **synclines**. Folding is most easily seen where road or other cuts have been made in hillsides or from satellite views over large areas of the earth's surface (Figure 6.25).

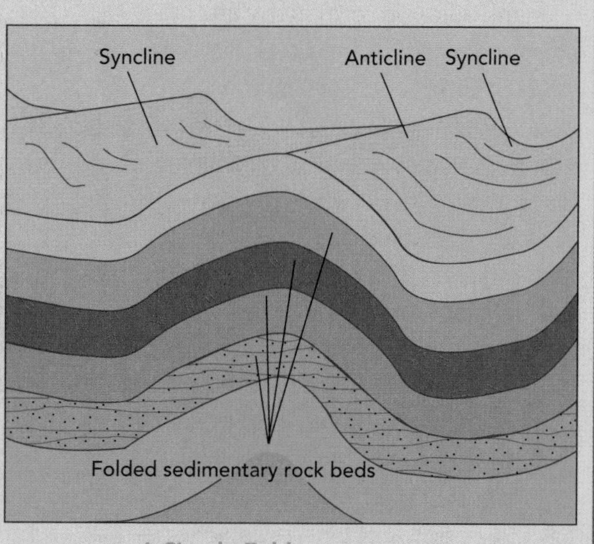

Figure 6.24 A. Simple Fold

Under severe stress, folds can become so distorted that they fall over to create **recumbent folds**, as shown in Figure 6.26a and b. Where compression continues, rocks are so distorted that thrust faulting occurs as shown by Figure 6.26c and d. Recumbent folds and thrust faulting are common in the Alps, Himalayas, and Rocky Mountains.

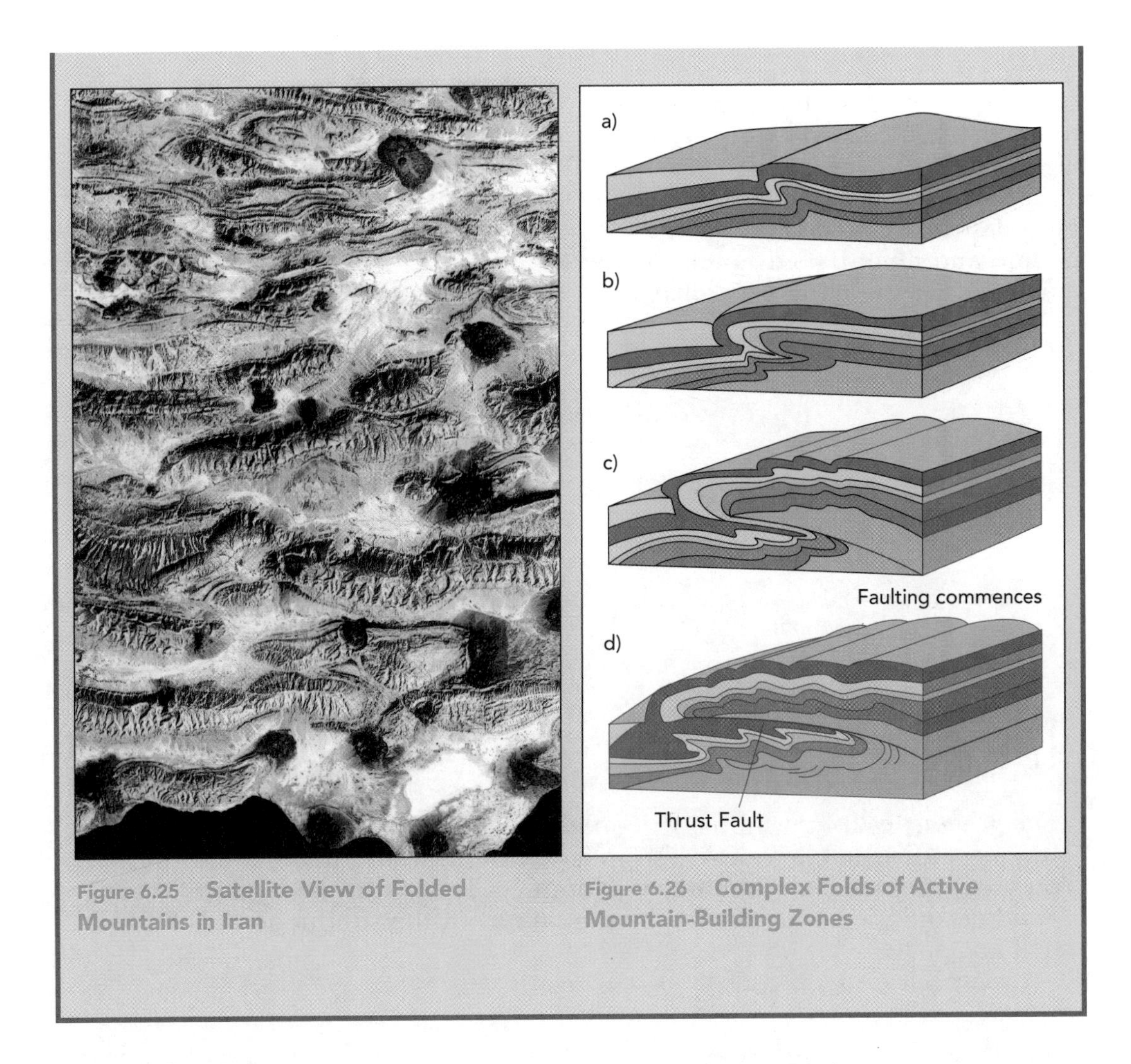

Figure 6.25 Satellite View of Folded Mountains in Iran

Figure 6.26 Complex Folds of Active Mountain-Building Zones

Earthquake Activity

Major fault lines mark the outer edges of descending plates that are moving into the asthenosphere. Faults also develop within the mountains created along collision zones. They combine with folding activity to distort and uplift the rocks trapped between two converging plates.

Earthquakes, associated with faulting and folding activity along converging plate boundaries, are frequent, violent, and destructive, since the plates are moving together with tremendous force.

Figure 6.27 shows an interesting observation about earthquakes at subduction zones that puzzled geologists. Earthquake foci start out near the earth's surface at the trenches, and then gradually descend along straight lines angled into the asthenosphere, beneath the island arcs. With the development of the plate tectonic theory, earth scientists realized these lines of earthquakes mark the edges of the descending oceanic plates as they are slowly pulled into the asthenosphere.

Japan is an active seismic zone typical of a converging plate boundary. A great number of earthquakes occur in the vicinity of this country each year. The 1923 Kanto/Tokyo earthquake, one of the most destructive quakes, killed 156 000 people and demolished much of the metropolitan area of Tokyo-Yokohama.

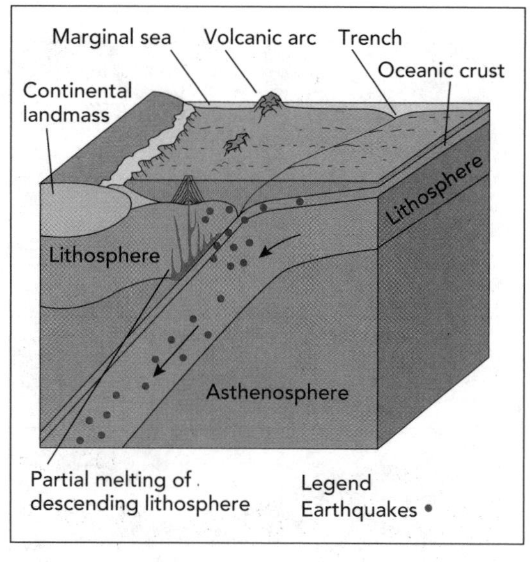

Figure 6.27 Cross-Section of Earthquakes at Subduction Zones

It's a Fact...

The greatest eruption in historic times was that of the Krakatau volcano in Indonesia in 1883, a volcano supplied with andesitic magma and located on a converging plate boundary. Mount St. Helens in the United States is another andesitic volcano, fed by magma from the subduction of the Juan de Fuca plate.

QUESTIONS

18. What would you say to a friend who asked you about moving to a place that was located on a converging plate boundary?
19. Explain why metamorphic rocks are formed at converging plate boundaries.
20. a) Describe the differences between the magmas erupted at diverging and converging plate boundaries.
    b) Explain how these differences account for the volcanic features and types of eruptions experienced at each of these plate boundaries.
21. a) Complete a comparison chart to show the differences between collision and subduction zones of converging plate boundaries.
    b) Explain why such differences occur.
22. a) Why might a cooperative effort by nations around the Pacific Rim help to reduce the loss of life and property in the region due to tectonic activities?
    b) What form might such cooperation take?
    c) What nations would likely be leaders and innovators in this effort? Why?

6·6 Plate Interiors: Zones of Inactivity

With the notable exception of hot spots, the interiors of tectonic plates are normally very quiet in geological terms. The centres of plates are either oceanic basins underlain by basaltic rocks with an average age of 60 million years, or continents composed largely of granitic rocks with an average age of 650 million years. Unlike ocean floor rocks, the light continental rocks are not consumed at subduction zones, accounting for the comparatively young age of the rocks of the ocean floors.

The Ocean Basins

Basalt, the rock type found beneath the oceans, is denser and younger than granitic continental rocks. Away from the mid-oceanic ridges, the ocean floors gradually slope downward into huge plains or basins. Here, the rocks of the oceanic plate have cooled down and, consequently, are thicker and denser than at the ridge (Figure 6.28). These denser, older parts of the plate sink deeper into the underlying asthenosphere. With the exception of ocean trenches, the deepest ocean basins occur where the rocks are the oldest.

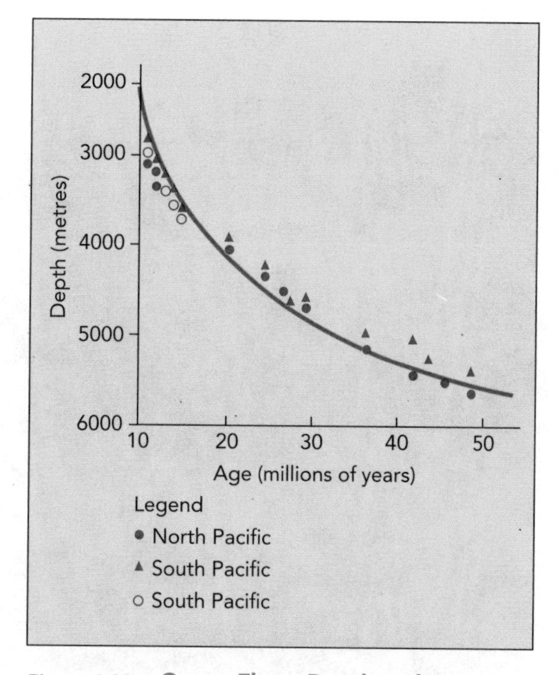

Figure 6.28 Ocean Floor: Depth and Age Comparison The average depth of the Pacific Ocean plotted against the age of the oceanic crust

The Continents

The oldest, most stable parts of continents occur where ancient rocks are found at the surface or overlain by flat-lying sedimentary rocks. Such areas, called **cratons** (Figure 6.29), make up the nuclei of the continents. They have not experienced mountain-building activity for hundreds of millions or billions of years and are almost earthquake free. Many rocks of cratons were originally the roots of mountain chains that were exposed over hundreds of millions of years by erosion. The Canadian Shield of North America is one of the major cratons of the world and is flanked by younger rocks to the east, west, and south, as shown in Figure 6.30.

The continents are not entirely composed of cratons. Their edges are often plate boundaries and are active zones of mountain building. It is through such activity that the sizes of continents are thought to increase by **accretion**. In this process, as mountain building subsides, erosion reduces the mountain chains and exposes the hard rocks at their roots. This adds new, stable areas to the original craton and increases the total area of continental crust. This, at least, is the basic hypothesis that earth scientists are researching.

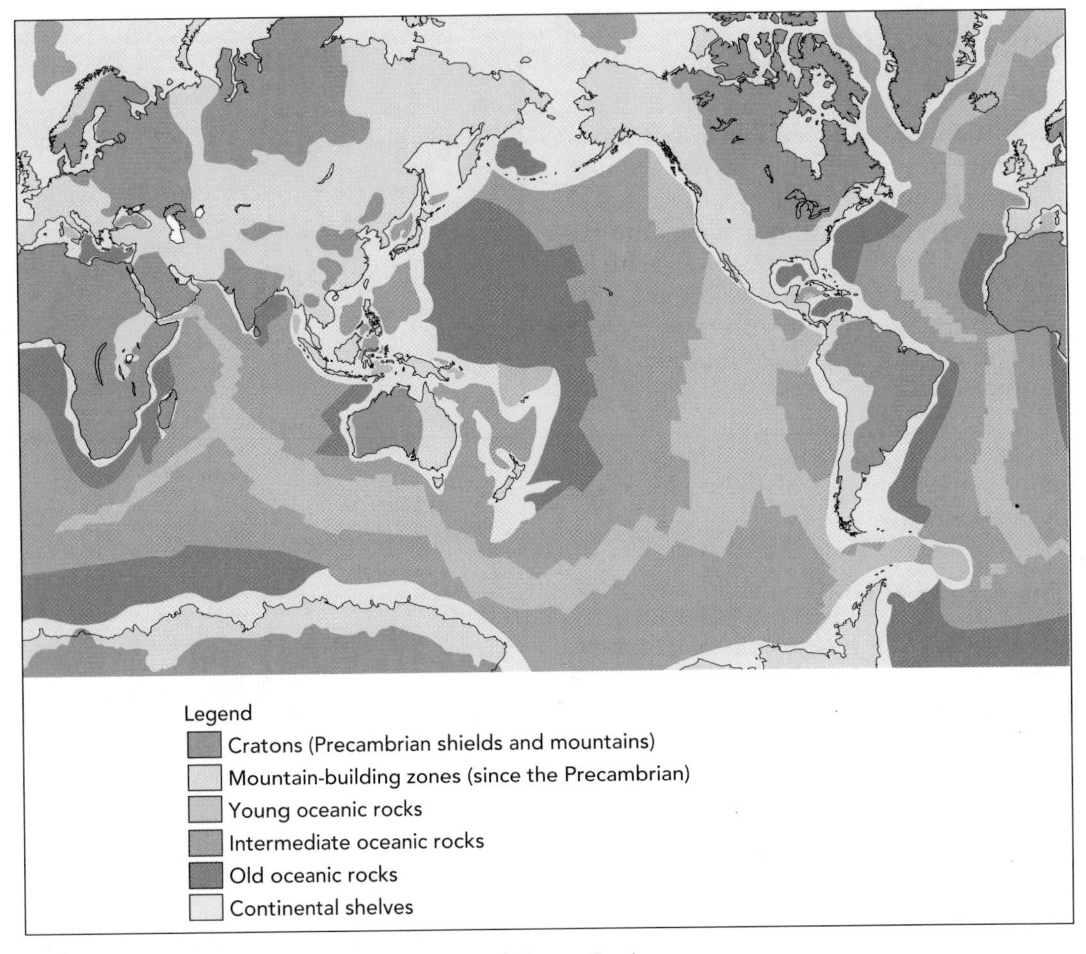

Figure 6.29 **World: Cratons, Continents, and Ocean Basins**

The great ages of continental rocks result from the fact that they are buoyant compared to the denser rocks of the lower lithosphere and mantle. Continental crust is not forced into the mantle in large amounts at subduction zones. It tends, instead, to float on the denser mantle, something like slag on molten iron.

Continents: Their Assembly and Break-Up

The observation of the close fit of the coastlines of Africa and South America led to the idea of drifting continents, first suggested by Alfred Wegener, a German meteorologist, in the early twentieth century. Although he gathered a great deal of evidence to back up his hypothesis, most of the scientific world was unconvinced. The great obstacles to acceptance of this hypothesis were the lack of a force great enough to move whole continents and an explanation of how continents could "drift" through the solid crust of the ocean basins.

New discoveries about the ocean basins after 1945 led to the development

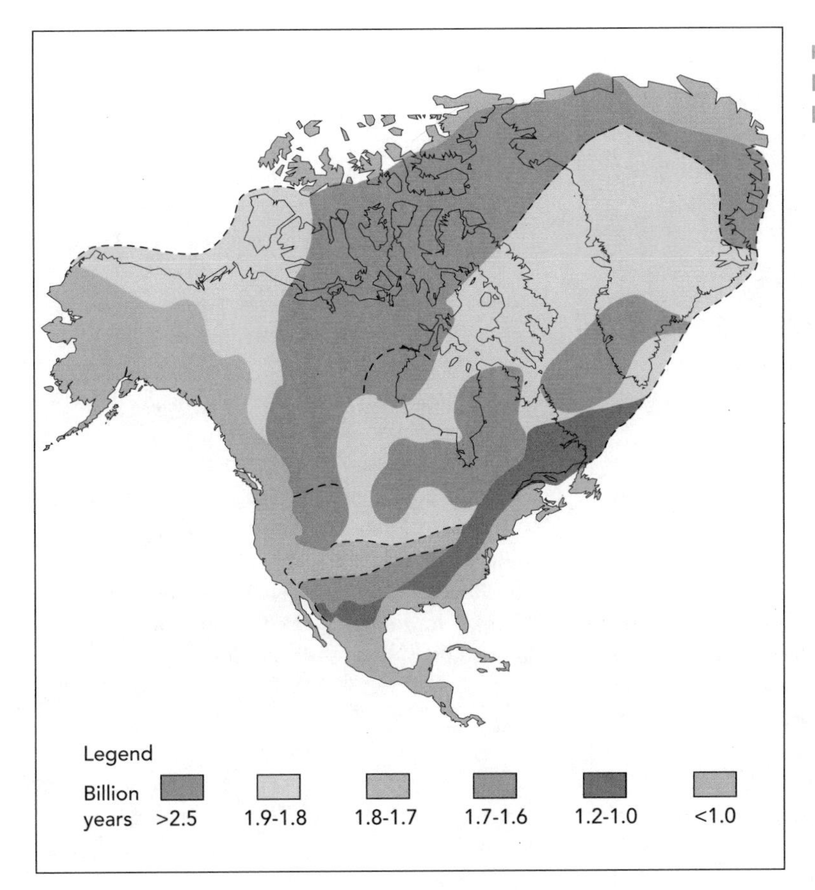

Figure 6.30
Foundation Rocks of North America

of the theory of plate tectonics. This theory was based on revolutionary ideas like sea-floor spreading, subduction zones, and moving plates. Evidence to support these ideas helped to overcome obstacles to the acceptance of Wegener's idea of moving continents. One of Wegener's ideas that gained much support was that all of the world's continents were once collected into a single **supercontinent**, which he called Pangea.

Paleomagnetism and other techniques are now being used to retrace the paths of the continents millions of years into the past.

The reasons for the break-up of continents are still being debated by earth scientists. A recent explanation suggests that supercontinents, such as Pangea, act as thermal caps, slowing the escape of heat energy from the mantle below them. Consequently, temperatures under the supercontinental "blanket" increase, producing a super hot spot several hundred metres high. This swelling causes cracks or rifts to develop in the supercontinent, splitting it apart. In effect, the swelling at the hot spot acts like a hill. Once the splitting occurs, the various fragments of the supercontinent begin to slip down the "hillside", away from the hot spot. Since the African continent has remained in the same position far longer than the other continents, earth scientists felt it was a good test of this hypothesis. Satellite observations of the height of Africa seem to show that it has risen about 400 m

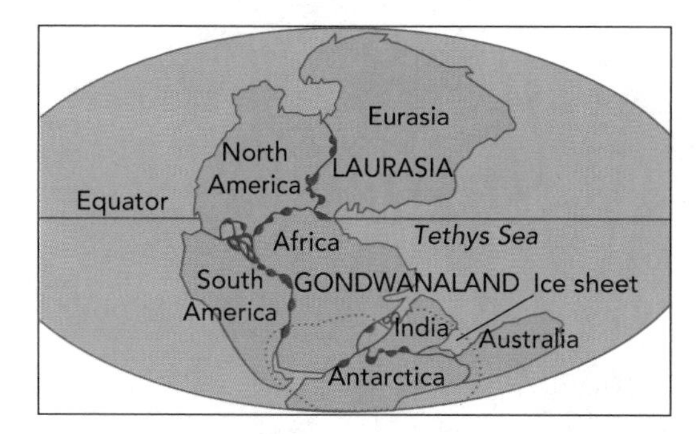

Figure 6.31 **The Supercontinent, Pangea, With Its Two Major Subdivisions of Laurasia and Gondwanaland**

since the break-up of the last supercontinent, Pangea, 200 million years ago.

On the other hand, where the crust of the earth was cooler, the denser material acted like a trough. Gravity then pulled the continents into the trough where they would be welded together along their collision boundaries. Recent research suggests that supercontinents similar to Pangea may have assembled and broken up at least six different times over the course of the earth's history. This is still a research frontier which is currently under investigation by earth scientists using the latest computer models and satellite data.

QUESTIONS

23. a) Based on Figure 6.29, draw a map of one continent showing its central craton and fringing active mountain belts.
    b) Africa is the only continent that does not display this pattern. Using Figure 6.1, suggest why this is so.
24. a) Using an outline map of the present world, attempt to reassemble the continent of Pangea.
    b) Describe the difficulties you encountered and why they might have arisen.
    c) Wegener and other earth scientists used the continental shelves in their reassemblies of Pangea. Why would this result in a more accurate map?
25. Draw a cartoon or sketch to illustrate the reasons for the break-up of Pangea, as suggested by a recent hypothesis.

It's a Fact . . .

- The oldest ocean basin rocks are about 180 million years old, about 4 percent of the earth's 4.6 billion years. The oldest continental rocks reach ages of 3.96 billion years, a rock record spanning almost 83 percent of the earth's history.
- The oceanic crust averages 5 to 8 km in thickness, while the continental crust averages 35 km and ranges from 10 to 70 km.
- Although oceans cover about 70 percent of the earth's surface, the oceanic basins actually make up only about 55 percent of the crust. The continental shelves — submerged areas of continental crust — combine with the continents to make up the other 45 percent of the earth's surface.

Review

- The theory of plate tectonics deals with the processes changing the earth's lithosphere.
- Plates are the rigid, moving sections of the lithosphere that float on the asthenosphere.
- Folding, faulting, and vulcanism are major tectonic processes.
- Plate boundaries are zones of intense tectonic activity. The three types of plate boundaries are diverging, transform fault, and converging.
- Plates are created and move apart at diverging boundaries, in association with volcanic and faulting processes.
- Plates slide past one another at transform fault boundaries, with associated faulting processes.
- Plates move together at converging plate boundaries, accompanied by folding, faulting, and volcanic processes.
- Tectonic processes at converging boundaries are violent and destructive.
- Plate interiors are relatively free of tectonic activities.
- The world's continents were part of the supercontinent of Pangea, which broke up 200 million years ago.

Geographic Terms

plate
asthenosphere
convection current
plate tectonics
tectonics
folding
faulting
vulcanism
hot spot
converging plate boundary
diverging plate boundary
transform plate boundary
collision zone
subduction zone
mid-ocean ridge
sea-floor spreading
magnetic reversal
viscosity
pillow lava
shield volcano
rift valley
transform fault
normal fault
reverse fault
fault scarp
strike-slip fault
graben
horst
block mountain
tilted block mountain
subduction
andesitic magma
continental volcanic arc
island arc
composite volcano
volcanic neck
cinder cone
dyke
sill
laccolith
batholith
anticline
syncline
recumbent fold
craton
accretion
supercontinent

Explorations

1. Research the basic ideas of continental drift proposed by Alfred Wegener in the early 1900s. Summarize the evidence Wegener used to justify his hypothesis and identify one major reason why his hypothesis failed to gain much support with the scientific community at the time.
2. a) Prepare a list of the most violent and destructive tectonic events over the past century or so.
   b) Plot the locations of these events on a world map.
   c) Describe the pattern of locations of these events.
   d) Explain how these events are related to plate boundaries and tectonic processes.
   e) Use this research to draw up a world map outlining zones of high, medium, and low tectonic risk.
   f) What use might such a map be to:
      i) someone deciding where to move in the world?
      ii) an insurance company?
      iii) an engineering company responsible for the locating of nuclear power plants?
3. How might a knowledge of plate tectonics theory and the movements of continents be useful to mining companies exploring for minerals, such as oil and gas, or metals?

Related Careers

- oceanographer
- physical oceanographer
- computer programmer
- geologist
- research technician
- cartographer
- geochemist
- prospector
- petrologist
- engineer
- marine biologist
- biologist
- seismologist
- biogeographer
- surveyor
- chemist
- geophysicist
- petroleum geologist
- hazard researcher
- civil engineer

Energy From Above

UNIT

4

The sun, a nuclear furnace located 150 000 000 km from the earth, provides the planet with a source of energy from above. This energy, called solar radiation, is responsible for the earth's climate and weather systems and for the ocean currents. It provides the earth with heat and light. It triggers the water cycle and allows for plant growth to occur. The five chapters in this unit examine the nature of this energy from above and explain the physical systems that are dependent on this energy input.

As you work through this unit, consider these questions:

- What effect does the earth's atmosphere have on solar radiation?
- What systems on earth are dependent upon the input of solar radiation?
- How would the earth's physical systems change if the amount of solar radiation that we receive was altered?
- In what ways are the various systems that depend upon the receipt of solar radiation interrelated?

CHAPTER

7

SOLAR RADIATION

OBJECTIVES:

By the end of this chapter, you will be able to:

- understand the nature of solar radiation;
- understand the characteristics of the layers of the atmosphere;
- appreciate the impact the earth's atmosphere has on solar radiation;
- identify several factors that will influence the amount of energy available for use on the earth's surface;
- summarize the ways solar energy is used by systems in the natural environment;
- recognize the actual and potential impact of humans on the earth's energy balance.

CHAPTER 1: The Nature of Physical Geography
CHAPTER 2: Earth: Its Place in the Universe
CHAPTER 3: The Earth in Motion
CHAPTER 4: The Earth's Interior
CHAPTER 5: The Earth's Crust
CHAPTER 6: The Lithosphere in Motion: Plate Tectonics
CHAPTER 7: Solar Radiation
CHAPTER 8: Climate
CHAPTER 9: Weather
CHAPTER 10: The Hydrosphere and the Hydrologic Cycle
CHAPTER 11: Natural Vegetation and Soil Systems
CHAPTER 12: Denudation: Weathering and Mass Wasting
CHAPTER 13: Distinctive Landscapes: Humid and Arid Environments
CHAPTER 14: Distinctive Landscapes: Glacial, Periglacial, and Coastal Environments
CHAPTER 15: Natural Hazards: Disrupting Human Systems
CHAPTER 16: The Disruption of Natural Systems
CHAPTER 17: Fragile Environments

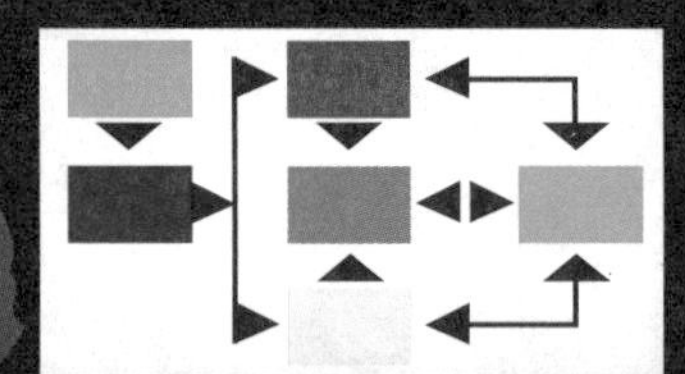

Introduction

The sun is the source of the energy that the earth receives from "above". With few exceptions, this energy input, which we call **solar radiation**, is responsible for the earth's biological and climatic systems. Animal life, plant growth, weather systems, and ocean currents all depend upon the heat and light derived from the daily input of solar radiation. In this section of the text, we will examine the characteristics of solar radiation and describe how the earth's biological and climatic systems make use of "energy from above".

7·1 The Source

Our sun is a star of average size and intensity, which releases vast amounts of electromagnetic energy. This energy is usually visualized or thought of as travelling in the form of waves. The source of this energy is the nuclear reaction involving the fusion of hydrogen to helium which is constantly occurring in the sun's core. Hydrogen makes up approximately 72 percent of the sun, while helium comprises another 27 percent. Figure 7.1 illustrates the characteristics of the different zones of the sun.

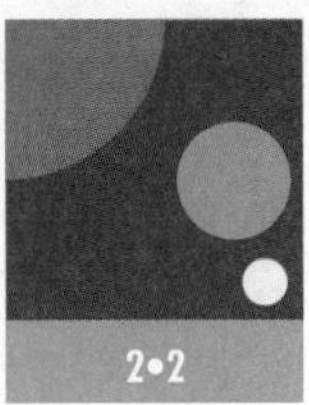

All objects will emit or give off energy in the form of **electromagnetic waves** if they are hotter than absolute zero. As the temperature of the emitting body increases, the amount of radiation released increases. That is, the greater the temperature, the greater the amount of radiation. Also, as the temperature of the emitting body increases, the wavelengths of the radiation decrease in size. That is, the greater the temperature, the shorter are the wavelengths of the radiation.

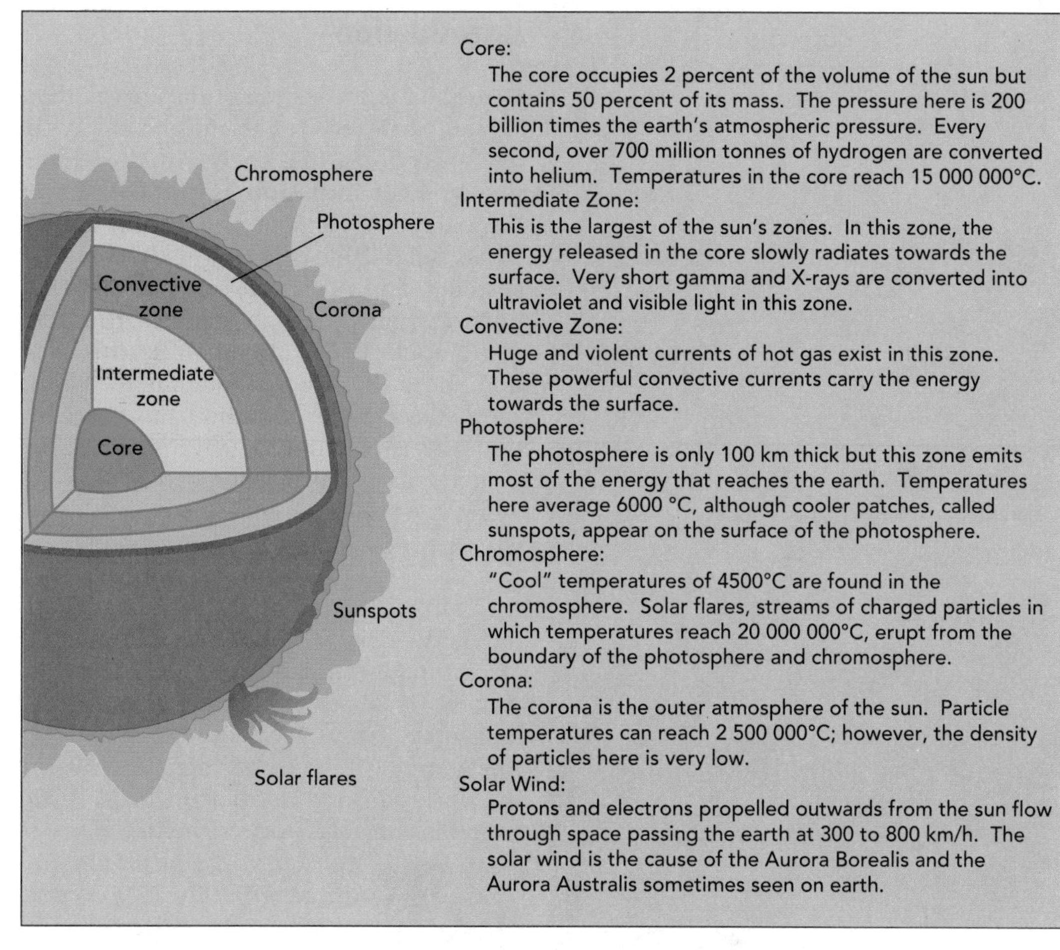

Figure 7.1 The Sun — Source of Our Energy From Above

It's a Fact...

The shortest waves on the electromagnetic spectrum are gamma rays which have a wavelength of less than 0.00 001 micrometres.
(1000 micrometres = 1 mm)

The sun, whose surface temperature is approximately 6000°C, emits a great deal of energy, primarily in short wavelengths. The earth, whose surface temperature is quite cool compared to the sun, emits much less energy and in much longer wavelengths than the solar radiation emitted by the sun. Figure 7.2 summarizes the radiation amounts and wavelengths emitted from the sun and earth.

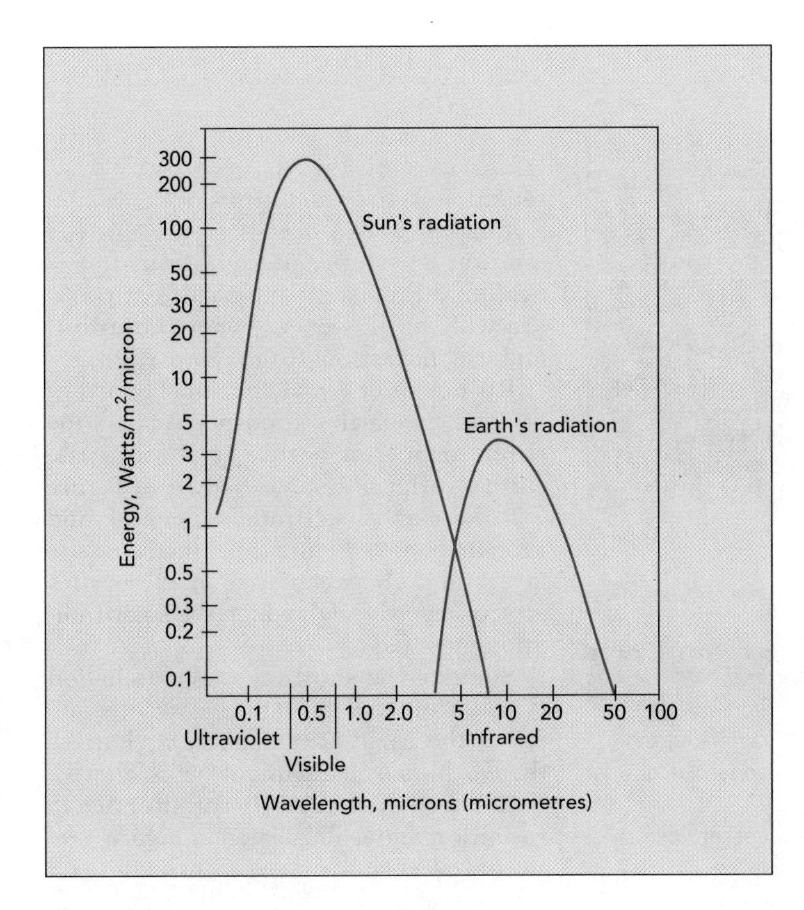

Figure 7.2
Solar and Terrestrial Radiation Spectrums
(A micron or micrometre equals one one-thousandth of a millimetre. A watt is a measurement of the rate at which energy flows and is equal to one joule per second.)

The Trip to Earth

The solar radiation released from the sun is emitted in all directions. A very small fraction of this solar radiation reaches the outer limits of the earth's atmosphere having travelled the 150 million kilometres from the sun at the speed of light (300 000 km/s). It takes approximately 500 s for the solar radiation to reach the outer limits of the earth's atmosphere.

The **solar constant** is a term used to describe the amount of solar energy received at the very outer limit of the earth's atmosphere. It is equal to approximately 1400 W/m^2.

It is estimated that only 0.00 000 005 percent of the energy released by the sun actually reaches the outer limits of the earth's atmosphere. Nevertheless, this small percentage represents a large amount of energy that fuels the earth's biological and climatic systems.

If the amount of solar energy that the earth receives each day is compared to other energy outputs or uses, the magnitude of the solar radiation that we receive from the sun each and every day becomes even clearer. (See Figure 7.3.)

Daily solar energy received by the earth	1
World production of energy in 1990	0.1
Strong earthquake	0.01
Daily heat flux from the earth's interior	0.0001
Major hurricane	0.0001
Daily output of the world's largest hydro-electric dam	0.0000001
Average summer thunderstorm	0.00000001

Note: The magnitude of the daily solar radiation reaching the outer limits of the earth's atmosphere has been arbitrarily set at 1.

Figure 7.3 Comparative Magnitudes of the Daily Input of Solar Energy With Various Other Energy Sources

QUESTIONS

1. Examine Figure 7.2. State three differences that exist between the solar and terrestrial radiation spectrums.
2. Why is the solar radiation system considered an "open cascading system"? (Refer back to Chapter 1, page 8, if necessary.)
3. Rank the following from the item emitting its maximum radiation in the longest wavelength to the item emitting its maximum radiation in the shortest wavelength:
   - an element on an electric stove turned on to high
   - a sandy beach on a hot summer's day
   - a human body sitting in a classroom
   - the sun
4. How far away would a star be if it required four years for the radiation released by that star to reach the earth?
5. It has been estimated that every second the sun releases the energy equivalent of 100 billion hydrogen bombs. Why does the earth only receive a small fraction of this energy?
6. Attempt to display the information in Figure 7.3 in a visual format.

7·2 The Earth's Solar Radiation Balance

We have traced the sun's shortwave solar radiation to the outer limits of the earth's atmosphere. We are interested in the amount of energy or radiation that is available for use at the earth's surface. We will call this energy "**net radiation**" and use the symbol **Rn** to represent it.

Rn does not equal the solar constant because the earth's atmosphere alters the solar radiation before it reaches the earth's surface. Also, as shown in Figure 7.2, the earth itself radiates energy and influences **Rn**. Figure 7.4 illustrates the layers and characteristics of the earth's atmosphere through which the solar radiation must pass.

Some of the shortwave solar radiation is able to travel through the various layers of the earth's atmosphere and strike the earth's surface without being altered. This energy is called direct **shortwave radiation** and is illustrated in Figure 7.5.

However, you can see from Figure 7.5 that most of the solar radiation is not able to travel directly through the atmosphere without being altered. Although the value for direct shortwave radiation in Figure 7.5 is a generalization and can differ dramatically for different locations and conditions, the direct shortwave radiation at any one location is usually only 20 to 30 percent of the value of the solar constant.

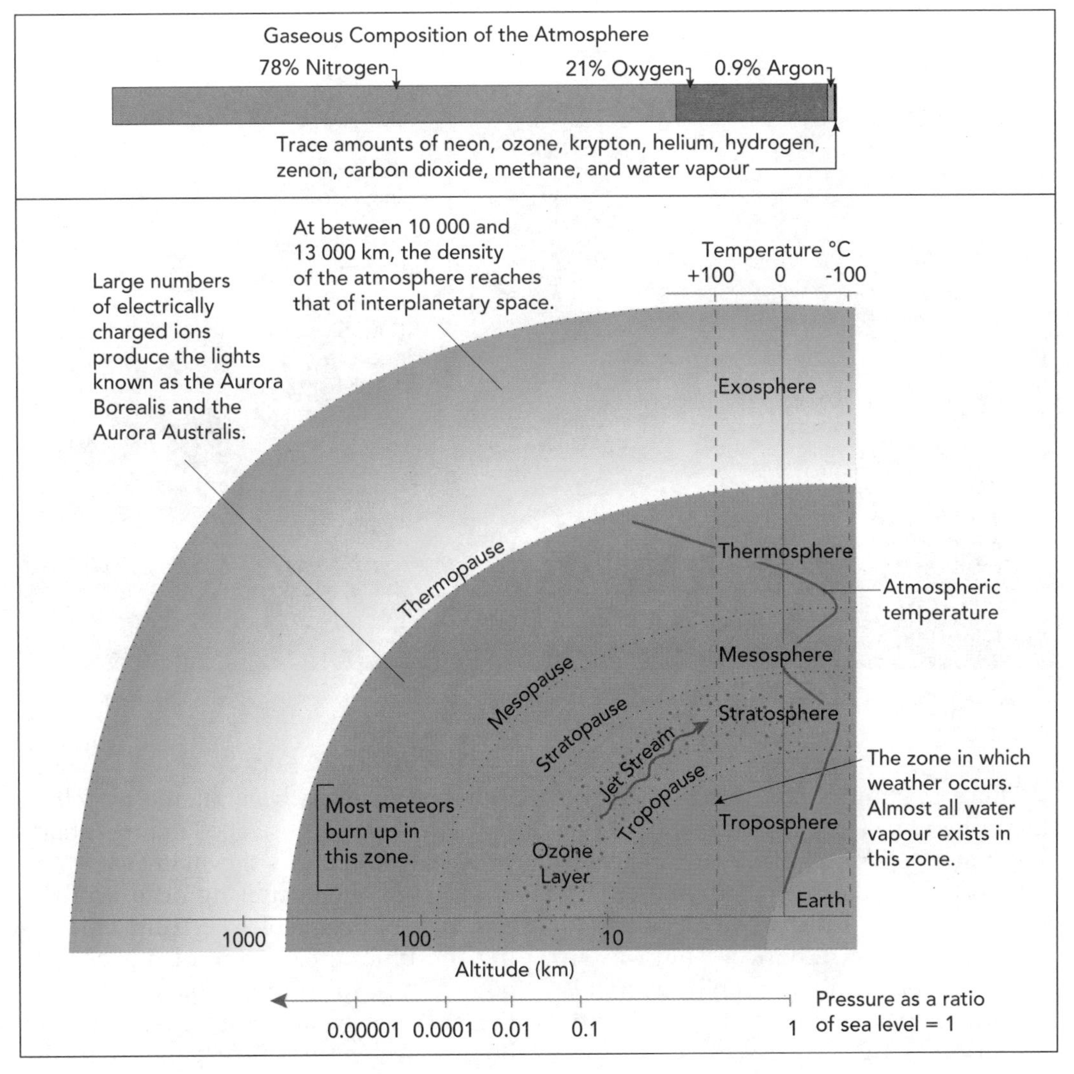

Figure 7.4 **The Characteristics of the Layers of the Earth's Atmosphere**

It's a Fact...

The atmosphere is less transparent to longwave radiation than it is to shortwave radiation. This fact sets up a natural "greenhouse effect" in the atmosphere. The amount of longwave radiation leaving the earth "system" is thus less than the amount of shortwave radiation entering the system. Without this natural greenhouse effect, the earth's average temperature would be -33°C instead of its present +15°C.

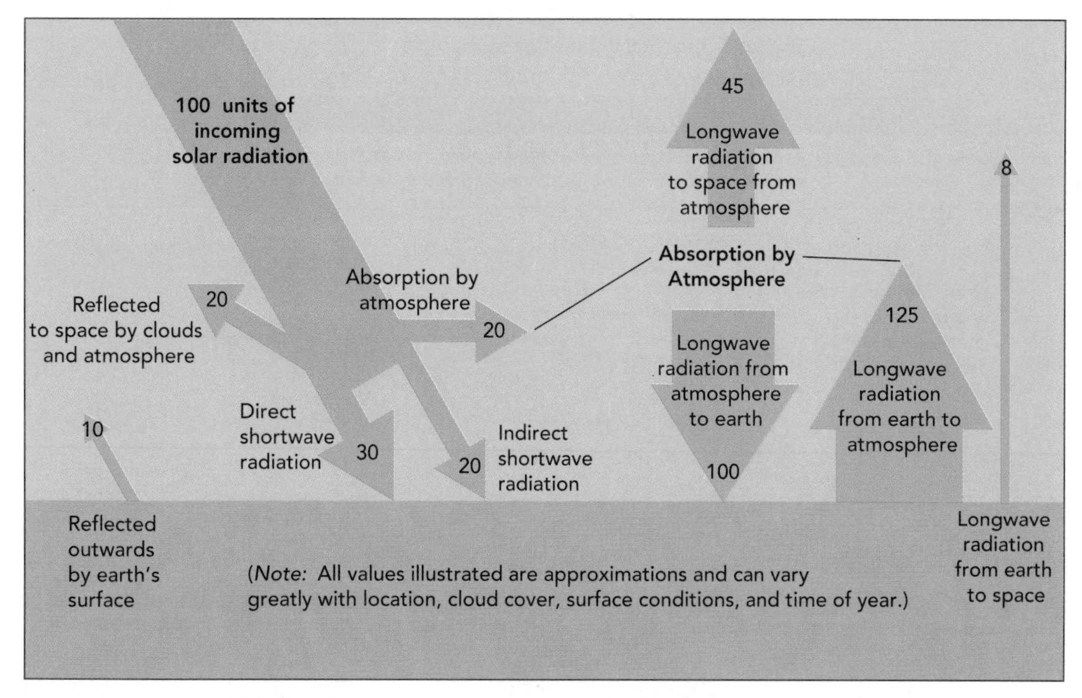

Figure 7.5 **The Radiation Balance at the Earth's Surface**

There are many components in the atmosphere, including dust, pollen, pollution, water vapour, and atmospheric gases such as carbon dioxide (CO_2), ozone (O_3), and oxygen (O_2), which have the ability to affect shortwave radiation. One way in which the solar radiation can be altered is by reflection. Similar to the process of reflection is the process of scattering. When scattering occurs, the wavelength of the solar radiation is changed. The solar radiation that has been altered by reflection and scattering is called indirect shortwave radiation and is shown in Figure 7.5.

Much of the scattering occurs in the blue part of the visible spectrum, enhancing the blue appearance of our sky. At other times, for example during sunsets, the shortwave radiation is scattered in other wavelengths corresponding to the orange, yellow, and red bands of the visible spectrum.

It's a Fact...

The visible spectrum lies between 0.4 and 0.7 mircrometres. These are the wavelengths that the human eye can "see". The maximum wavelength for solar radiation is 0.49 micrometres. This wavelength corresponds to the blue section of the visible spectrum. Our sky thus appears to us as blue.

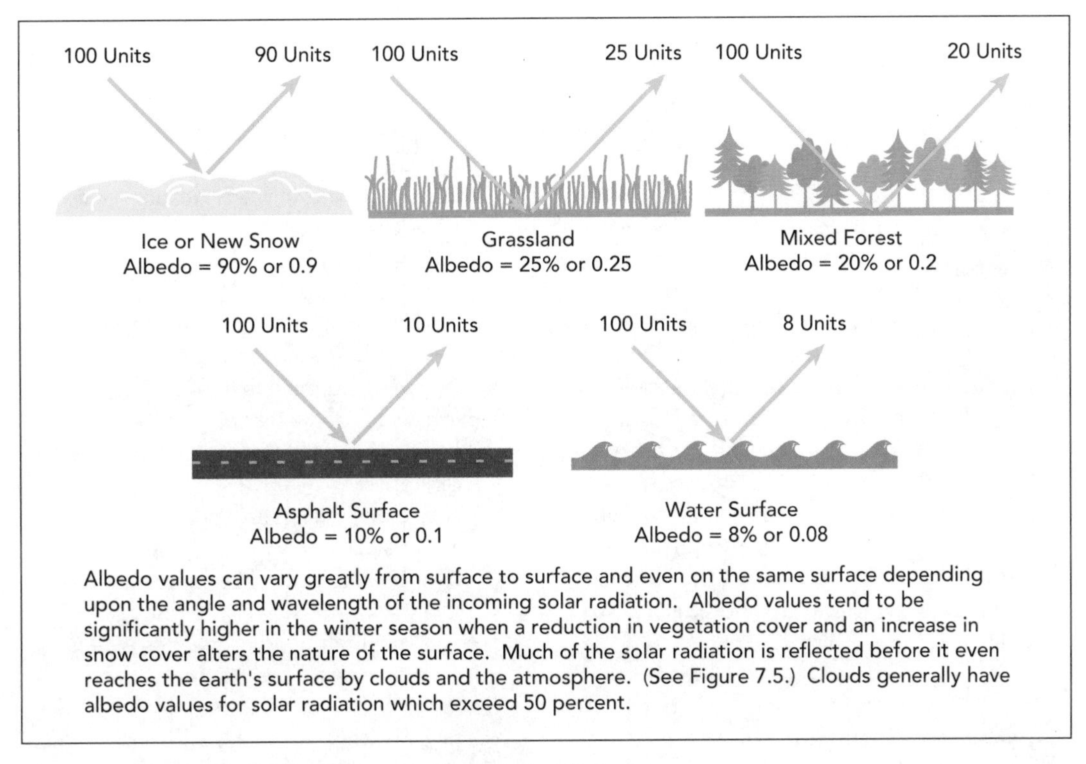

Figure 7.6 **Typical Albedo Values for Various Earth Surfaces**

However, not all of the incoming shortwave radiation is available for use at the earth's surface. Some of this incoming shortwave radiation is reflected or bounced off the earth's surface. The amount that is reflected is called **albedo** and is dependent upon the angle of the incoming radiation, the texture of the surface receiving the radiation, the colour of the surface receiving the radiation, and the wavelength of the radiation itself. At low angles, with smoother surfaces, and with lighter colours, the albedo is higher. A field of ice will therefore have a high albedo value, whereas a black asphalt road will have a low albedo value. The effect that the wavelength of the incoming radiation has on the albedo varies greatly from surface to surface. Some surfaces' albedos will peak at a particular wavelength but decline for both longer and shorter wavelengths. Figure 7.6 illustrates some typical albedo values for earth surfaces. In Figure 7.5, 10 percent of the shortwave radiation is being reflected or bounced off the earth's surface.

Some of the solar radiation on its path through the earth's atmosphere is absorbed or "captured" by particles and molecules. Clouds, dust, water vapour, carbon dioxide, oxygen, and ozone can capture or absorb solar radiation. Due to their molecular structures, certain components of the atmosphere are particularly adept at absorbing certain wavelengths. For example, ozone, which is concentrated in the stratosphere (see Figure 7.4), absorbs most of the very short, ultraviolet wavelengths of solar radiation, those below 0.3 micrometres. Upon receiving this added energy, the absorbing body will heat up and become warmer than its surroundings. It will then radiate energy to its cooler surroundings. Because the atmosphere is not nearly as hot as the sun, the energy emitted by it is called **longwave radiation**. This longwave radiation is emitted in all directions. When it is radiated down to the earth's surface, it represents an additional energy input for use at the earth's surface. This energy is called longwave incoming radiation. Some of the absorbed energy is radiated outwards by the atmosphere and does not add to or take away from the quantity of radiation available at the earth's surface. (See Figure 7.5.)

Finally, the earth itself absorbs a portion of the radiation that reaches it, heats up, and radiates longwave radiation back into the atmosphere. Most of this radiation is re-absorbed by the atmosphere, while a smaller amount escapes the earth's atmosphere and reaches outer space. Carbon dioxide found in the atmosphere is very efficient at absorbing longwave radiation in the 13.1 to 16.9 micrometre range, while water vapour absorbs much of the radiation between 5.3 and 7.7 micrometres. This longwave radiation leaving the earth's surface is shown in Figure 7.5. The longwave radiation exchanges, or fluxes, are complex due to the cycle of continuous absorption and re-emittance between the earth and the atmosphere.

At night, when there is no input of solar radiation into the system, **Rn** is determined simply from the longwave terms and is almost always negative.

QUESTIONS

7. Consider two locations in the same city. Location A is in clean air, while Location B is located under a pollution haze caused by industrial activity. In which location would each of the following terms have the greatest value? Draw a labelled diagram to help explain your answer.
   - direct shortwave radiation
   - indirect shortwave radiation
   - longwave incoming radiation
8. Explain each of the following observations:
   - People in hot climates tend to wear white clothes.
   - A field of ice is not melting on a sunny winter's day.
   - An asphalt driveway becomes sticky on a sunny summer's day.
   - The temperature drops more on a clear night than on a cloudy night.
9. Redraw Figure 7.5 to illustrate the effect that a major volcanic eruption would have on the values of direct shortwave radiation, indirect shortwave radiation, and incoming longwave radiation.
10. Why is it difficult to exactly define the outer limits of the earth's atmosphere?

7·3 Variation in Rn Values Over Space and Time

The radiation reaching the earth's surface varies greatly from place to place. We noted earlier that cloud cover and albedo values greatly affect **Rn**. Chapter 8 discusses the influence that latitude has on **Rn** and explains the important climatic effects that result from the variation in net radiation values received at the earth's surface.

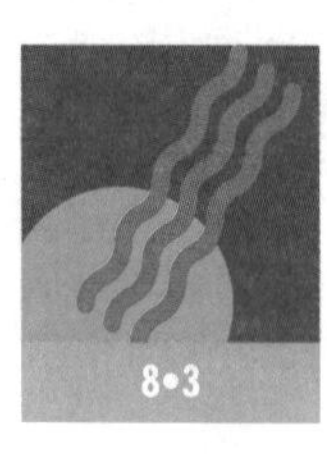

The radiation available at the earth's surface has not always been constant. Volcanic eruptions can spew vast amounts of ash into the atmosphere. At times in the past, such eruptions have reduced the amount of incoming solar radiation and thus cooled the earth by reducing **Rn**. A collision of a large meteor with the earth might also have the same effect. Some theories suggest that such events as a meteor collision or volcanic activity have in the past significantly changed **Rn**, resulting in the extinction of entire species.

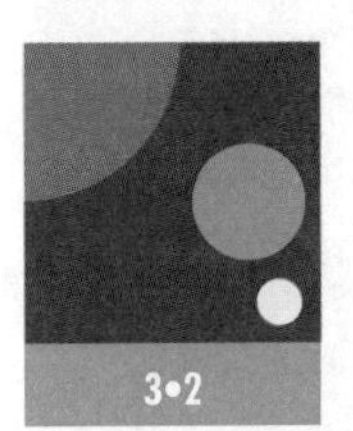

The amount of the sun's energy that we receive can fluctuate. In Chapter 3, we referred to the Milankovitch Cycles which suggest that changes in incoming solar radiation of up to 5 percent can occur when the variations in the tilt of the earth's axis correspond to certain variations in the earth's orbit. Many scientists are now convinced that the Milankovitch Cycles are the key to explaining why the earth experiences periods of glaciation.

In the twentieth century, human activities have altered and are continuing to alter the earth's radiation balance in many planned and unplanned ways. Frost protection for crops and solar heating techniques are examples of planned alterations of the radiation balance. These techniques are illustrated in Figures 7.7 and 7.8.

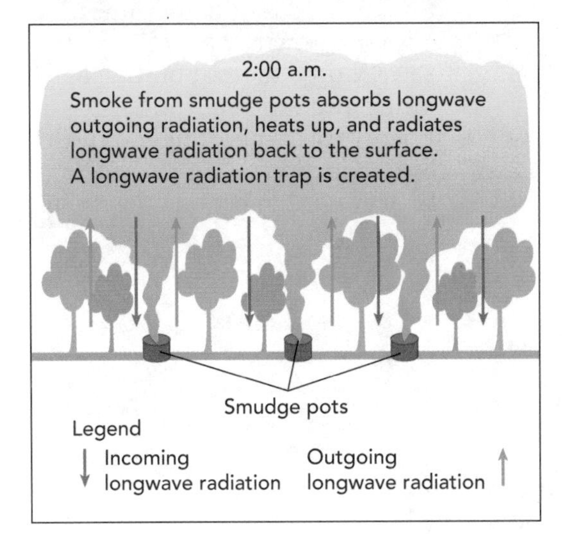

Figure 7.7 The Planned Alteration of the Radiation Balance for the Purpose of Frost Protection

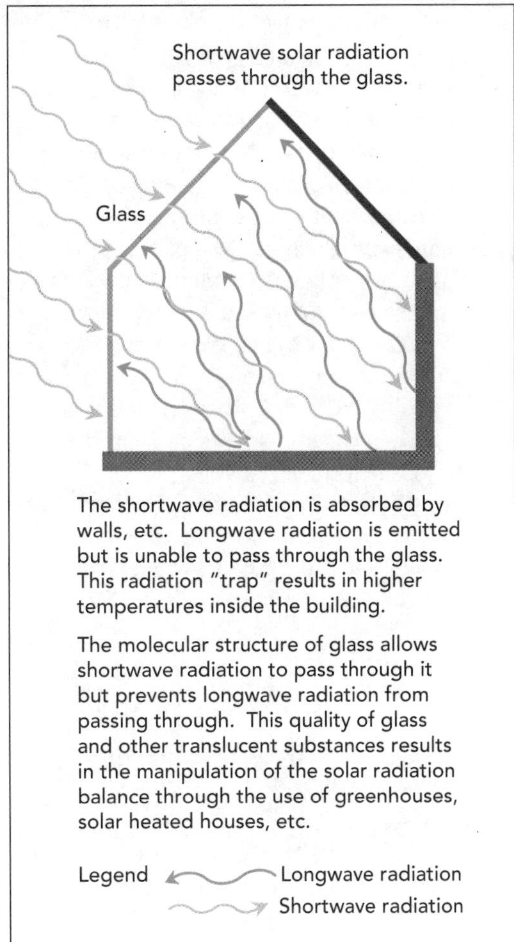

Figure 7.8 The Planned Alteration of the Radiation Balance for the Purpose of Solar Heating

The greenhouse effect and the depletion of the earth's ozone layer have brought about unplanned alterations in the radiation balance which have the potential to alter and threaten life on earth. These important environment issues are outlined in Chapter 16.

QUESTIONS

11. Choose either Figure 7.7 or 7.8 and explain the diagram in words that a ten-year-old student would understand.
12. The Inuit who lived in the Arctic traditionally used "glasses" made of bone or wood with narrow horizontal slits to protect their eyes from "snow blindness". The glasses protruded two or three centimetres outwards from the eyes. What would cause "snow blindness" and why might these glasses have been effective in preventing it?
13. Football and baseball players put black cork or powder under their eyes on sunny days. Why do they do this?
14. Why do you feel immediately cooler on a hot summer's day when a cloud passes in front of the sun? Refer to Figure 7.5 in your answer.

7·4 The Energy Balance

We have traced our "energy from above", solar radiation, from its origin in the sun, through outer space, through the earth's atmosphere to the surface of the earth. As this energy is converted and consumed on earth, the sun provides more every day of every year. This energy system is "open" because we will continue to be supplied with solar radiation for at least another five billion years. Figure 7.5 indicates that the earth receives more radiation than it gives off. The excess radiation available at the earth's surface, **Rn**, is used by the earth in one of four ways. The **energy balance** equation summarizes how the earth makes use of **Rn**.

The energy balance equation reads:

Rn = Ph + E + G + H

where,

Rn = Net radiation, the amount of energy available for use at the earth's surface

Ph = Energy used for photosynthesis or plant growth

E = Energy used for evaporation and transpiration

G = Energy used in heating the earth by conduction

H = Energy used in heating the air by convection

It's a Fact...

Conduction is the process that occurs when heat is transferred from molecule to molecule. It involves no movement of particles. Convection is the process that occurs when heat is transferred through the movement of substances or particles. Little conduction but a great deal of convection occurs in the atmosphere. Little convection but a great deal of conduction occurs in the earth itself.

The sizes of the variables in the energy balance vary from location to location. For example, more energy is being used to evaporate water over a warm ocean than over a dry desert. Also, more energy is being used by plants for photosynthesis in a rainforest than in an Arctic tundra area.

The energy balance equation stresses once again the importance of the energy that we receive from above. Plant growth, the water cycle, and the warming of our earth and atmosphere are all dependent on solar radiation. Figure 7.9 illustrates the subsystems on earth which are dependent upon **Rn**. The following chapters examine these subsystems.

QUESTIONS

15. Chapter 16 discusses the "Greenhouse Effect" which involves an increase in the amount of radiation available at the earth's surface. How might each of the following be affected by the "Greenhouse Effect": cloud cover, vegetation growth, sea levels, your own life?
16. Examine Figure 8.9 on page 141. This map shows the amount of radiation available at the earth's surface on an annual basis. Using Figure 8.9 and the energy balance equation, explain why Arctic areas are generally drier and colder and contain less plant life than the mid-latitude and equatorial areas on earth.

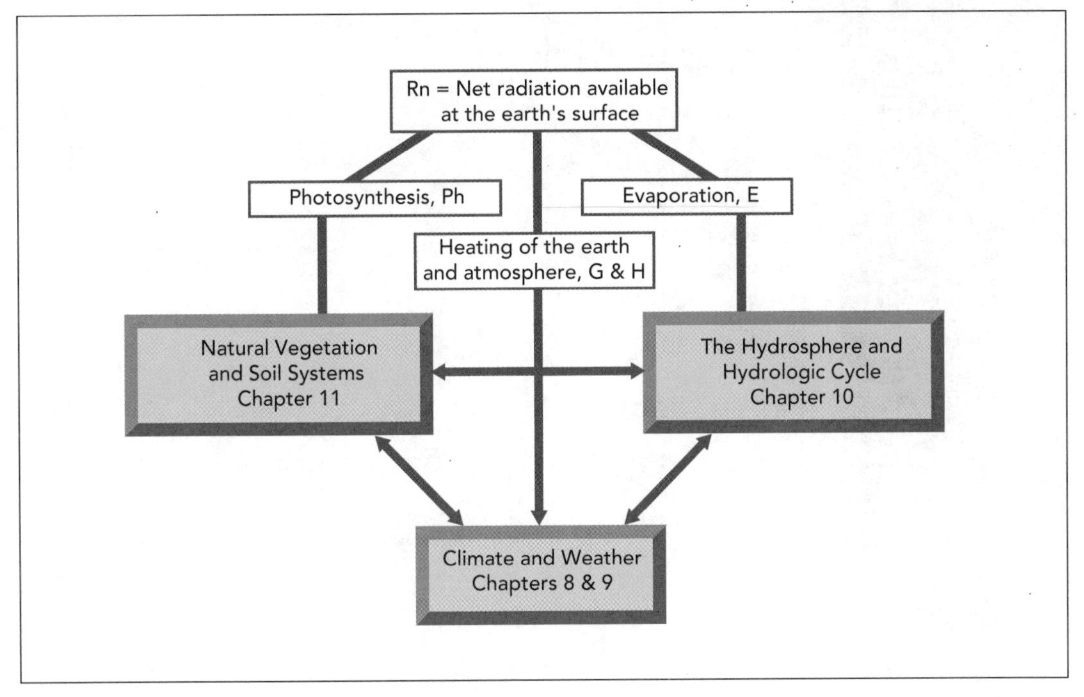

Figure 7.9 The Systems Influenced by the Amount of Radiation Available at the Earth's Surface

Review

- Because of its great heat, the sun gives off tremendous amounts of shortwave radiation.
- Part of the solar radiation travels directly through the earth's atmosphere to reach the earth's surface.
- Most solar radiation is affected by the atmosphere, either being reflected, scattered, or absorbed.
- Some of the radiation that does reach the earth's surface is reflected back to the atmosphere; this is referred to as albedo.
- The solar radiation that reaches the earth's surface warms it up and the earth radiates longwave radiation back to the atmosphere.
- The amount of energy available for use at the earth's surface is the net radiation.
- Humans have the ability to modify the net radiation available at the earth's surface in many planned and unplanned ways.
- The energy is used by the natural systems of the planet to support life.

Geographic Terms

solar radiation
electromagnetic wave
solar constant
net radiation (**Rn**)
shortwave radiation
albedo
longwave radiation
energy balance

Explorations

1. Many societies have used the sun as a religious symbol. Why do you think that they have done this?
2. Why was this chapter in the book placed before Chapters 8, 9,10, and 11? (See Figure 7.9.)
3. What procedures do you use to manipulate solar radiation in your own house or apartment?
4. Describe an experiment or demonstration that you could use to illustrate to a friend how the solar radiation balance works. (Figures 7.5, 7.6, 7.7, and 7.8 might help you to think of ideas.)

CHAPTER

8

CLIMATE

CHAPTER 1: The Nature of Physical Geography

CHAPTER 2: Earth: Its Place in the Universe

CHAPTER 3: The Earth in Motion

CHAPTER 4: The Earth's Interior

CHAPTER 5: The Earth's Crust

CHAPTER 6: The Lithosphere in Motion: Plate Tectonics

CHAPTER 7: Solar Radiation

CHAPTER 8: **Climate**

CHAPTER 9: Weather

CHAPTER 10: The Hydrosphere and the Hydrologic Cycle

CHAPTER 11: Natural Vegetation and Soil Systems

CHAPTER 12: Denudation: Weathering and Mass Wasting

CHAPTER 13: Distinctive Landscapes: Humid and Arid Environments

CHAPTER 14: Distinctive Landscapes: Glacial, Periglacial, and Coastal Environments

CHAPTER 15: Natural Hazards: Disrupting Human Systems

CHAPTER 16: The Disruption of Natural Systems

CHAPTER 17: Fragile Environments

OBJECTIVES:

By the end of this chapter, you will be able to:

- understand the relationship between weather and climate;
- understand important terms that are associated with the study of climate;
- appreciate the importance of classifying climatic conditions in various ways;
- recognize that the climatic patterns on earth are complex and that they result from a wide variety of interacting forces which include the earth's basic motions, the earth's surface features, and the arrangement of the earth's land masses and water bodies;
- identify the importance of the sun in powering climatic systems;
- appreciate that our understanding of climate and weather is still developing and that many theories have yet to be proven to be true;
- appreciate the extreme variability and complexity of the earth's climates;
- examine ways that humans both influence and are influenced by climate and weather.

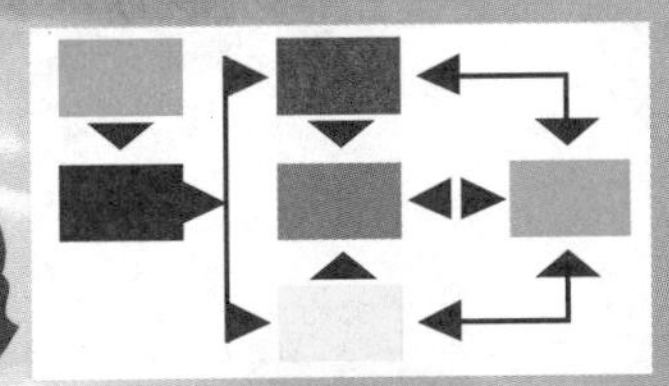

Introduction

The earth receives massive amounts of energy — solar radiation — from the sun, and this energy powers the physical systems that give the planet its climatic characteristics. But, the climatic patterns are complex. This is due to the variety of ways in which the sun's energy is used and influenced by forces on the earth. In this chapter, you will identify the forces that act as climatic controls and explore the ways these controls shape global climate patterns. You will recognize that humans, too, can influence the climate in many ways.

8·1 Climate and Weather

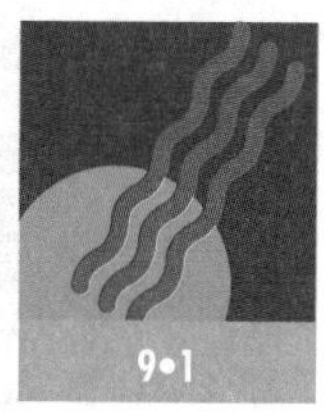

Perhaps no systems in physical geography affect people more directly than those associated with climate and weather. **Weather** refers to the short-term, usually daily, characteristics of the atmosphere. Wind, humidity, cloud cover, visibility, temperature, precipitation, and air pressure are the variables that comprise our weather. **Climate** takes into account the long-term characteristics of our atmosphere. When the weather at a location on the earth's surface is measured over many years, an average can be obtained. This average represents the climate of an area.

In a few locations on earth, the terms "weather" and "climate" are almost interchangeable. This means that the weather does not significantly change from day to day and, therefore, the climate represents an accurate summary of the weather conditions. If one knew the climate of these locations, then one could accurately predict the weather. Consider the equation:

(4 + 4 + 4 + 4)/4 = 4. Since all of the terms in the numerator are "4", then the average is also "4". In other words, weather equals climate. However, in the equation: (2 + 4 + 5 + 5)/4 = 4, the average is only the same as one of the digits in the numerator. In other words, three-quarters of the terms in the numerator (weather) are different from the average (climate). In most locations on earth, climatic statistics are remarkably different from the day-to-day atmospheric characteristics. Studying only the climatic statistics provided by travel agents is a hazardous way to predict the actual weather you would experience at a holiday resort.

For most locations in the world, the day-to-day weather is very changeable, while the climate, averaged over many years, is relatively stable and unchanging. In this chapter, we will first examine the complex and varied climates that exist on earth.

8·2 Climatic Classification Systems

There are thousands of different climatic stations throughout the world, each one unique in its own way. However, in examining climatic data, similarities emerge which form a basis by which we can classify climates. We classify climates in order to identify patterns that exist on the earth's surface. The identification of patterns or common climatic characteristics allows us to study and isolate those factors that are controlling our climate. Many climatic classification systems exist, most of which are based on the twin variables of temperature and precipitation. Examine Figure 8.1. This figure represents a simple way in which to classify climates.

Most climatic classification systems are far more complex than the one illustrated in Figure 8.1. The most famous classification system was developed by Dr. Wladimir Köppen in 1918. (See Figure 8.2.) This classification system uses a hierarchical series of letters to group together areas of the world that have similar climates. A less mathematical, more descriptive system of climatic classification remains popular in many texts today. (See Figure 8.3.)

The climatic systems outlined in Figures 8.1 through 8.3 illustrate the remarkable diversity that exists in the world's climates. This diversity is the result of many factors or climatic controls that interact with one another. Some of these controls are global in their impact, such as the patterns of winds and ocean currents, while others exert their influence on a continental or regional scale, for example, landforms and the influence of bodies of water. Still other controls influence the climates of limited areas, exerting a local influence. This chapter explains the various factors that influence the earth's climates.

It's a Fact...

The sunniest place in the world is the eastern Sahara desert, which has a mean annual sunshine total of 4300 hours, 97 percent of the possible total.

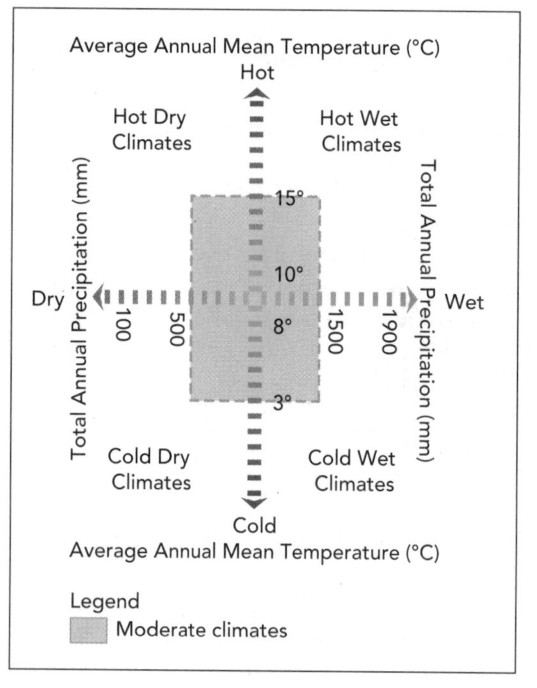

Figure 8.1 A Simple Climatic Classification System

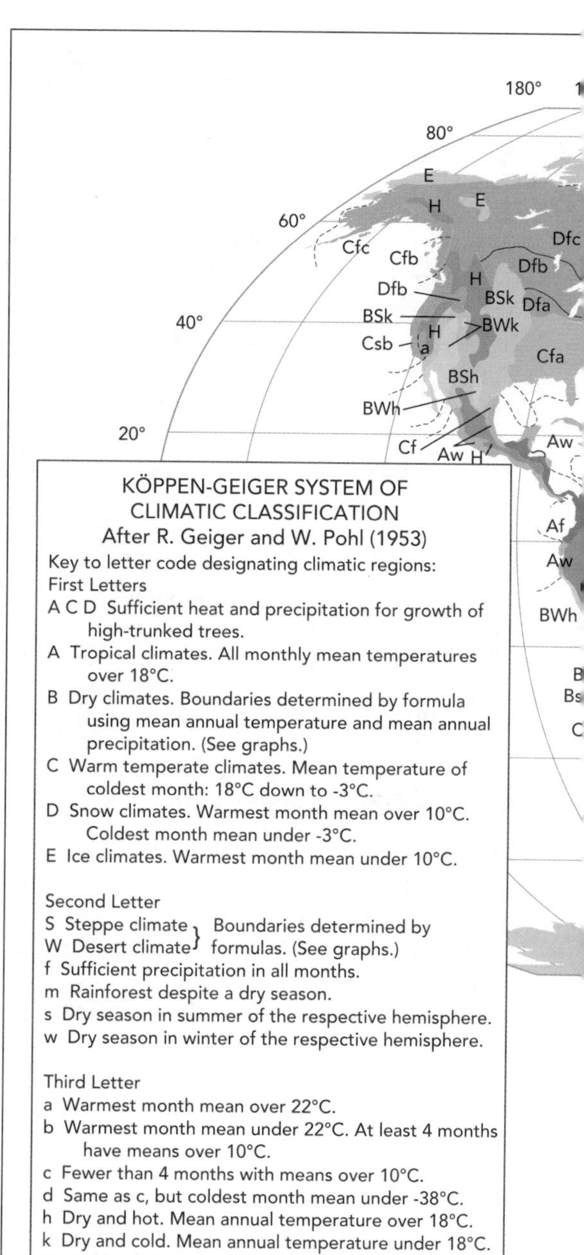

KÖPPEN-GEIGER SYSTEM OF CLIMATIC CLASSIFICATION
After R. Geiger and W. Pohl (1953)

Key to letter code designating climatic regions:

First Letters

A C D Sufficient heat and precipitation for growth of high-trunked trees.
A Tropical climates. All monthly mean temperatures over 18°C.
B Dry climates. Boundaries determined by formula using mean annual temperature and mean annual precipitation. (See graphs.)
C Warm temperate climates. Mean temperature of coldest month: 18°C down to -3°C.
D Snow climates. Warmest month mean over 10°C. Coldest month mean under -3°C.
E Ice climates. Warmest month mean under 10°C.

Second Letter

S Steppe climate } Boundaries determined by
W Desert climate } formulas. (See graphs.)
f Sufficient precipitation in all months.
m Rainforest despite a dry season.
s Dry season in summer of the respective hemisphere.
w Dry season in winter of the respective hemisphere.

Third Letter

a Warmest month mean over 22°C.
b Warmest month mean under 22°C. At least 4 months have means over 10°C.
c Fewer than 4 months with means over 10°C.
d Same as c, but coldest month mean under -38°C.
h Dry and hot. Mean annual temperature over 18°C.
k Dry and cold. Mean annual temperature under 18°C.
H Highland climates.

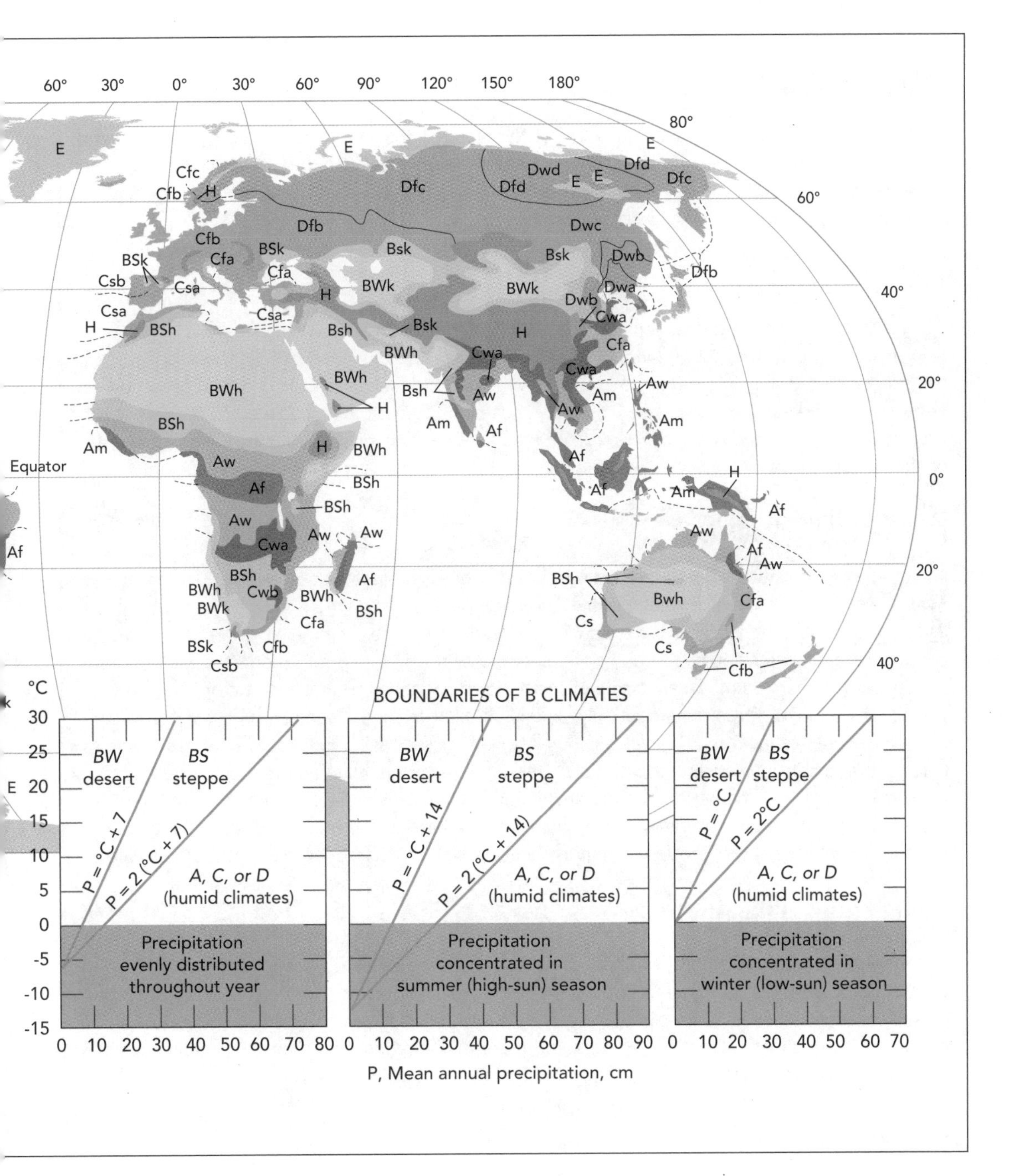

Figure 8.2 World Map of Climates According to the Kôppen-Geiger-Pohl System

Tropical Climates

These climates are usually found within 15° of the equator and are characterized by hot temperatures throughout the year. In *tropical wet* climates, significant amounts of precipitation occur throughout the year, while in *tropical wet-dry* climates, a distinct dry season can be recognized. Monsoon climates can be included in this latter subdivision.

Desert Climates

Where precipitation is extremely limited, desert conditions occur. *Hot weather deserts* such as the Sahara can be easily distinguished from *cold weather deserts* found in high latitudes. Specific climatic factors can give rise to desert conditions in various world locations. Cold ocean currents flowing along the western edges of the Americas and Africa combine with very stable air to produce *west-coast deserts* such as the Atacama in Peru and the Namib in Namibia and Angola. Rainshadow areas on the leeward side of many of the world's great mountain ranges can result in desert conditions.

Mid-latitude Climates

Great variety highlights mid-latitude climates. Warm, dry summers and cool, wet winters are characteristic of *Mediterranean climates* which occur in the Mediterranean, California, Chile, South Africa, and parts of western Australia. *West-coast climates* are found in areas on the windward side of mountain ranges and are characterized by heavy amounts of precipitation and moderate temperatures throughout the year. On the leeward side of these mountain ranges, drier climates called *mid-latitude steppe* can be found. These climates usually experience a large range of temperatures between winter and summer. *Humid subtropical climates* generally experience rainfall throughout the year. They experience warm to hot summers and cool winters. They are most often found on the southeastern sides of the continents. *Humid continental climates* are similar to humid subtropical climates but are found farther north and, therefore, experience more severe winters with significant snowfall amounts. Cold winters and warm summers with precipitation throughout the year characterize humid continental climates.

High Latitude Climates

Because of their extreme latitudes, these climates are characterized by extremely cold winters and short, cool summers. *Continental/subarctic/climate* areas experience very cold, long winters and receive a minimal amount of precipitation. *Marine subarctic climates* are noted for their wet, windy conditions. They experience milder winters and more precipitation than their continental subarctic cousins. True *Arctic climates* are found north of 70° and are so cold they experience no real summer. They are also very dry and in most areas can be classified as cold weather deserts.

Highland Climates

Variations in altitude can result in all the above climates being experienced on a single mountain near the equator. Great differences over short distances are characteristic of *highland climates* and, therefore, it is very difficult to accurately classify the climates of mountainous areas.

Figure 8.3 **A Qualitative Descriptive Climatic System**

QUESTIONS

1. List what you consider to be three advantages and three disadvantages of the climate experienced in the area in which you live.
2. Figures 8.4 and 8.5 display climatic data for four world locations. Figure 8.4 shows two **climographs**, while Figure 8.5 shows two **hythergraphs**. For each location, choose two different classification systems (Figures 8.1 through 8.3) and classify the climate at that location according to the climatic classification system being used.
3. Compare climographs and hythergraphs. What are the advantages and disadvantages of each of these methods used to display climatic data?
4. Why would the study of world climates be complicated if we did not classify the climates in various ways?
5. Identify three ways in which the climate of your area influences the economic activities found there.

8·3 Climatic Controls: The Variations in Solar Radiation Inputs

The nature and importance of solar radiation were considered in Chapter 7. The amount of solar radiation available for use at the earth's surface varies greatly from location to location and from season to season. These variations lead to dramatic differences in world climates. An important factor that contributes to variations in solar radiation is the latitude of places on the earth's surface.

Due to the curvature of the earth, solar radiation received at higher latitudes is less intense than at the equator. This is because the same amount of solar radiation is spread out over an ever greater area of the earth's surface, as Figure 8.6 illustrates.

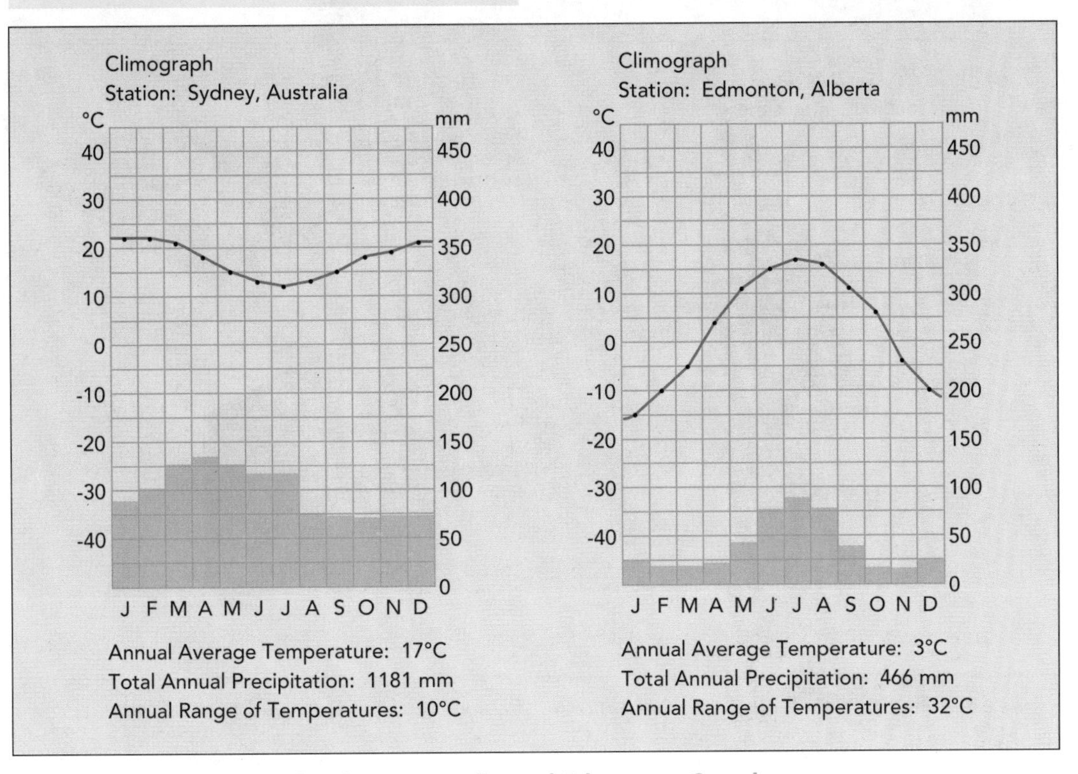

Figure 8.4 Climographs of Sydney, Australia, and Edmonton, Canada

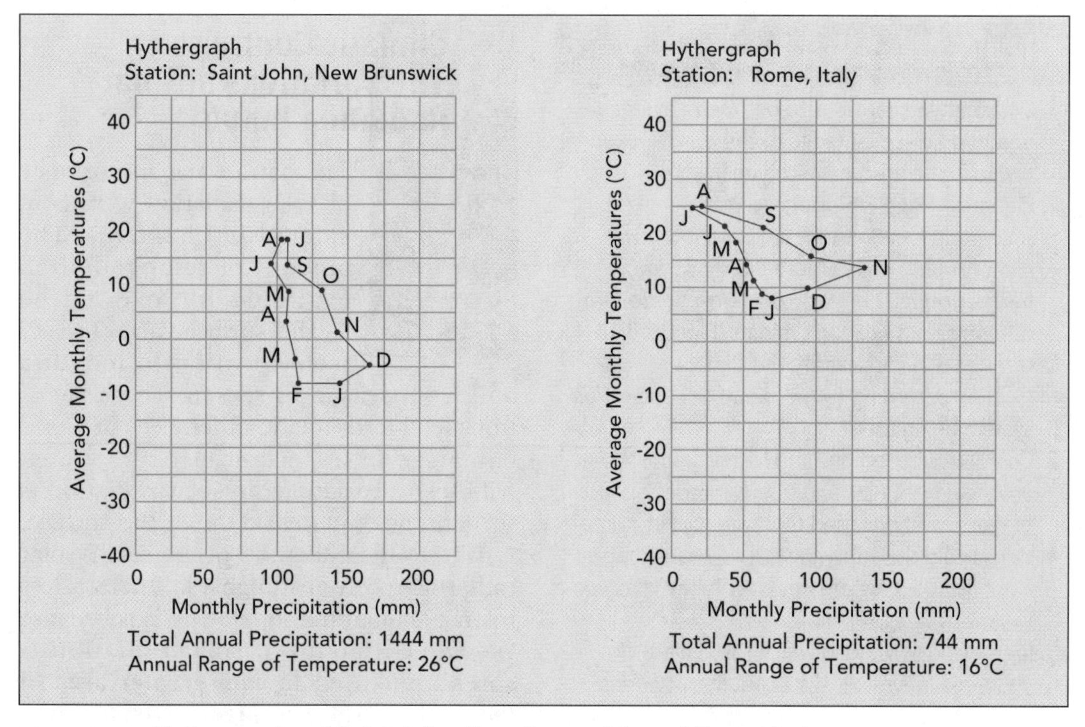

Figure 8.5 **Hythergraphs of Saint John, New Brunswick, and Rome, Italy**

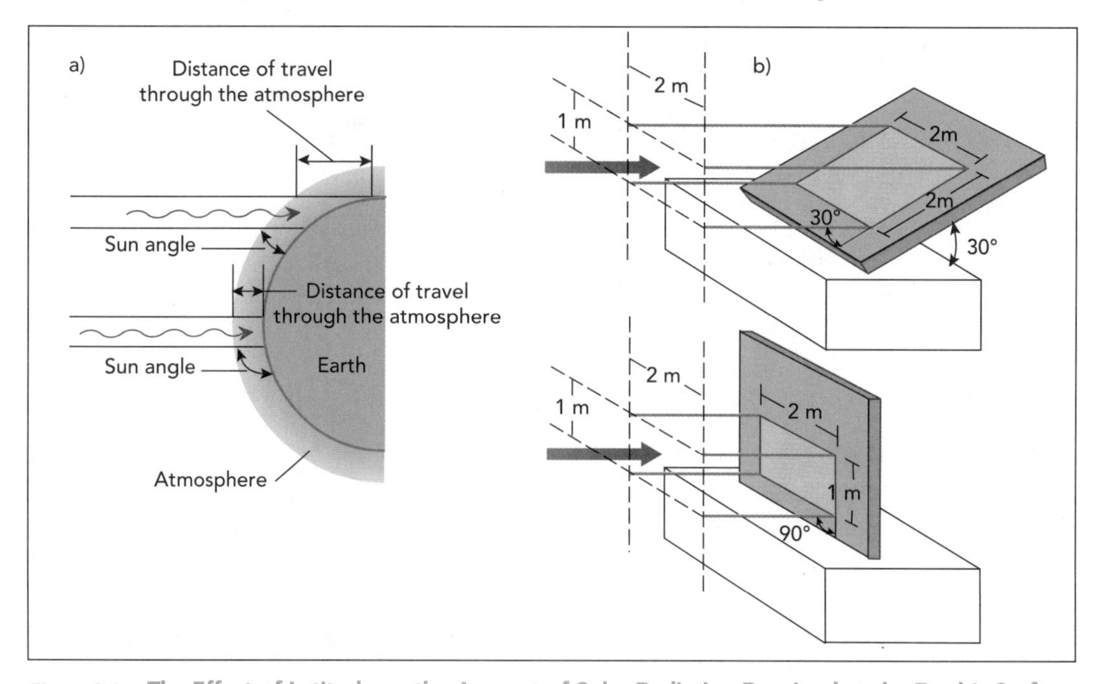

Figure 8.6 **The Effect of Latitude on the Amount of Solar Radiation Received at the Earth's Surface**

As a result, temperatures and evaporation rates usually decrease as latitude increases. Other factors come into play as well. The angle of the incoming solar radiation becomes more oblique at higher latitudes, and so albedo values tend to increase. In addition, Figure 8.6 illustrates that as you move away from the equator, the thickness of the atmosphere through which solar radiation must pass increases. This leads to more absorption, scattering, and reflection of solar radiation at these latitudes, which further reduces the amount of incoming solar radiation received at higher latitudes. Figure 8.7 illustrates the relationship between the sun's altitude, the thickness of the atmosphere through which solar radiation must pass, and the intensity of solar radiation that would be received at the earth's surface.

The variation in the amount of incoming solar radiation caused by the apparent motion of the sun north and south of

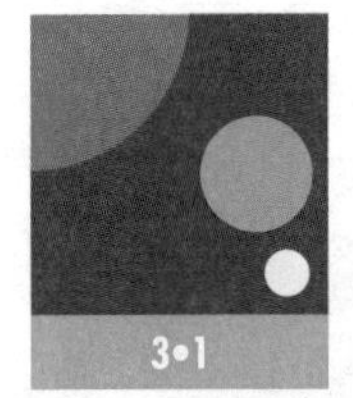

the equator is also a significant factor in explaining the differences in climates experienced on earth. On June 21, the sun is directly overhead on the Tropic of Cancer, 23.5°N. The northern hemisphere is experiencing summer, characterized by warm temperatures and long days. On December 21, the sun is directly overhead on the Tropic of Capricorn, 23.5°S. The southern hemisphere at this time is experiencing summer, while winter is beginning in the north. Figure 8.8 illustrates the variation in the length of day and thus in incoming solar radiation for the northern hemisphere. Remember, however, as you study Figure 8.8 that the amount of net radiation available for use at the earth's surface is not simply a function of the number of daylight hours.

As a result of the latitude and the season of the year, places on the earth's surface experience great differences in the

Sun's Altitude	Distances Rays Must Travel Through Atmosphere*	Radiation Intensity on a Surface Perpendicular to Rays	Radiation Intensity on a Horizontal Surface
90	1.00	78	78
80	1.02	77	76
70	1.06	76	72
60	1.15	75	65
50	1.31	72	55
40	1.56	68	44
30	2.00	62	31
20	2.92	51	17
10	5.70	31	5
5	10.80	15	1
0	45.00	0	0

* Expressed in atmospheres where 1.00 is the thickness of the atmosphere when the sun is directly overhead. A Transmission Coefficient of 78 percent is used. This means that it is assumed that 78 percent of the radiation entering the earth's atmosphere will reach the planet's surface.

Figure 8.7
The Relationship Between the Sun's Altitude and the Thickness of the Earth's Atmosphere

	0°	10°	20°	30°	40°	50°	60°	70°	80°	90°
Jan.	12:07	11:35	11:02	10:24	9:37	8:30	6:38	0:00	0:00	0:00
Feb.	12:07	11:49	11:21	11:10	10:42	10:07	9:11	7:20	0:00	0:00
Mar.	12:07	12:04	12:00	11:57	11:53	11:48	11:41	11:28	10:52	0:00
Apr.	12:07	12:21	12:36	12:53	13:14	13:44	14:31	16:06	24:00	24:00
May	12:07	12:34	13:04	13:38	14:22	15:22	17:04	22:13	24:00	24:00
June	12:07	12:42	13:20	14:04	15:00	16:21	18:49	24:00	24:00	24:00
July	12:07	12:40	13:16	13:56	14:49	15:38	17:31	24:00	24:00	24:00
Aug.	12:07	12:28	12:50	13:16	13:48	14:33	15:46	18:26	24:00	24:00
Sept.	12:07	12:12	12:17	12:23	12:31	12:42	13:00	13:34	15:16	24:00
Oct.	12:07	11:55	11:42	11:28	11:10	10:47	10:11	9:03	5:10	0:00
Nov.	12:07	11:40	11:12	10:40	10:01	9:06	7:37	3:06	0:00	0:00
Dec.	12:07	11:32	10:56	10:14	9:20	8:05	5:54	0:00	0:00	0:00

Figure 8.8 **Hours and Minutes of Sunlight Experienced on the 15th of Each Month vs. Degrees of Latitude in the Northern Hemisphere**

amount of solar radiation received. These differences in turn directly affect the climates experienced on earth. Figure 8.9 summarizes and illustrates the variations in the amount of solar radiation received at the earth's surface. In general, the solar radiation values decrease from the equatorial regions to the poles. However, the pattern is complex due to factors such as albedo and cloud cover that can distort the expected pattern.

The major climatic effects of these variations in the amount of incoming solar radiation are felt in the areas of temperature, air pressures, and wind and ocean currents. Figure 8.10 illustrates the effect that latitude has on temperatures. The imbalance in heat that exists between latitudinal zones is responsible for triggering global winds and ocean currents.

QUESTIONS

6. Imagine living above the Arctic Circle and experiencing 24 hours of daylight in the summer and 24 hours of darkness in the winter. What would be five advantages and five disadvantages of these unusual days and nights?
7. Figure 8.8 indicates that, during the northern summer, locations above the Arctic Circle (66.5°N) experience 24 hours of daylight. What factors reduce the impact or effectiveness of this constant supply of solar radiation?
8. Using columns 3 and 4 of Figure 8.7, draw a diagram that illustrates the difference between radiation intensity on a surface perpendicular to the sun's rays as opposed to the radiation intensity on a surface horizontal to the earth.
9. How would the pattern of solar radiation received at the earth's surface (Figure 8.9) be altered if the earth were a cube instead of a sphere? Redraw Figure 8.6 to illustrate your answer.

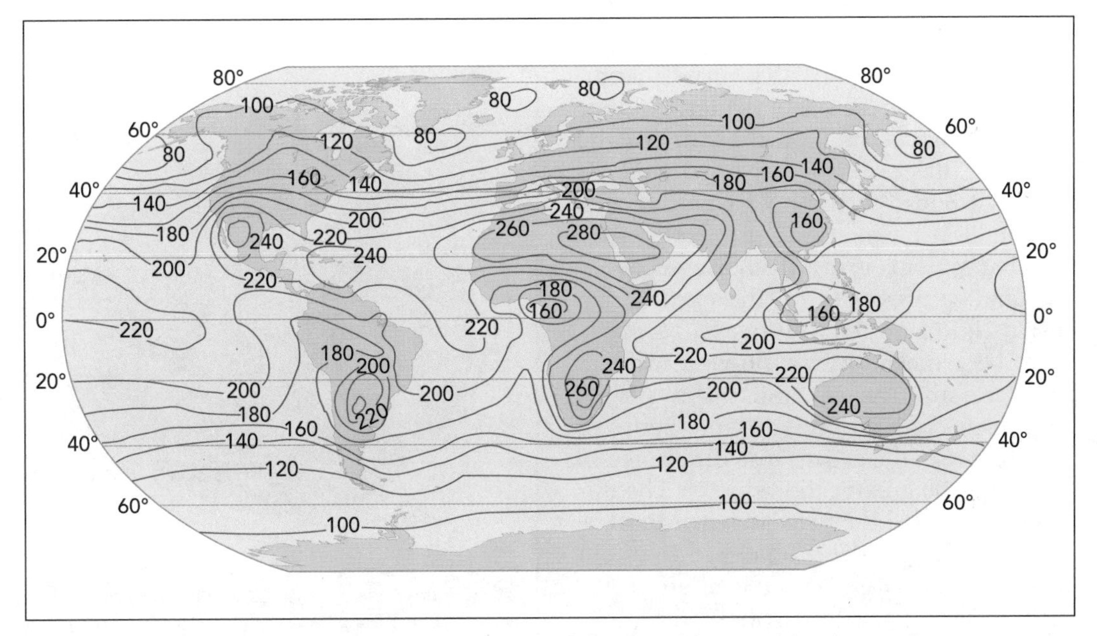

Figure 8.9 Solar Radiation Amounts Across the Globe Highly generalized map of mean annual solar radiation received at the earth's surface. Units are watts per square metre.

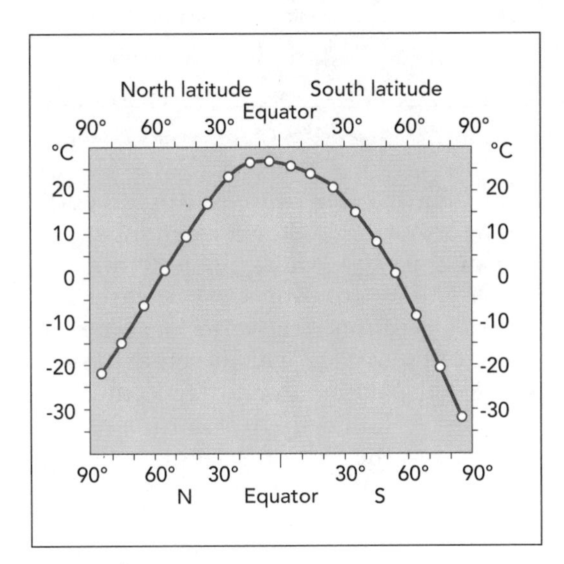

Figure 8.10 Mean Surface Air Temperature by Latitude Zones

8·4 Climatic Controls: The Global Wind Systems

A second global influence on climate is the earth's wind systems.

Hot air is less dense than cool air. This is due to the activity of the air molecules themselves. As heat is introduced into an air mass, the air molecules absorb energy and their activity or movement increases, leading to fewer air molecules per unit volume of atmosphere. Less dense, warm air tends to rise. This is why a hot air balloon rises. In addition, hot or warm air is buoyant; this means that because of the increased molecular activity, the air is able to support or hold a greater amount of other atmospheric components, such as water vapour. Areas of warm, rising air have low atmospheric pressures.

By contrast, the number of air molecules per unit volume of atmosphere in cold air is greater. This leads to heavier or more dense air which, due to gravity, has a tendency to sink to the earth's surface creating high pressure areas. Further, cold air is less buoyant than warmer air, as the air molecules are less active, so cold air masses are less able to hold water vapour and other atmospheric components.

Wind systems result from the necessity to equalize pressures on the earth's surface. Rising air in low pressure areas must be replaced by air flowing in from high pressure areas. Winds always blow from zones of high pressure into zones of low pressure; the greater the difference in pressure, the greater the velocity. The difference is called the **pressure gradient**.

Theoretically, the global wind system would involve two giant cells of air movement, one in each hemisphere. The cells would be triggered by hot air rising at the equator and cold air flowing down from the polar regions to replace it. A low pressure area would exist at the equator as air rises while high pressure would exist at the polar regions as cold air falls and presses down on the earth's surface. Figure 8.11 illustrates this theoretical global wind system.

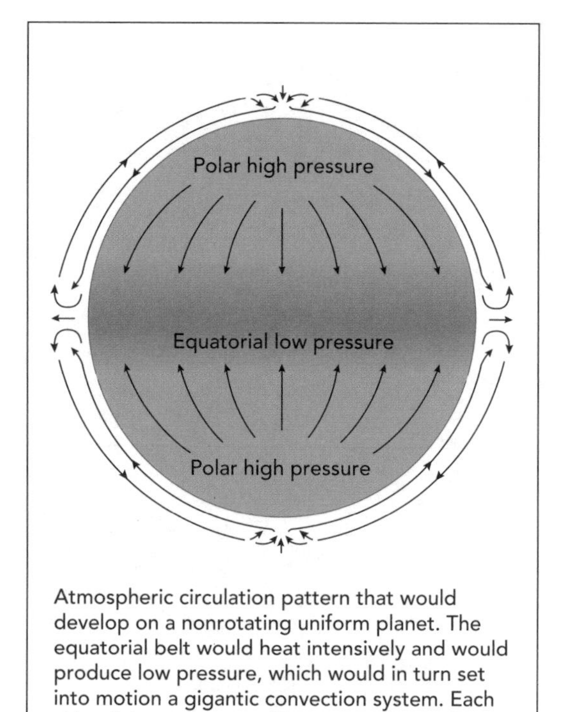

Atmospheric circulation pattern that would develop on a nonrotating uniform planet. The equatorial belt would heat intensively and would produce low pressure, which would in turn set into motion a gigantic convection system. Each side of the system would span one hemisphere.

Figure 8.11 Theoretical Global Wind System

The idealized or theoretical wind system shown in Figure 8.11 is distorted by many forces that are at work in the real world. The most important of these forces is called the Coriolis force. Due to the spinning of the earth on its axis, atmospheric movement in the northern hemisphere is deflected to the right while movement in the southern hemisphere is deflected to the left. This occurs because the speed of the earth's rotation at the equator is much greater than at the poles.

Consider an airborne object at the north or south pole that has been aimed at a location on the equator directly south or north of its own location. As the object travels towards the equator, the earth will be rotating west to east. The object's position relative to locations on the earth's surface will therefore appear to be deflected west or to the right of its motion in the northern hemisphere and west or to the left of its motion in the southern hemisphere. Figure 8.12 illustrates this effect. Note that the deflection to the right or left is an apparent one caused by the rotation of the earth. From outer space, the object would be seen to travel in a straight line.

We have noted that air rises in the equatorial regions and begins to flow at high altitudes north and south. Unlike

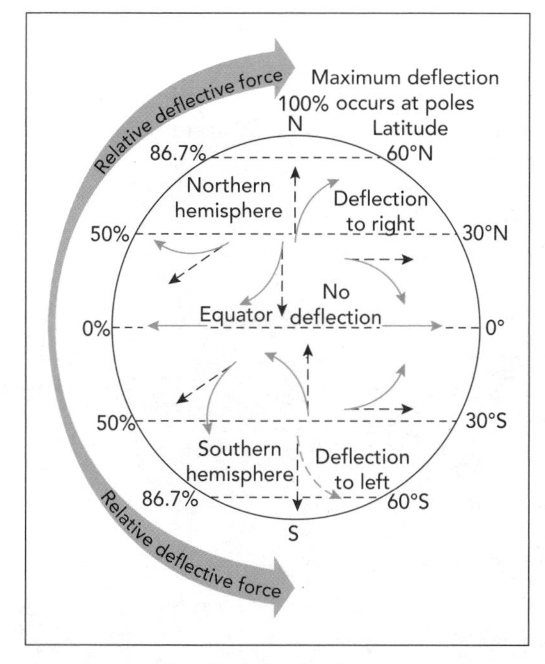

Figure 8.12 **The Coriolis Force**

our theoretical model (Figure 8.11), these upper atmospheric winds flowing away from the equator meet winds flowing towards the equator at the same altitude, and air is forced downwards at this point of convergence. This sinking air forms high pressure zones on the earth's surface at approximately 30°N and 30°S. These areas of high pressure are called the **subtropical highs**. They are caused by the convergence of air more so than by the thermal qualities of the air. As this air descends to the earth's surface, it spreads out to form two important and well-known surface wind systems. The air flowing towards the equator becomes the northeast and southeast trade winds, while the air flowing towards the poles becomes the westerly winds. The westerlies flow poleward and meet surface winds flowing out of the high pressure zones at the poles. The convergence of these two surface wind systems causes air to be forced upwards, creating low pressure zones. These low pressure zones are called the **subpolar lows**. As a result of the flow of these various winds, three distinctive cells of air movement exist in each hemisphere. These cells are illustrated in Figure 8.13. Figure 8.13 also shows the effects that the Coriolis force exerts on the surface winds associated with these three cells.

Although surface wind systems have been studied and documented for many years, only recently has detailed information concerning high altitude or upper atmospheric wind systems been gathered. The most important of these high altitude winds are the planetary **jet streams** that race around our planet at heights of between 9 and 12 km and at speeds that can reach 450 km/h. The jet streams develop in a broad area of upper level westerly winds that flow around our planet in both hemispheres from about 25° to almost the poles. Their path is wavy or meandering in nature and they mark the division between cold polar air and warm tropical air. The waves or meanders in the jet streams are fundamentally important in both the development of **cyclones**, or low pressure systems, and **anticyclones**, or high pressure systems. These cyclones and anticylones dominate the weather of much of the mid- and high latitude zones on earth. The jet streams and their meanderings appear also to be fundamentally important in the planetary heat exchange, the process by which surplus energy in the equatorial regions is moved to the areas of energy deficit at the poles.

In addition to the broad westerly flow of winds mentioned above, a second major high altitude group of winds is an easterly flow in the tropical areas on either side of the equator. The upper atmospheric flow of winds is shown in Figure 8.14.

It should be noted that Figures 8.13 and 8.14 show the global wind patterns in relationship to the equator. In fact, due to the tilt of the earth's axis and the apparent

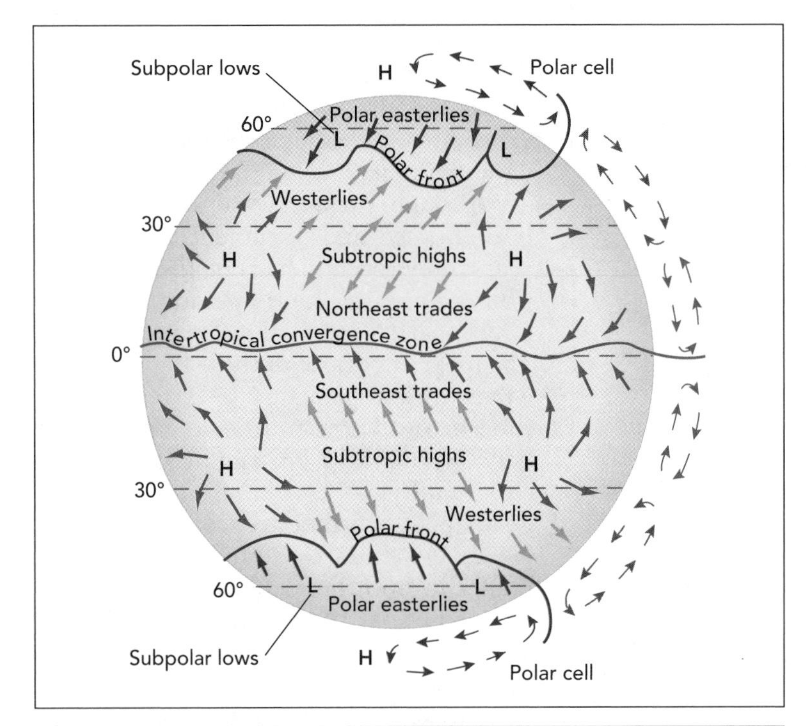

Figure 8.13 **Schematic Arrangement of Winds and Pressure Belts**

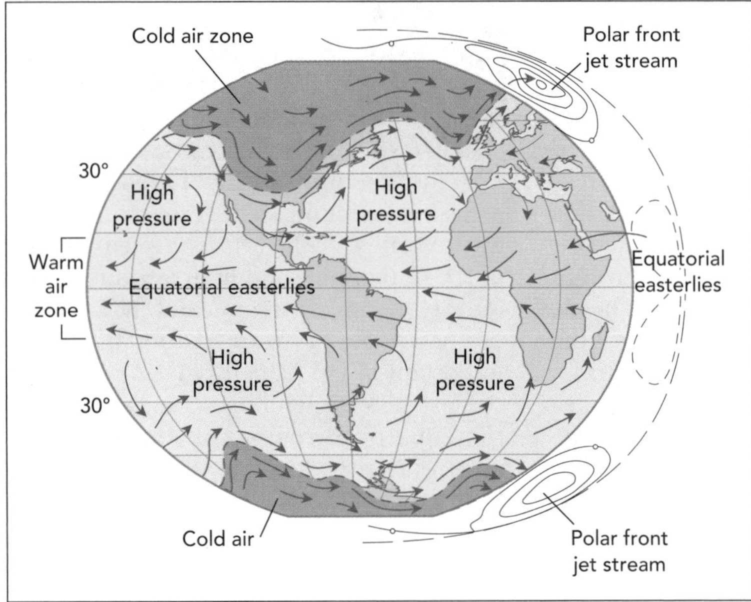

Figure 8.14 **Upper Atmospheric Flow of Winds**

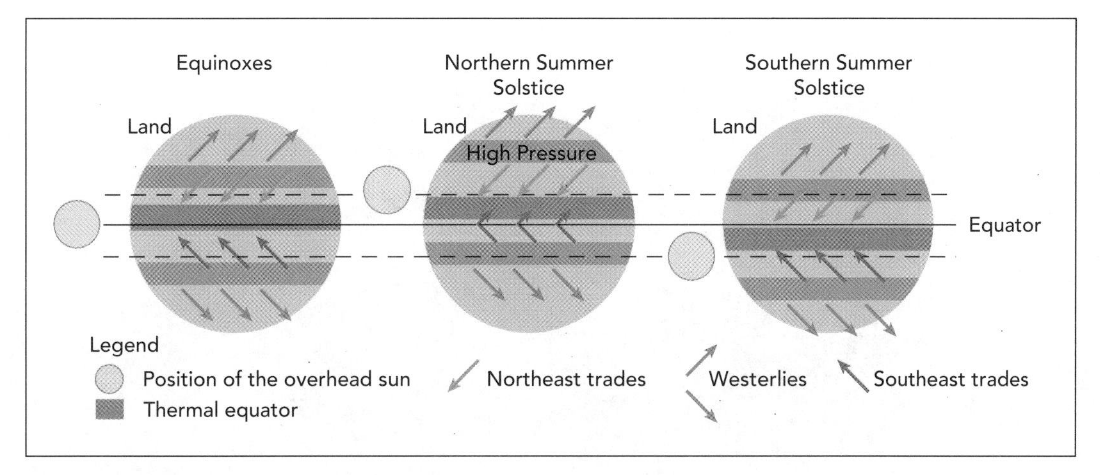

Figure 8.15 **The Influence of the Thermal Equator on the Earth's Wind Systems**

movement of the sun north and south of the equator, the global wind belts shift their positions from season to season. When discussing global wind patterns and pressure zones, the thermal equator or the area of greatest heat must be considered. The thermal equator ranges from 12°N to 8°S of the actual equator and results in the wind and pressure shifts illustrated in Figures 8.15 and 8.16.

Figure 8.13 assumes a uniform surface and thus it omits the effect that differential heating can have on wind systems and patterns. Land masses heat up and cool down much more quickly than do water bodies. As the land heats up, it heats the air above the land, which tends to rise.

This creates low pressure zones over land masses. As land masses cool down, the air above the land is cooled, resulting in high pressure zones. The differential heating of land and water bodies is discussed in detail later in this chapter. When the influence of differential heating is added to Figure 8.13, we get the actual global surface wind and pressure patterns illustrated in Figure 8.17.

We have examined the global wind systems in detail because they exert a fundamental influence on the world's climates. The wind systems move air masses from one area of the globe to another. The

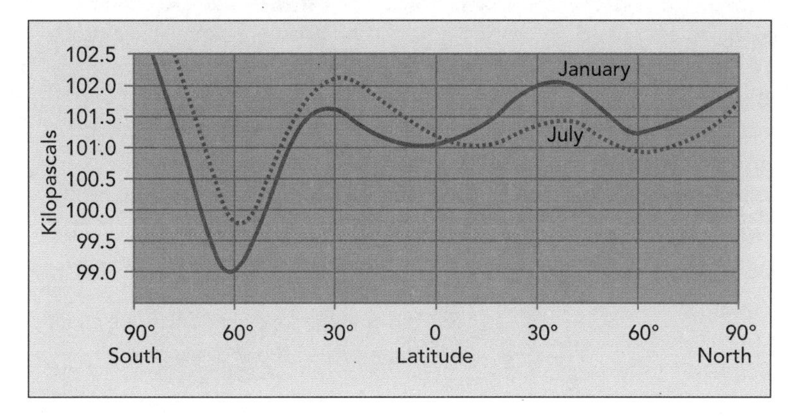

Figure 8.16 **Changes in Sea-Level Air Pressure vs. Latitude**

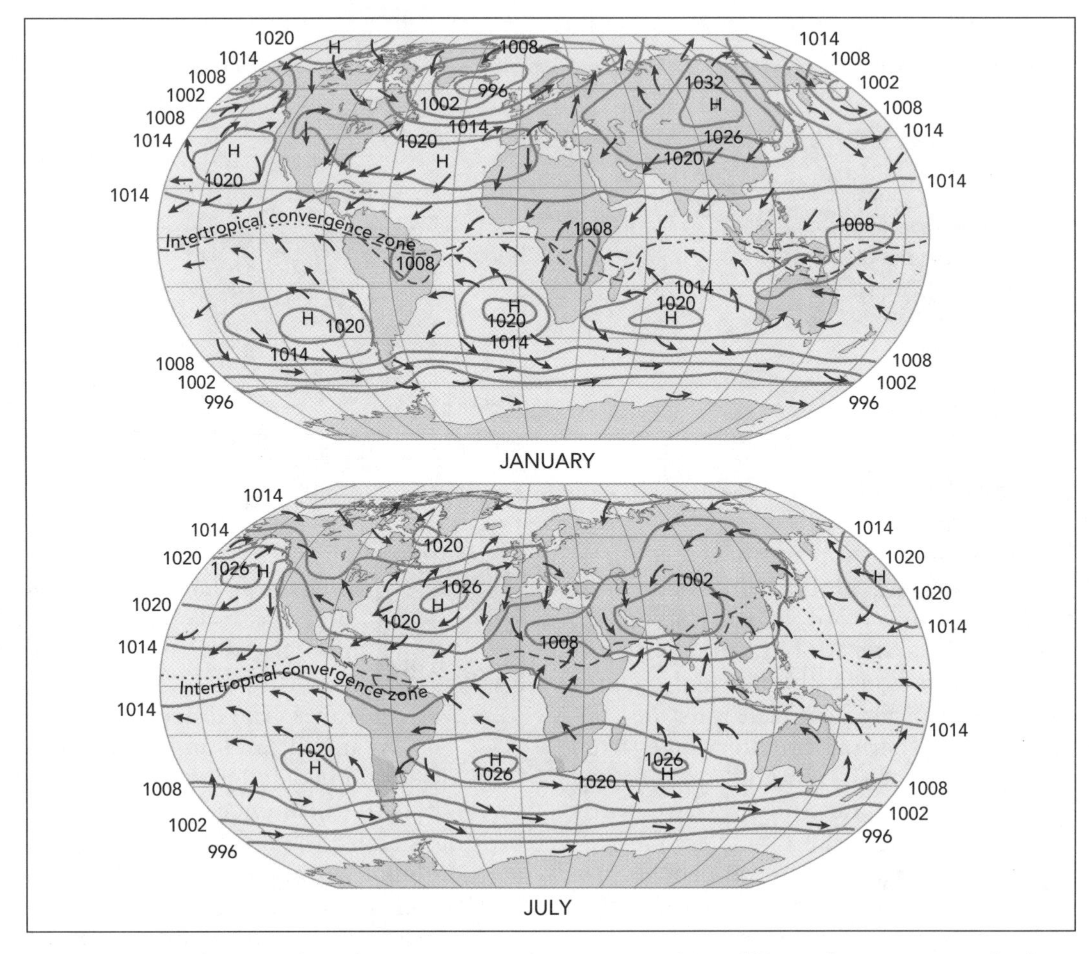

Figure 8.17 Surface Winds and Pressure Systems (Measured in Millibars) for January and July

moisture and temperature characteristics of these air masses have a direct influence on the climates of the areas over which they pass. In attempting to understand and explain both climate and weather conditions, we must look at the prevailing winds that are influencing a location. The importance of wind systems will become clearer as we discuss other climatic controls in the following sections.

QUESTIONS

10. What are some activities that are enhanced by windy days? What are some activities that are impeded by windy days?
11. Why is there a variation in the extent of the thermal equator north (12°) and south (8°) of the actual equator?
12. a) The "prevailing wind" refers to the wind direction which occurs most often in an area. What is the prevailing wind in your community?
    b) Explain why the following decisions might be influenced by the prevailing wind:

- You are deciding where to build your new house.
- A planner is deciding where to locate a new steel mill.
- Airport officials are deciding how to align their new runways.

13. If you were sailing from England to Australia, how would the global wind systems influence your route and speed?

8·5 Climatic Controls: Ocean Currents

Flowing through the world's oceans are huge rivers. These rivers do not flow downhill, nor do they have any banks, nor are they contained by any valleys. They are defined by their temperatures and chemical make-ups that are distinct from the water through which they are flowing. They are driven by temperature differences and by the wind. They are as complex and three-dimensional as the global wind systems and, like the wind systems, they are fundamentally important in global heat exchanges and in influencing the climates of land masses along which they flow. Figure 8.18 illustrates the major surface ocean currents, their temperature characteristics, and their direction of flow.

The currents move vast quantities of water. The Gulf Stream alone carries approximately 30 million cubic metres of water per second, an amount more than five times the total flow of all the freshwater rivers in the world combined! Many climatic effects that are associated with our ocean currents are understood and well documented. The moderating effect that the Gulf Stream and North Atlantic drift have on the British Isles and Scandinavia has been understood for hundreds of years. Figure 8.19 is a small-scale map that displays January temperatures for London, England; Bergen, Norway;

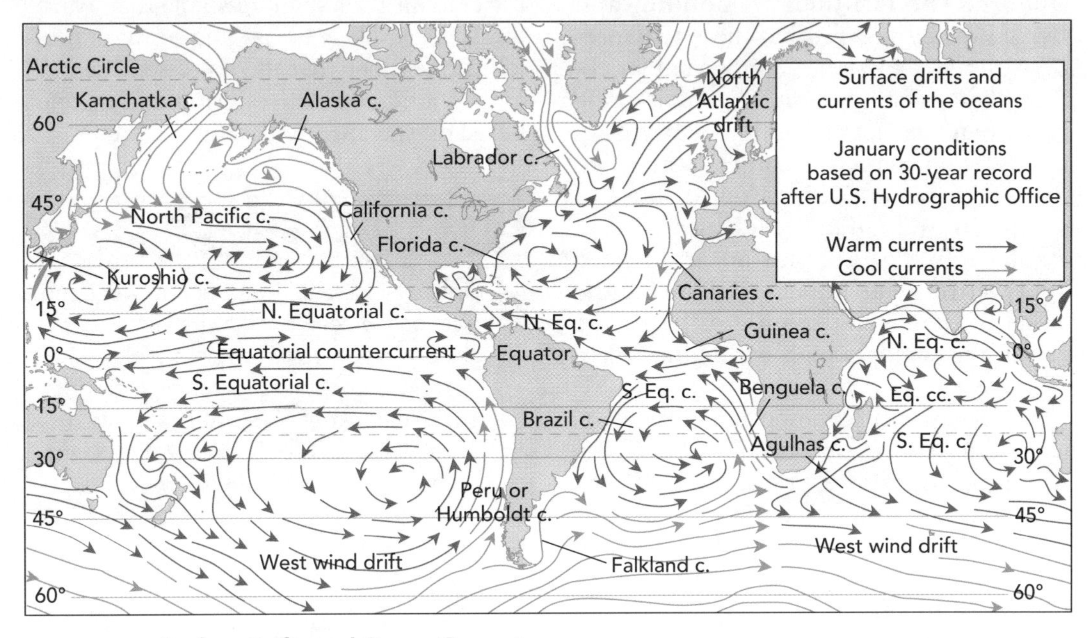

Figure 8.18 Surface Drifts and Ocean Currents

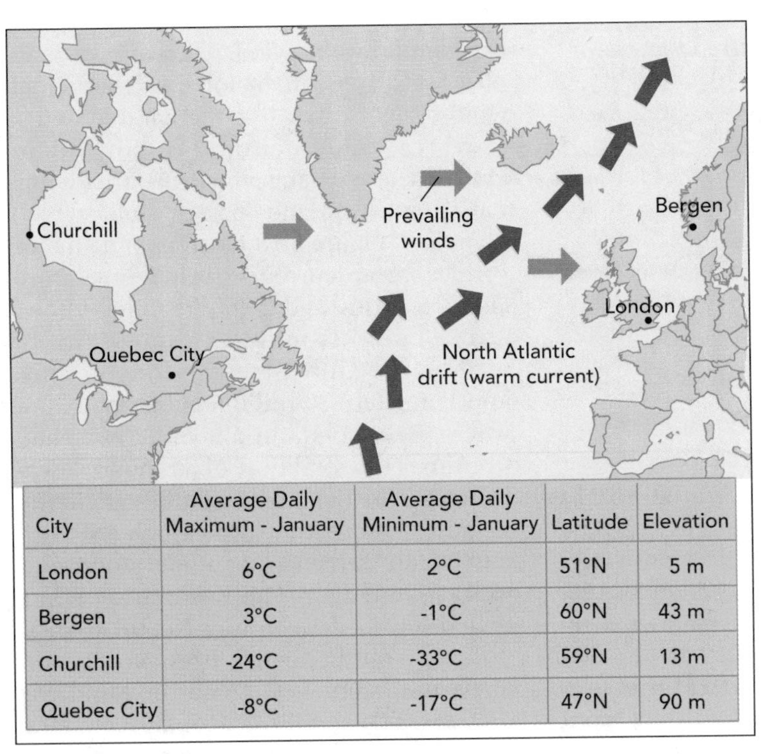

City	Average Daily Maximum - January	Average Daily Minimum - January	Latitude	Elevation
London	6°C	2°C	51°N	5 m
Bergen	3°C	-1°C	60°N	43 m
Churchill	-24°C	-33°C	59°N	13 m
Quebec City	-8°C	-17°C	47°N	90 m

Figure 8.19
Average Maximum and Minimum Daily January Temperatures for Four Selected Climatic Stations

Churchill, Manitoba; and Quebec City, Quebec. The latitudes of London and Bergen would suggest that these two locations should experience a much colder climate than in fact they do. The moderating effect that the North Atlantic drift has on London and Bergen becomes clear when their climates are compared to those of Churchill and Quebec and locations at similar latitudes that are not influenced by a warm ocean current. Enhancing the effect of the North Atlantic drift are the prevailing winds which are indicated in Figure 8.19.

When a warm ocean current sweeps by a coastline, the result is usually an increase in temperature and precipitation. This occurs as a result of the warm, moist air which originates above the warm current. When a cold current encounters a coastline, the affected land usually experiences dry and cool conditions. Cold air cannot evaporate as much water as warm air, and the cold air itself is more stable than warm air. It is less likely to rise and result in rainfall. Remember that the adjectives "warm" and "cold" are relative terms and are not based on specific temperatures.

QUESTIONS

14. Compare Figure 8.17 with Figure 8.18. Comment on the degree to which the pattern of global winds matches the pattern of global currents. Suggest reasons for the similarities and/or differences that exist in the patterns.
15. Examine Figure 8.18. From what direction(s) do most of the cold currents flow? From what direction(s) do most of the warm currents flow? Why would this be so?
16. If you like to swim in warm water, why should you visit the beaches in Florida instead of those in California? Be specific.

8·6 Climatic Controls: Water Bodies and Continents

The pattern and extent of land masses and water bodies on the earth's surface have a fundamental influence on the world's climates.

Water bodies provide sources of moisture for the land masses of the world. In general, locations near water bodies experience wetter climates than locations found at a distance from large bodies of water. An important exception to this rule occurs at high latitudes where the air temperatures are very cold. At these latitudes, the cold air is unable to evaporate and hold very much water vapour; therefore, cold weather deserts occur. A second exception occurs in areas that experience offshore winds for much of the year. These winds that blow from the land to the water are dry, and thus even coastline locations can experience desert conditions. A third exception, discussed in the previous section, is associated with the cool air resulting from a cold ocean current which does not evaporate nor hold very much water vapour. In addition, this air is generally stable and therefore does not easily rise and produce rainfall.

A second important effect that water bodies have on climates involves their ability to influence the temperatures of the land masses adjacent to them. Land masses receiving solar radiation convert the energy into heat that tends to be concentrated in the top few centimetres of the earth. The solar radiation is unable to penetrate into the soil or rock, and although some heat is conducted downwards, the surface layer remains much hotter than the material found even a few centimetres below the surface. We have all felt the heat of the sand on a beach on a hot summer's day. Rock and soil have a low specific heat when compared to water. This means that it does not require as much energy to raise the temperature of one cubic centimetre of soil or rock as it does to raise the temperature of one cubic centimetre of water.

By contrast, solar radiation falling on a water body is able to penetrate to significant depths and thus the energy does not remain concentrated at the surface. In addition, convection currents found in large bodies of water are able to distribute the energy received from the sun downwards. Evaporation from a large body of water further cools the temperature of the surface layers. As a result of all of these conditions, water bodies do not heat up as much in the summer as land masses. They thus act as "air conditioners" to the land masses that are adjacent to them, keeping the temperatures of these locations lower than they would otherwise be. In the winter, the opposite effect occurs. Water bodies maintain their heat longer and do not cool off as much nor as quickly as land masses. In the winter, therefore, large water bodies act as "heaters" to the land that is adjacent to them, keeping these locations warmer than they otherwise would be. The overall effect is a moderating influence on the land adjacent to large bodies of water. Cooler summers and warmer winters result in coastal locations experiencing a lower range of temperatures than locations found further inland. This effect is particularly dramatic in mid-latitude zones that receive onshore winds. Figure 8.20 summarizes the temperature effects that large water bodies have on adjacent land masses.

The differential heating between land and water illustrated in Figure 8.20 results in differences in air pressure. These differences in air pressure in turn

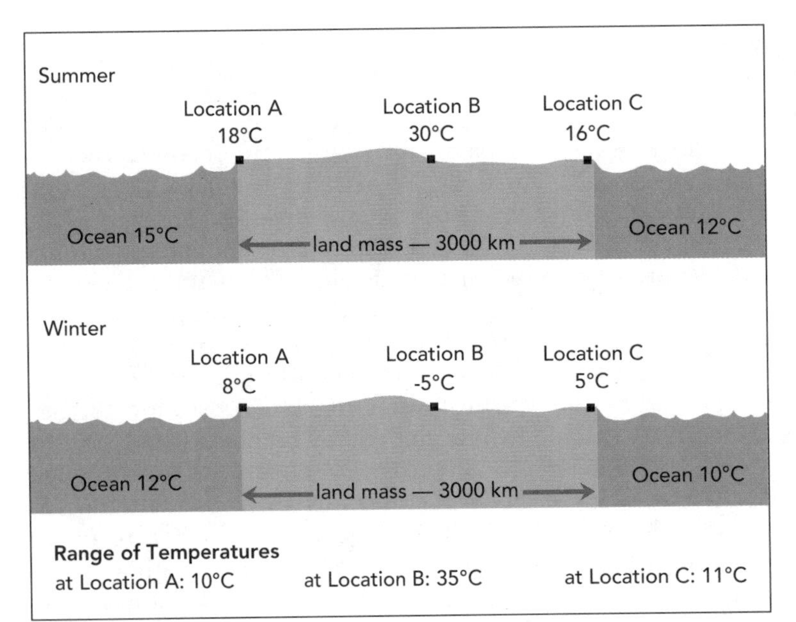

Figure 8.20
The Effect of Large Bodies of Water on the Temperatures Experienced by Adjacent Land Masses

result in the development of regional wind systems that can alter the global wind systems. We have already noted that hot air is less dense than cold air and has a tendency to rise, creating a low pressure area. Surface winds blow towards a low pressure area to replace the rising air. Over land masses in the summer, low pressure areas develop and winds blow inland towards the low pressure system. In the winter, the land masses cool off to a much greater degree than surrounding water bodies. High pressure areas develop over the land mass as cold, dense air sinks to the earth's surface. Winds blow offshore out of the high pressure system. The largest land mass in the world is Asia and, therefore, this effect is most noticeable there and results in the distinctive wind system know as the **monsoons**. Figure 8.21 displays climate graphs which illustrate the effects of the seasonal wind reversals associated with the monsoons.

QUESTIONS

17. Why does the water temperature in a swimming pool not get as hot as that of the concrete patio surrounding it?
18. Although the monsoon effect is most noticeable in Asia, locations in Africa and Australia are also affected by this differential heating of land and water. Use the climate statistics in an atlas to identify five locations other than those in Figure 8.21 which experience monsoon winds.
19. If all the land masses in the world were joined in one supercontinent, the monsoon effect would be dramatic. Draw sketch maps for winter and summer which would illustrate the monsoon effect on this supercontinent.
20. Describe an experiment or demonstration which would illustrate that soil and water have different heating abilities.

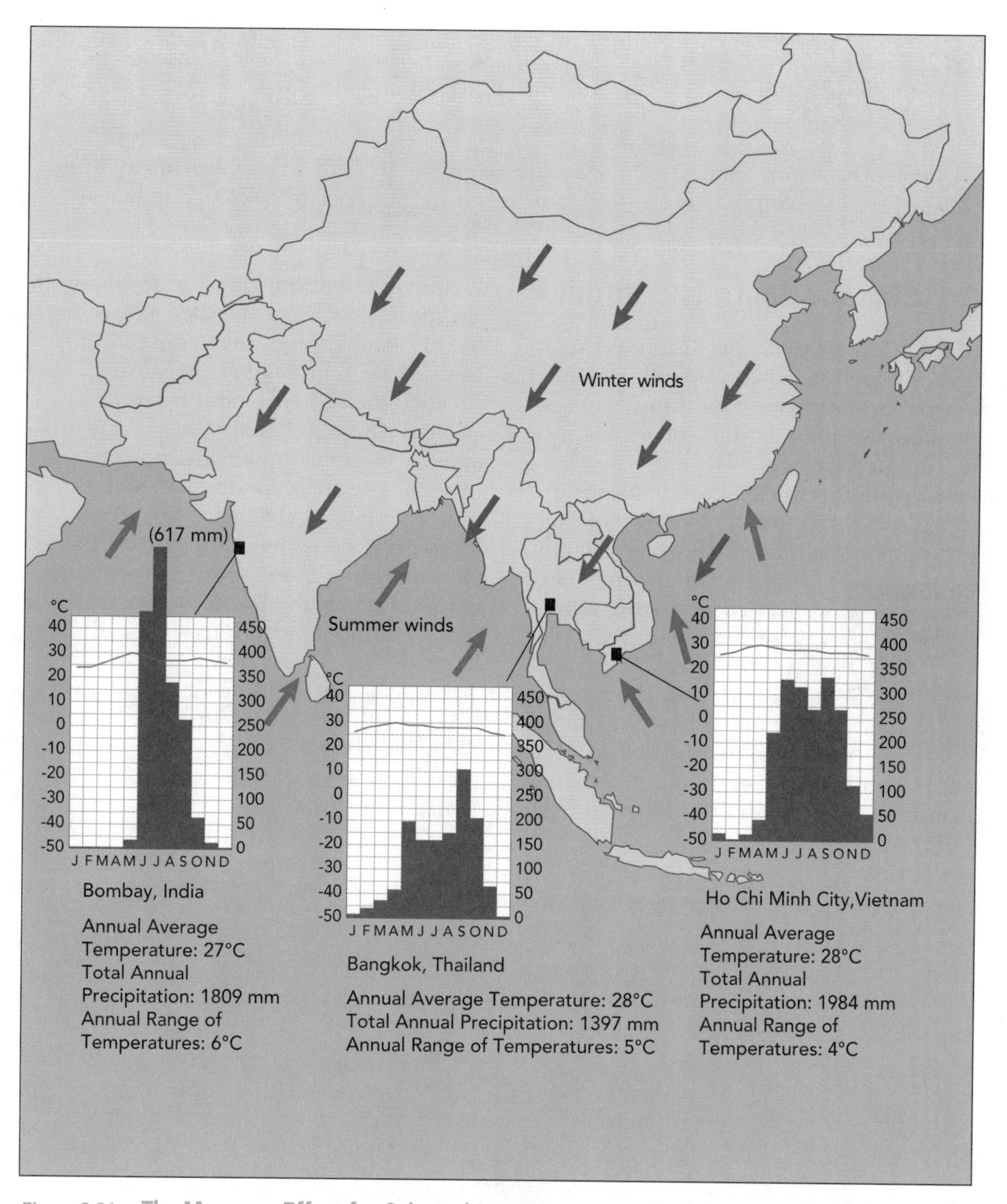

Figure 8.21 The Monsoon Effect for Selected Locations Note the dramatic effects on the precipitation patterns for Bombay, Bangkok, and Ho Chi Minh caused by the differential heating of the land masses and water bodies which results in the seasonal reversal of winds.

It's a Fact...

Due to the monsoon circulation, Cherrapunji in India receives on average 2922 mm of precipitation in June, but only 5 mm of precipitation in December!

8•7 Climatic Controls: Altitude

Altitude has a direct effect on the temperatures experienced at a location. Under normal atmospheric conditions, air temperatures within the troposphere decrease as altitude increases. This decrease in air temperatures with increased elevation is called the **environmental lapse rate** and on average equals 6.4°C for every 1000 m.

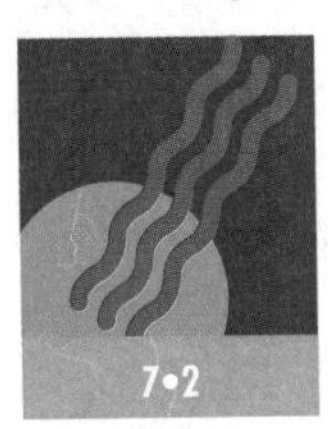

Solar radiation is not converted into heat until it is absorbed by a body. The earth itself, as discussed in Chapter 7, absorbs a significant amount of the incoming solar radiation. The earth, therefore, heats up and warms the atmosphere from below by radiating longwave radiation to the atmosphere. In addition, heat is distributed upwards to the atmosphere by warm convection currents of air generated at the surface. This is called the **sensible heat flux**.

Near the earth's surface, the atmosphere is denser, containing more air molecules, water vapour, dust, pollen, and so on per cubic centimetre than it does at higher altitudes. As compared to higher elevations, therefore, the atmosphere near sea level is able to absorb more solar radiation and also more longwave radiation from the earth and convert it into heat. The effect that altitude has on temperatures is illustrated in Figures 8.22 and 8.23. Figure 8.23 displays the temperatures for ten stations located at a similar latitude but at different altitudes.

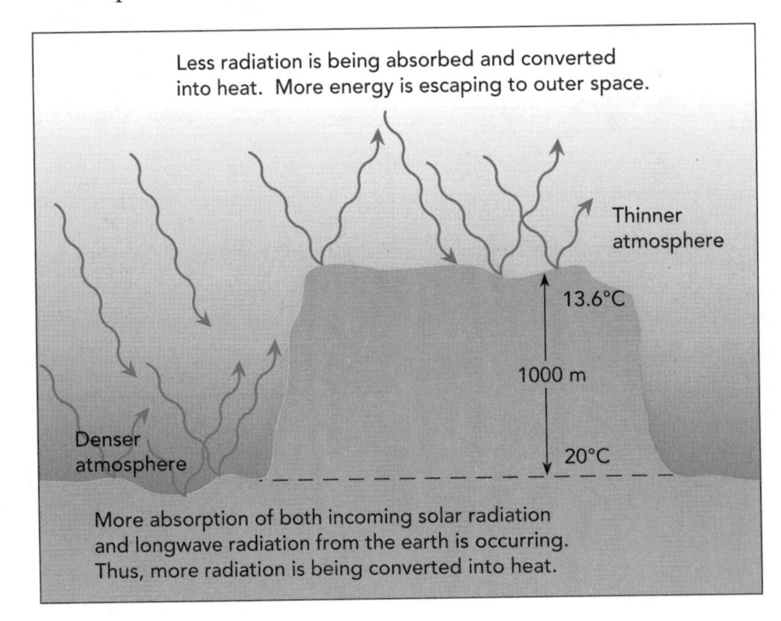

Figure 8.22
The Effect of Altitude on Temperature

Station	Latitude	Altitude Above Sea Level	Average Annual Daily Maximum Temperature	Average Annual Daily Minimum Temperature
Guayaquil, Ecuador	2°S	6 m	31°C	20°C
Padang, Indonesia	0°	7 m	31°C	23°C
Douala, Cameroon	4°N	8 m	29°C	23°C
Mogadishu, Somali	2°N	12 m	30°C	24°C
Manaus, Brazil	3°S	44 m	32°C	24°C
Kisangani, Zaire	0°	418 m	30°C	20°C
Kisumu, Kenya	0°	1148 m	28°C	18°C
Rubona, Rwanda	2°S	1706 m	25°C	14°C
Kabale, Uganda	1°S	1871 m	23°C	10°C
Quito, Ecuador	0°	2879 m	22°C	8°C

Figure 8.23 The Effect of Altitude on Temperature Note that as altitude increases, average temperatures generally decrease.

QUESTIONS

21. Describe a personal experience that illustrates that temperatures usually decrease with altitude.
22. Explain how it is possible for mountains located at the equator to be snow covered throughout the year.
23. Graph the temperatures displayed in Figure 8.23 in a manner that illustrates the influence altitude has on temperature.
24. Assuming stable air conditions and a normal environmental lapse rate, what would the temperature be outside of an airplane flying at 8000 m if the surface temperature was 20°C?
25. Draw a sketch diagram showing the location of an imaginary place that would experience extremely cold average monthly temperatures. Be sure to take into account all the climate controls discussed in this chapter that could influence temperature. Indicate these climatic controls on your sketch map.

8·8 Climatic Controls: Mountains

Because temperatures usually decrease with elevation, mountains have a direct effect on temperatures. Mountains also act as barriers to air masses moving around the earth's surface. This influence is particularly dramatic when mountain ranges block warm, humid air from moving farther inland.

When an air mass meets a mountain range, the air is forced to rise. As air rises, the effect of gravity is reduced and the air expands, becomes less dense, and cools. In fact, the amount of heat in the air remains the same; however, because of decreased pressure, the air now occupies a greater volume, and its temperature decreases. The rate at which a rising air mass cools is called the **dry adiabatic lapse rate** which for dry air equals approximately 10°C per 1000 m of vertical rise. The word "adiabatic" means that no heat is lost from the air mass itself.

As an air mass rises and air molecules are spread farther apart, its ability to hold water vapour decreases. At some height, the ability of the air to hold water vapour matches the amount of water vapour in the air. This is the condensation point — the temperature at which air becomes saturated. If air temperatures fall below the condensation point, the air will begin to release water vapour in the form of fog, dew, clouds, rain, or snow. Above this elevation, the air temperature decreases according to the **wet adiabatic lapse rate** which, on average, is 3°C per 1000 m. The wet adiabatic lapse rate is less than the dry adiabatic lapse rate because a lot of the available energy is released into the air through the condensation process, thus slowing the rate of cooling.

As an air mass crosses the mountains and falls down the leeward side, increased atmospheric pressures cause the air molecules to compact closer and temperatures within the air mass rise.

Adiabatic lapse rates are associated with air that is either rising or falling. These rates should be distinguished from the environmental lapse rate that refers to the rate at which temperatures decrease with height in stationary air. Figure 8.24 illustrates the difference between the adiabatic lapse rates and the environmental lapse rates.

Dry air descending the leeward sides of mountains gives rise to distinctive regional winds throughout the world. In North America, the dry, warm wind flowing down the east slopes of the Western Cordillera is known as the chinook, while in the lee of the European Alps it is known as the fohn.

To summarize: global climate patterns arise due to the influences of solar radiation inputs, global wind systems, ocean currents, and the arrangement of the continents and oceans. These large global patterns are modified and changed due to differences in altitude and the presence of mountain ranges. The influence of all the factors noted above results in the world temperature and precipitation patterns illustrated in Figures 8.25 and 8.26.

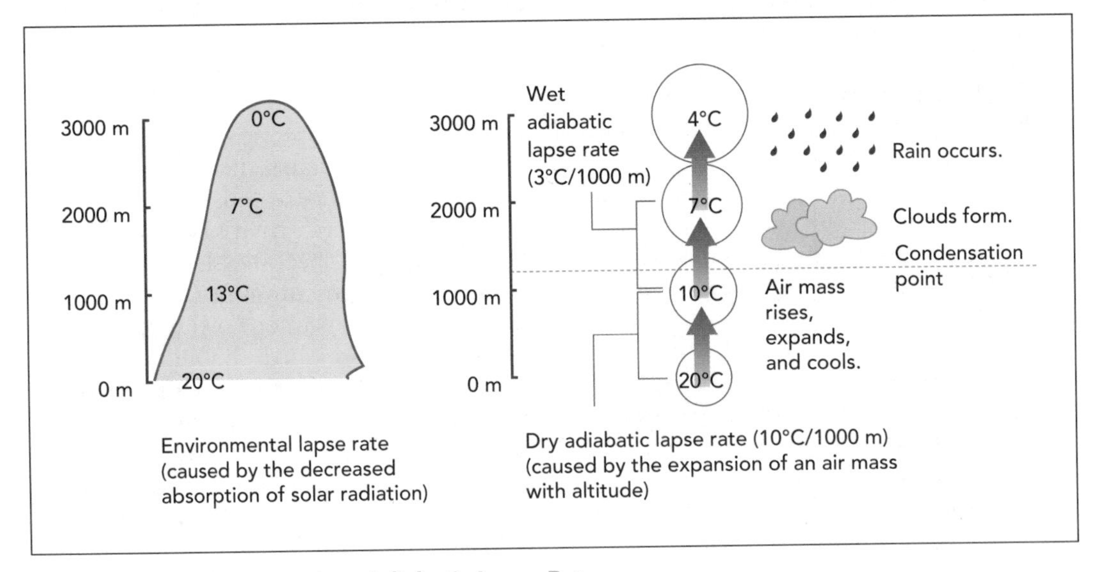

Figure 8.24 **Environmental vs. Adiabatic Lapse Rates**

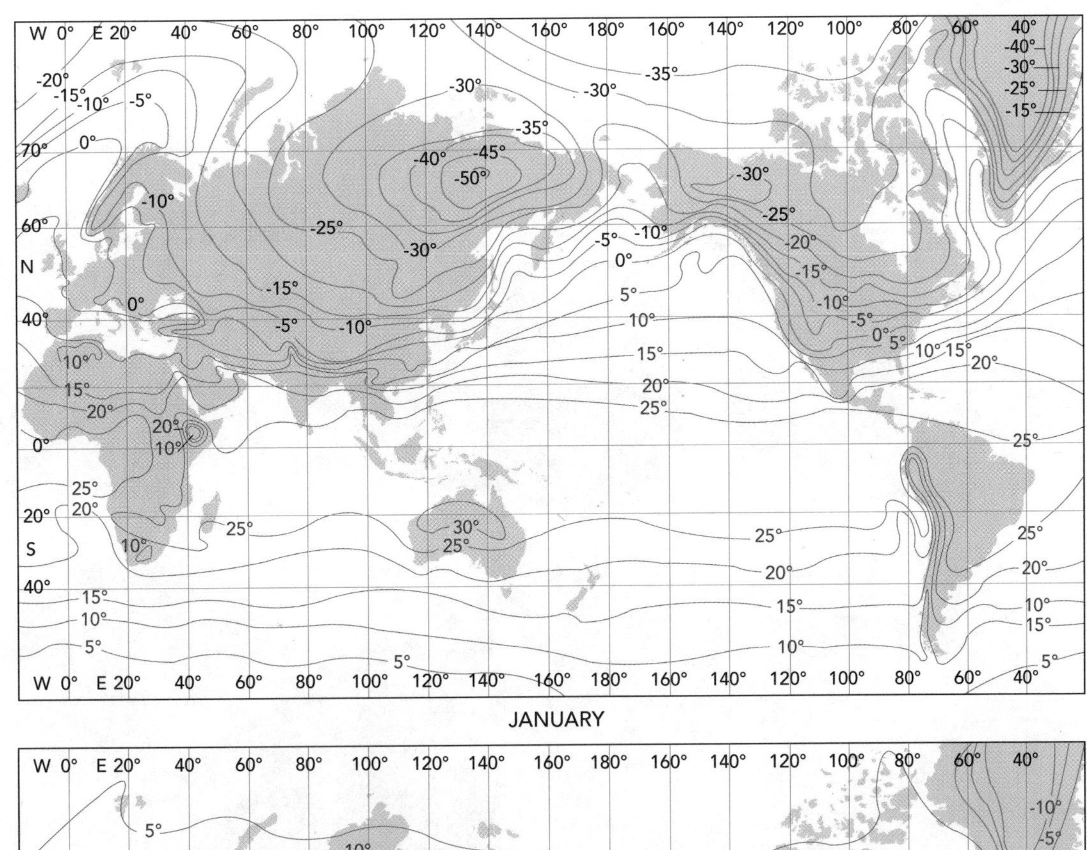

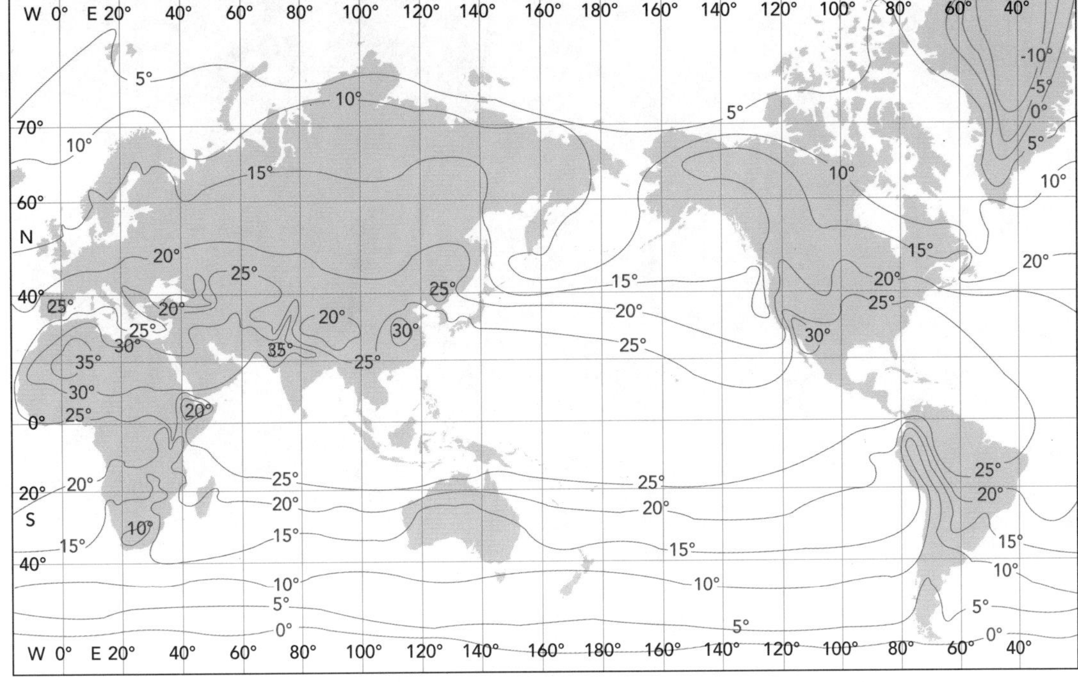

Figure 8.25 Average Global Temperatures The blue isotherms show temperatures of 0°C and lower: the red isotherms show temperatures above 0°C.

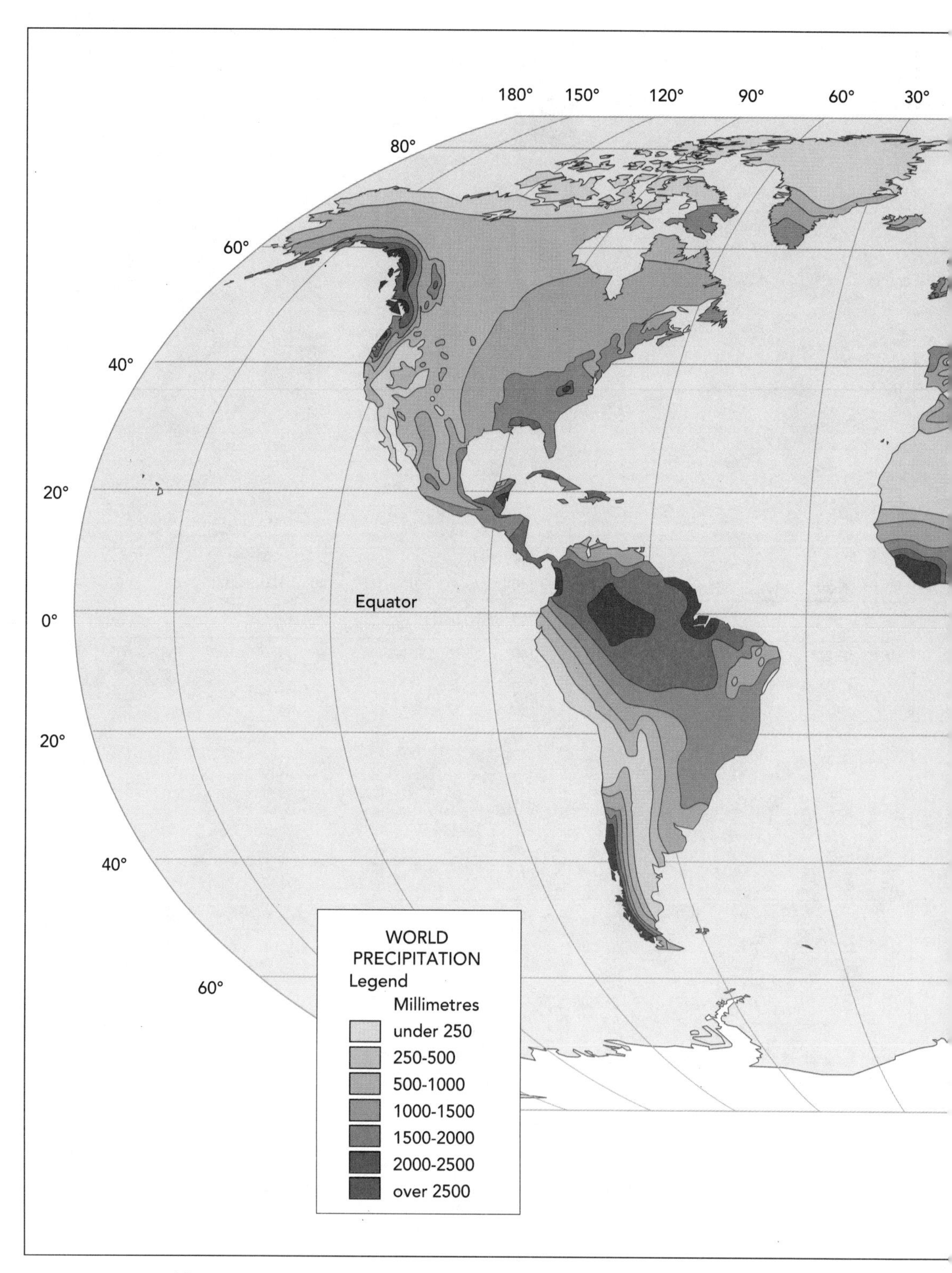

Figure 8.26 World Mean Annual Precipitation Patterns

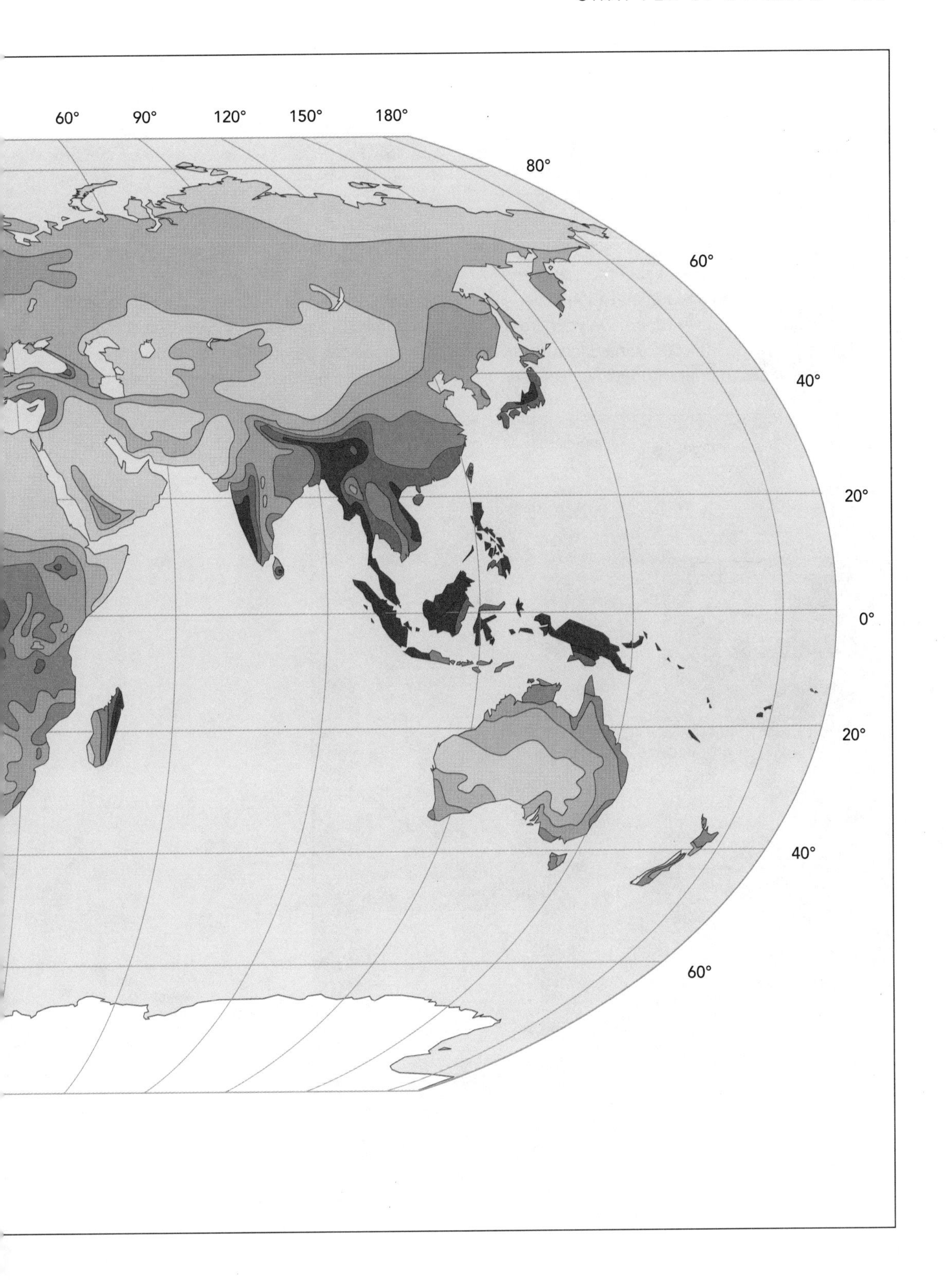
60°
90°
120°
150°
180°
80°
60°
40°
20°
0°
20°
40°
60°

QUESTIONS

26. Figure 8.27 shows a sketch map of Japan with precipitation data for stations on the east and west coasts of the country. Explain the pattern of precipitation experienced at these two locations.
27. Using an atlas, sketch a cross-section to represent western Canada. Label the ocean and mountains, put on a rough elevation scale, and include Vancouver Island. On your cross-section, indicate the approximate locations and elevations of Prince Rupert, Vancouver, Calgary, and Regina. Use symbols to indicate the mean annual precipitation values shown below for each location and explain the differences between the precipitation amounts experienced at each location.

Prince Rupert	–	2414 mm
Calgary	–	437 mm
Vancouver	–	1068 mm
Regina	–	398 mm

28. The range of temperature is the difference in degrees between a location's coldest monthly average temperature and that location's warmest monthly average temperature.

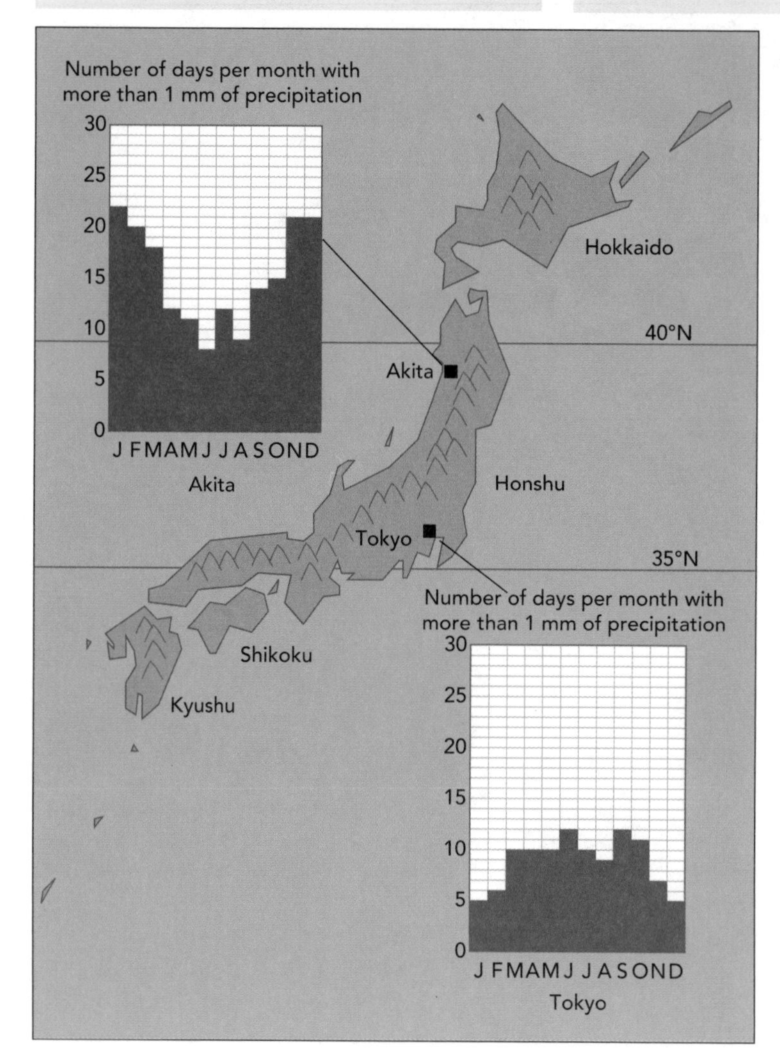

Figure 8.27
Selected Climatic Stations in Japan

Explain why the range of temperature for Vancouver is 15°C, while for Regina the range is 33°C.

29. Decide if the following statements are true or false and explain your answers.
    a) A baseball will travel farther in warm, humid air than in cold, dry air.
    b) A long-jump record is more likely to be set at 2000 m than at sea level.
    c) Canadian agriculture would benefit if the Western Cordillera were flattened.
    d) If you have forgotten your sleeping bag, it is wise to pitch your tent beside a large body of water when camping in November in Canada.

8•9 Climatic Controls: Local Influences on Climate

Many influences on an area's climate are localized in their effects, altering the climate over an area of only a few square kilometres to a few hundred square kilometres in extent.

Local Winds

As noted previously, winds are a response to differences in air pressure that in turn can be caused by differential heating; that is, one area of the earth's surface being heated to a different degree than an adjacent area. Figure 8.28 illustrates four types of local winds that are generated due to the differential heating and cooling of the earth's surface.

Urban Areas

An ever greater percentage of the world's people are living in urban areas. Cities of over one million people are now commonplace, and in many parts of the world megalopolises exist, supercities covering thousands of square kilometres. The urban environment can influence climates in many ways. Increased chemical and **particulate matter** pollution over urban areas can alter the radiation balance resulting in an increase in indirect and scattered solar radiation and a significant increase in the longwave radiation exchanges. The net effect is an increase in the amount of radiation being received at the earth's surface, resulting in higher temperatures. Adding to this effect is the reduction in photosynthesis and evapotranspiration rates experienced in the concrete world of the modern city. More radiation is thus utilized to produce heat. A **heat island** is present at most times in our urban areas. Measurements taken in and around London, England, indicate that temperatures in downtown London are on average 2°C to 3°C warmer than those of the surrounding countryside. Heat islands of 7°C to 10°C have been measured. Heat islands in turn can result in the wind system illustrated in Figure 8.29.

The albedo of an urban area is markedly different from the forest or grassland that it replaced. This, too, can alter the radiation balance, resulting in urban climates that are distinct from those of rural areas only a few kilometres away.

The increase in particulate matter over urban areas also results in an increased incidence of fogs and even precipitation, especially in coastal cities. The particulate pollution serves as hygroscopic particles on which water vapour can easily condense. Central London experiences twice the number of fogs than does the surrounding countryside. When combined with rising air, caused by the heat island effect described above, this increase in particulate matter becomes even more influential.

The morphology or structure of a modern city can also influence the climate.

The forest of skyscrapers which characterizes the modern city can produce wind tunnels along downtown streets. This effect, which can be very dramatic, is caused by air being funnelled between the tall buildings in the downtown area. This effect is illustrated in Figure 8.30.

Figure 8.28
Local Winds Created by Differential Heating and Cooling

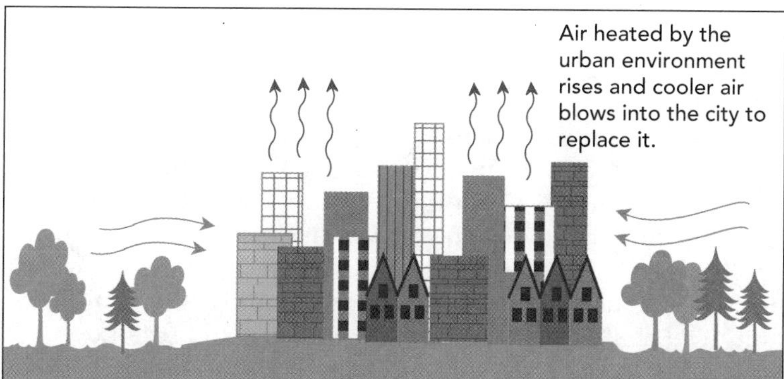

Figure 8.29 Urban Winds

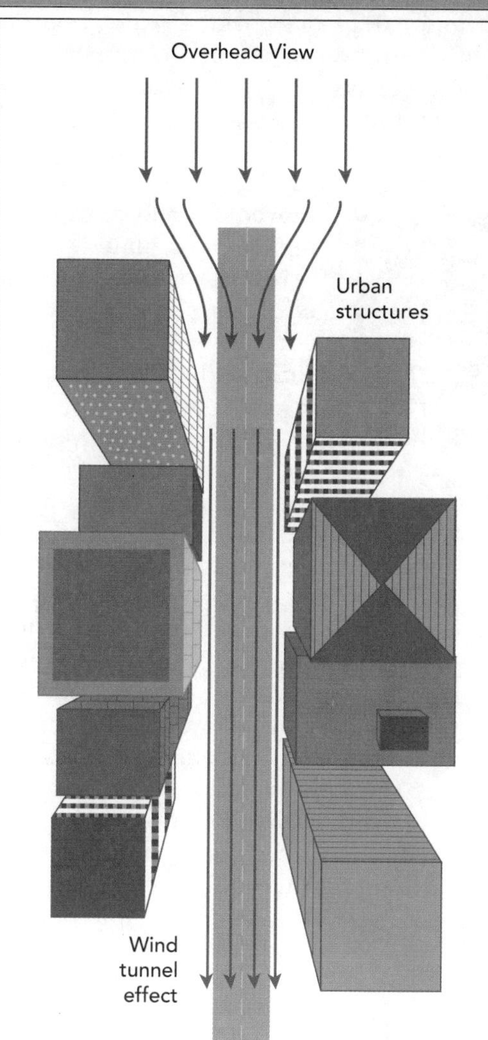

Figure 8.30 Wind Tunnel Effect in Urban Areas

Site and Landforms

A station's location in relationship to water bodies and landforms can give rise to dramatic climate differences among places on the earth's surface that are only a short distance apart. **Snow belts** are often found in the lee of lakes during the winter time, while the lakes themselves exert a moderating influence on climates at all times of the year. In low-lying areas, fog and dew will be more prevalent due to cold air drainage at night. Southern slopes in northern latitudes receive a great deal more solar radiation than northern slopes in the same vicinity. Differences in snow melt and in vegetation patterns between the two slopes can result.

QUESTIONS

30. Create a collage to illustrate all the ways in which cities can affect climates.
31. In what ways are land and sea breezes similar to monsoon winds? How are they different?
32. What effect might a stadium, such as Toronto's Skydome, have on local wind systems? Use sketch diagrams to help you answer.
33. From your own knowledge, explain two local climatic factors that influence your local area.

8•10 Climatic Change

Although at the beginning of this chapter we noted that climate, as distinct from weather, appears to be relatively stable and unchanging, there is increasing evidence to suggest that, when examined over centuries and especially over the geologic time scale, climatic patterns are dynamic and very changeable. The analysis of polar ice core samples, deep sea sediments, long-term pollen records, fossil evidence, sea level variations, glaciated landscapes, and even recorded histories, give evidence that the earth's climatic patterns are constantly changing and fluctuating. Five theories that have been put forth to explain climatic change are outlined below.

a) It has been suggested that very long-term fluctuations occur in the energy output of the sun as it revolves around the centre of the Milky Way galaxy. Such fluctuations would operate on a time scale of hundreds of millions of years. Scientists have suggested that the sun, due to its internal chemistry or due to the rotation of other systems in the galaxy, "flickers" approximately every 300 million years. They have attempted to link these "flickers" with past ice ages on earth.

b) Due to the variations in the orbit of our own planet around the sun, changes occur in the amount of solar radiation reaching the earth. These cycles were outlined by Milankovitch and are discussed in Chapter 3. Many researchers claim these changes in the input of solar radiation can be directly linked to glacial and interglacial periods. Others remain unconvinced of the link.

c) There is evidence to suggest that the earth has undergone dramatic periods of intense volcanic activity during which a great deal of ash has been pumped into the atmosphere. If enough material were put into the atmosphere, it is possible that incoming solar radiation could be reduced to the point that significant cooling of the earth's surface could result.

d) We have noted that the arrangement and extent of land and water bodies, the location of mountain ranges, and the elevation of land masses have a direct effect on climates. When these controls are looked at in terms of plate tectonics, it is clear that the constant movement and rearrangement of the plates would bring about fundamental climatic changes. These changes would occur on a time scale of millions of years.

e) Related to d) above is the apparent link that exists between plate tectonic activity and atmospheric carbon dioxide content. Scientists have noted that during periods of increased plate tectonic activity, excessive amounts of carbon dioxide might be released into the atmosphere from the deep trenches formed by spreading plates. Carbon dioxide is one of a group of gases known as "greenhouse gases". It has the ability to allow shortwave solar radiation to penetrate to the earth's surface, but it prevents the earth's longwave radiation from escaping to outer space. The suggestion is that as the CO_2 content in the atmosphere increases, the net radiation available at the earth's surface increases.

This results in warmer climates and increased plant growth. Evidence exists that the Mesozoic period on earth was a time of increased plate tectonic activity, increased levels of atmospheric CO_2, and warmer temperatures.

Many questions remain concerning the role carbon dioxide plays in influencing the earth's climate. From the analysis of deep sea cores, atmospheric carbon dioxide levels can now be approximated fairly accurately for the past 350 000 years. Figure 8.31 compares the variations in carbon dioxide with temperature changes. Very recently, researchers have linked the variations in atmospheric carbon dioxide content to changes in the earth's orbit. It appears that orbital variations precede changes in atmospheric carbon dioxide levels, that in turn trigger periods of glaciation and deglaciation. This area of scientific research is extremely important as humans are now putting large amounts of carbon dioxide and other greenhouse gases into our atmosphere. If carbon dioxide is fundamentally important in determining the earth's climate, then the human race might be bringing about changes in our climate over the next century that are unprecedented. Chapter 16 discusses the "Greenhouse Process" in detail.

The earth's atmosphere and climate have been evolving since the beginnings of our planet. The speculation that climate might be the unifying concept that helps scientists to link and understand the relationships among the atmosphere, lithosphere, and biosphere is a new and exciting area of research in physical geography.

QUESTIONS

34. Why is it difficult to research long-term climatic trends?
35. Examine Figures 8.32, 8.33, and 8.34. Each figure presents measurements that suggest that changes in the earth's climate have occurred. In your own words, summarize the data presented in each figure and state whether the data suggests a warming or cooling trend for the earth's climate at the present time.

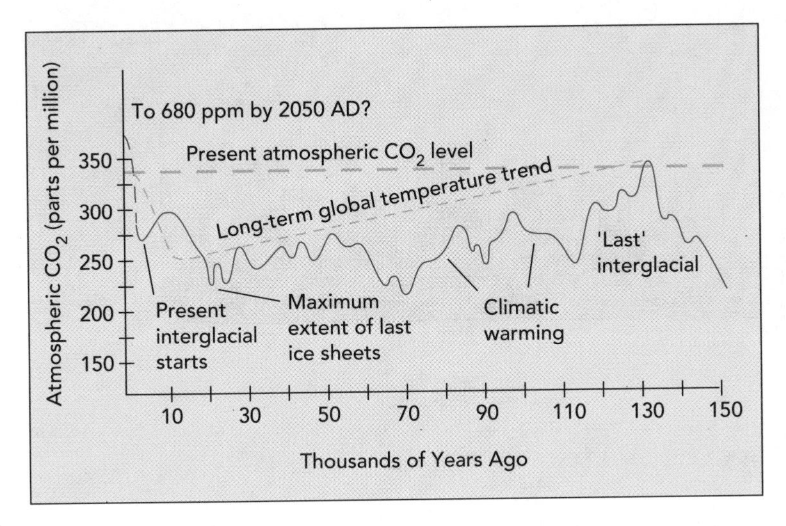

Figure 8.31
CO_2 and Temperature Changes for the Last 150 000 Years

36. By understanding climatic controls and the mechanisms that influence climates, we might be able to better plan for the future. Explain three reasons why the ability to predict climatic trends for the next century would be useful.

	% Advancing	% Stationary, Hidden by Snow, or Otherwise Uncertain	% Retreating
1920-9	7.6	6.4	86.0
1930-9	9.6	9.1	81.3
1940-9	10.6	8.9	80.5
1950-9	6.4	10.7	82.9
1960-9	10.9	35.6	53.5

Figure 8.32 **Glacier Movements in the Italian Alps (Decade Averages)**

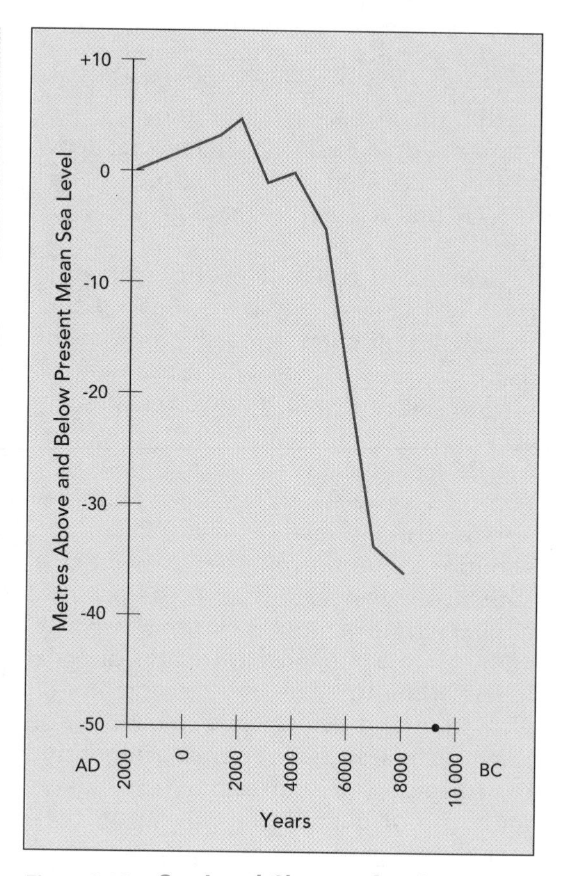

Figure 8.33 **Sea Level Changes for the Last 12 000 Years**

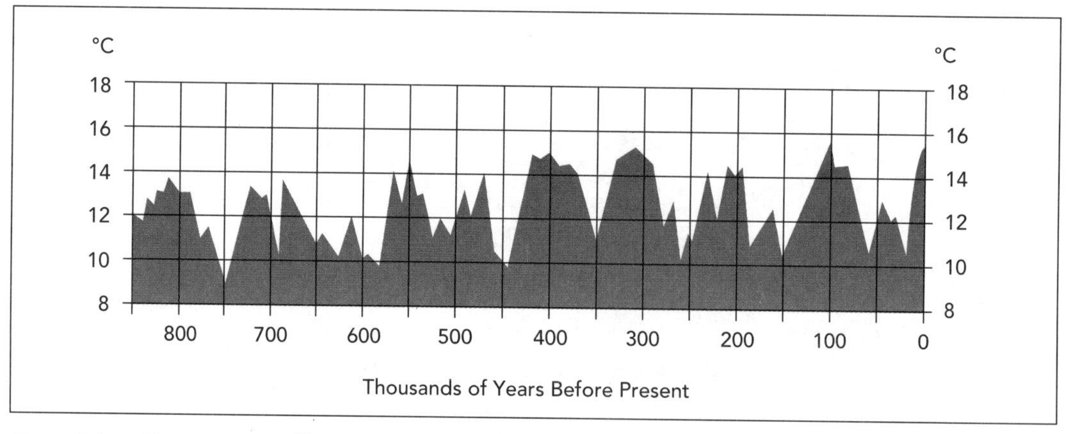

Figure 8.34 **Temperature Changes Over the Last 850 000 Years**

Review

- The term "weather" refers to short-term atmospheric conditions and "climate" to long-term characteristics of the atmosphere.
- Climatic classification schemes help to simplify the study of climate.
- Solar radiation is not received equally on all parts of the earth's surface; this imbalance in energy leads to pressure and wind systems.
- Wind systems are affected by a variety of forces, including the Coriolis force and the differential heating effect of land and water.
- Ocean currents are another mechanism that redistributes heat about the earth.
- Climates on land are influenced by nearby oceans and by the presence of mountain ranges.
- Some climatic factors are quite local in effect, such as the heat islands created by large urban areas.
- Climate, when viewed from the perspective of centuries, goes through important changes.
- Humans are significant agents of climatic change.

Geographic Terms

weather
climate
climograph
hythergraph
pressure gradient
subtropical highs
subpolar lows
jet stream
cyclone
anticyclone
monsoon
environmental lapse rate
sensible heat flux
dry adiabatic lapse rate
wet adiabatic lapse rate
particulate matter
heat island
snow belt

Explorations

1. What do you personally consider to be the ideal climate? Explain your answer.
2. Many areas of the world do not experience seasons as we understand the term. Use atlases or textbooks to identify three locations in the world where the climate would be divided into wet and dry seasons.
3. Eleven locations are marked on the world map in Figure 8.35. Figure 8.36 contains eleven sets of climatic statistics. List the locations shown on the world map and beside each location place the number of the climatic statistics that represents the climate experienced at that location. Attempt to match the locations and statistics yourself without referring to an atlas. For two of the locations, fully explain why you felt the graph matched the location.
4. List all the climatic controls described in this chapter. Explain how each climatic control influences, or does not influence, the climate of the area in which you live. To help you answer this question, obtain the climatic statistics for the area in which you live.

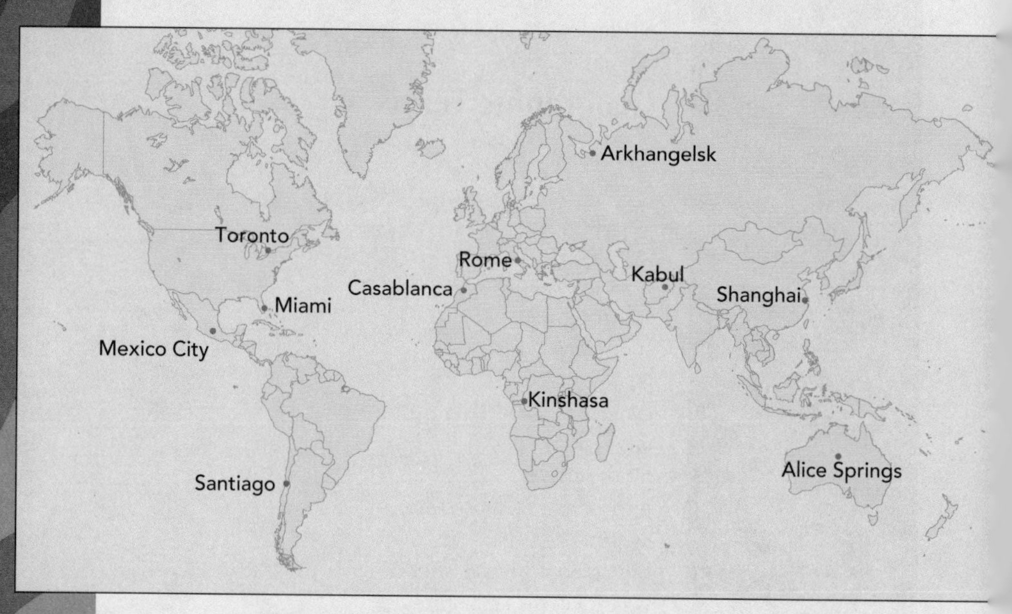

Figure 8.35 **Eleven Climatic Stations in the World**

Station		J	F	M	A	M	J	J	A	S	O	N	D	Yearly	Annual Temperature Range
1	Precipitation (mm)	43	33	28	10	15	13	8	8	8	18	31	38	*253	
	Temperature (°C)	29	28	25	20	15	12	12	14	18	23	26	28	**21	17
2	Precipitation (mm)	31	36	94	102	20	5	3	3	3	15	20	10	342	
	Temperature (°C)	-3	-1	6	13	18	22	25	24	20	14	7	3	12	28
3	Precipitation (mm)	48	58	84	94	94	180	147	142	130	71	51	36	1135	
	Temperature (°C)	4	5	9	14	20	24	28	28	23	19	12	7	16	24
4	Precipitation (mm)	53	48	56	36	23	5	0	3	8	38	66	71	406	
	Temperature (°C)	13	13	14	16	18	20	22	23	22	19	16	13	18	10
5	Precipitation (mm)	50	46	61	70	66	67	71	77	64	62	63	65	762	
	Temperature (°C)	-7	-6	1	6	12	18	21	20	16	9	3	-4	7	28
6	Precipitation (mm)	31	19	25	29	42	52	62	56	63	63	47	41	530	
	Temperature (°C)	-16	-14	-9	0	7	12	15	14	8	2	-4	-11	0	31
7	Precipitation (mm)	3	3	5	13	64	84	76	56	31	15	8	5	363	
	Temperature (°C)	21	20	18	15	12	9	9	10	12	15	17	19	15	12
8	Precipitation (mm)	135	145	196	196	158	8	3	3	31	119	221	142	1357	
	Temperature (°C)	26	26	27	27	26	24	23	24	25	26	26	26	25	4
9	Precipitation (mm)	13	5	10	20	53	119	170	152	130	51	18	8	749	
	Temperature (°C)	12	13	16	18	19	19	17	18	18	16	14	13	16	7
10	Precipitation (mm)	71	53	64	81	173	178	155	160	203	234	71	51	1494	
	Temperature (°C)	20	20	22	23	25	27	28	28	27	25	22	21	24	8
11	Precipitation (mm)	71	62	57	51	46	37	15	21	63	99	129	93	744	
	Temperature (°C)	8	9	11	14	18	22	25	25	22	17	13	10	16	17

Figure 8.36 **Selected World Climatic Statistics**

* Total Annual Precipitation in Millimetres

** Average Yearly Temperature in Degrees Celsius

CHAPTER

9

WEATHER

OBJECTIVES:

By the end of this chapter, you will be able to:

- identify forms of instrumentation and methods used to compile weather information;
- recognize the importance of the movement of air masses in shaping global weather patterns;
- distinguish between the forces influencing the weather in the mid-latitudes and in the equatorial regions;
- recognize the significance of frontal systems in the weather patterns of North America;
- describe and explain the development of a mid-latitude cyclonic storm;
- appreciate the immense power and destructive abilities of weather events such as hurricanes and tornadoes;
- appreciate the complexity of the atmosphere and the difficulties involved in accurately predicting the weather.

CHAPTER 1: The Nature of Physical Geography
CHAPTER 2: Earth: Its Place in the Universe
CHAPTER 3: The Earth in Motion
CHAPTER 4: The Earth's Interior
CHAPTER 5: The Earth's Crust
CHAPTER 6: The Lithosphere in Motion: Plate Tectonics
CHAPTER 7: Solar Radiation
CHAPTER 8: Climate
CHAPTER 9: Weather
CHAPTER 10: The Hydrosphere and the Hydrologic Cycle
CHAPTER 11: Natural Vegetation and Soil Systems
CHAPTER 12: Denudation: Weathering and Mass Wasting
CHAPTER 13: Distinctive Landscapes: Humid and Arid Environments
CHAPTER 14: Distinctive Landscapes: Glacial, Periglacial, and Coastal Environments
CHAPTER 15: Natural Hazards: Disrupting Human Systems
CHAPTER 16: The Disruption of Natural Systems
CHAPTER 17: Fragile Environments

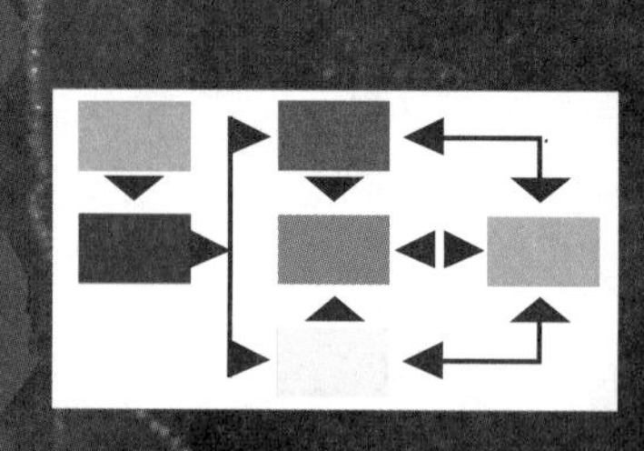

Introduction

Be still, sad heart! and cease repining;
Behind the clouds is the sun still shining.

Longfellow, "The Rainy Day"

It is remarkable how often weather is referred to in daily life. Weather affects us in different ways, directly by forcing us to wear suitable clothing and indirectly by shaping our moods. But what causes weather? Why is one day rainy and cool and the next one dry and warm? What creates weather patterns? These and other topics will be explored in this chapter.

9·1 Weather and Air Masses

The harvesting of crops, the playing of a golf game, the launching of a space shuttle, the choosing of a wardrobe — these are only four of the thousands of human activities that are directly influenced by the weather. Weather is perhaps the most dynamic of earth's physical systems. It often changes from minute to minute. Weather is defined as the condition of the atmosphere at a particular point in time. The components or elements that make up our weather include temperature, air pressure, humidity, precipitation, visibility, cloud cover, and wind. These components are constantly measured by thousands of weather stations located around the globe. Figure 9.1 illustrates some of the various instruments used to measure these components of weather.

Wind is one of the most important of the components of weather because our wind systems transport air masses across the earth's surface. An air mass is a three-dimensional parcel of air whose temperature and humidity characteristics are

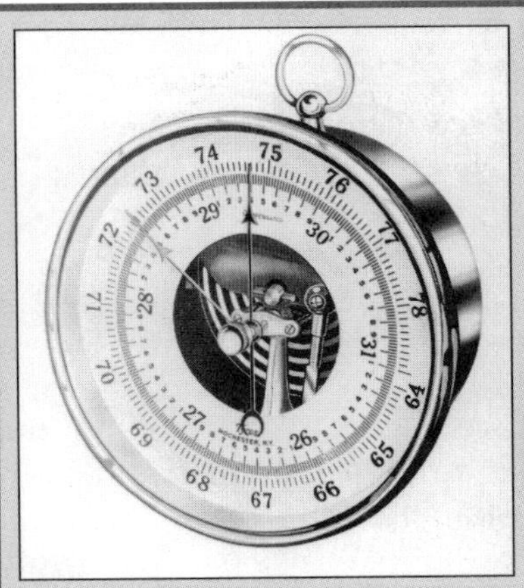

a) Barometer

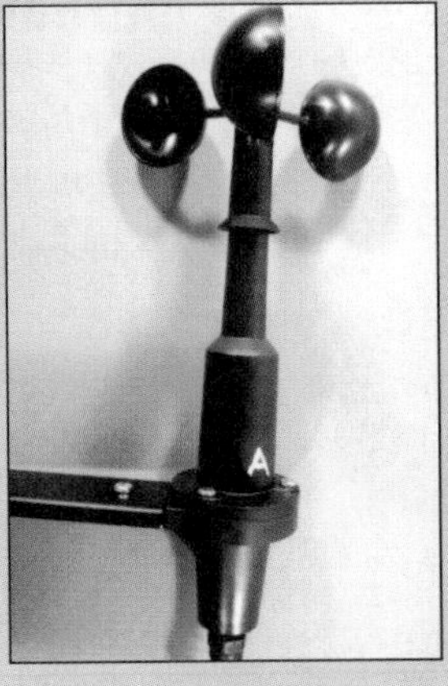

b) Anemometer

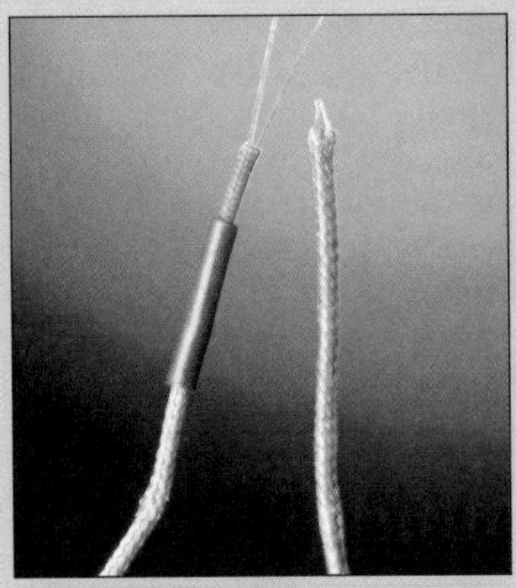

c) Thermocouples

Figure 9.1 **Instruments Used to Measure Weather Components**

a) The barometer is used to measure air pressure. The aneroid barometer consists of a sealed chamber which has been partly emptied of air. The lid of the chamber is a flexible diaphragm which moves up and down depending on the air pressure. The diaphragm is connected to a scale which can be calibrated in millibars, inches, or kilopascals. The instrument can also be calibrated in units of altitude to be used in an aircraft. It is then called an altimeter.

b) The anemometer is used to measure wind speed. When placed in the wind, the cups on the anemometer will begin to spin at a rate dependent on the wind speed. Sometimes the spinning of the cups is used to generate a small electrical current, the strength of which can be converted into the correct wind speed.

A hot wire anemometer uses a heated wire which is placed in the wind. The degree to which the wire cools off is reflected in the current running through the wire and this reduction in current can be used to measure wind speed.

A wind vane is often attached to the anemometer. The blade on the wind vane will align itself to the wind and thus indicate the wind direction.

c) Thermometers and thermocouples are used to measure temperature. Most thermometers measure temperatures by the expansion of a liquid in a calibrated tube. The tube has a metal end or bulb which conducts the heat to the liquid.

There are many types of thermometers. A maximum-minimum thermometer measures the highest and lowest temperatures which were reached between resets. This is a common measurement at most weather stations as the average between the maximum and minimum temperatures gives us the daily mean temperature.

A recording thermometer or thermograph gives us a continuous record of temperature changes. When the temperature of a "wet-bulb" thermometer, a thermometer whose bulb is kept wet, is compared to a dry-bulb thermometer, a different reading is obtained. The difference between the two readings can be used to obtain a measurement of relative humidity. A thermocouple consists of two different metals. One junction of the thermocouple is placed in a medium of known temperature, for example, ice water, while the other end is placed in the soil or air or fluid whose temperature you wish to measure. The electrical current which is set up can be measured and converted into degrees.

nearly homogeneous. The areas of origin give the air masses their characteristics. Depending upon the wind systems, air masses with quite different characteristics can invade an area of the earth's surface over a short period of time. Cold, dry air can be replaced by warm, humid air, resulting in dramatic changes in the weather that a location experiences. Figure 9.2 outlines the major air masses and their paths of movement that influence North America and summarizes the characteristics of those air masses.

Sometimes the various weather elements can combine to bring about conditions that are dangerous to human health. Figure 9.3 illustrates the results that occur due to the interchange between wind and temperature, while Figure 9.4 combines temperature with relative humidity to produce a "Comfort Index".

QUESTIONS

1. In your own words and based on your own feelings, describe "the perfect day" in terms of weather. Be sure to address all the components of weather noted on page 169. How often do these "ideal" conditions occur in your local area?
2. Using Figure 9.2 and Figure 8.17 (page 146), predict which air masses would have the most influence on the weather of the following places: Vancouver, Regina, Toronto, and Halifax. Organize your answer by examining both winter and summer conditions.
3. If you were setting up a weather station using the instruments described in Figure 9.1, where should you locate each of the instruments in order to obtain accurate readings?
4. Copy Figures 9.3 and 9.4 into your notebook. Add to each figure, at different spots on the index, three pieces of advice that you would give to joggers who wish to run outdoors.

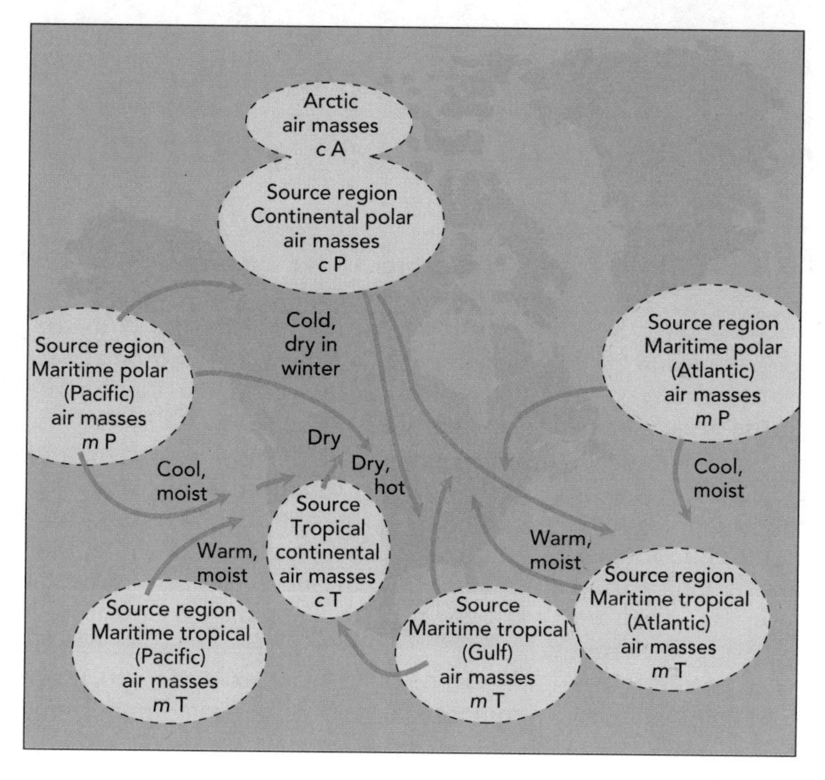

Figure 9.2
Air Masses Affecting North America

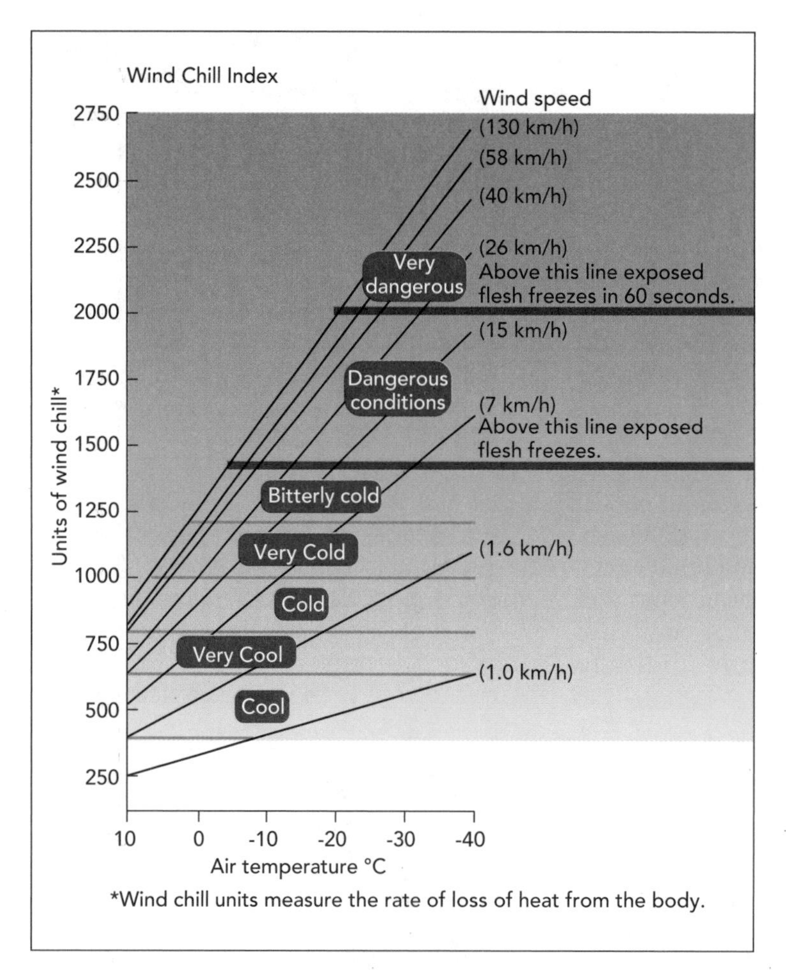

Figure 9.3
Wind Chill Index

9·2 Weather of the Mid- and High Latitudes

Weather is easy to understand and predict when one air mass dominates the day-to-day conditions of the atmosphere. The location in question will experience the weather associated with the dominant air mass. (See Figure 9.2.) There are locations where one air mass dominates the weather over a relatively long period of time, sometimes months. At these locations, weather is relatively unchanging and easy to predict, and the terms "climate" and "weather" are almost interchangeable. However, in much of the mid-latitude areas of the world, the weather is influenced by the constant movement of air masses of different characteristics, and by the meeting and interaction of these air masses.

The most dramatic and influential zone where air masses with different characteristics meet is called the **polar front**. A front is the boundary zone between two different air masses. Along the polar front, cold, dry, polar air comes in contact with warm, humid, sub-

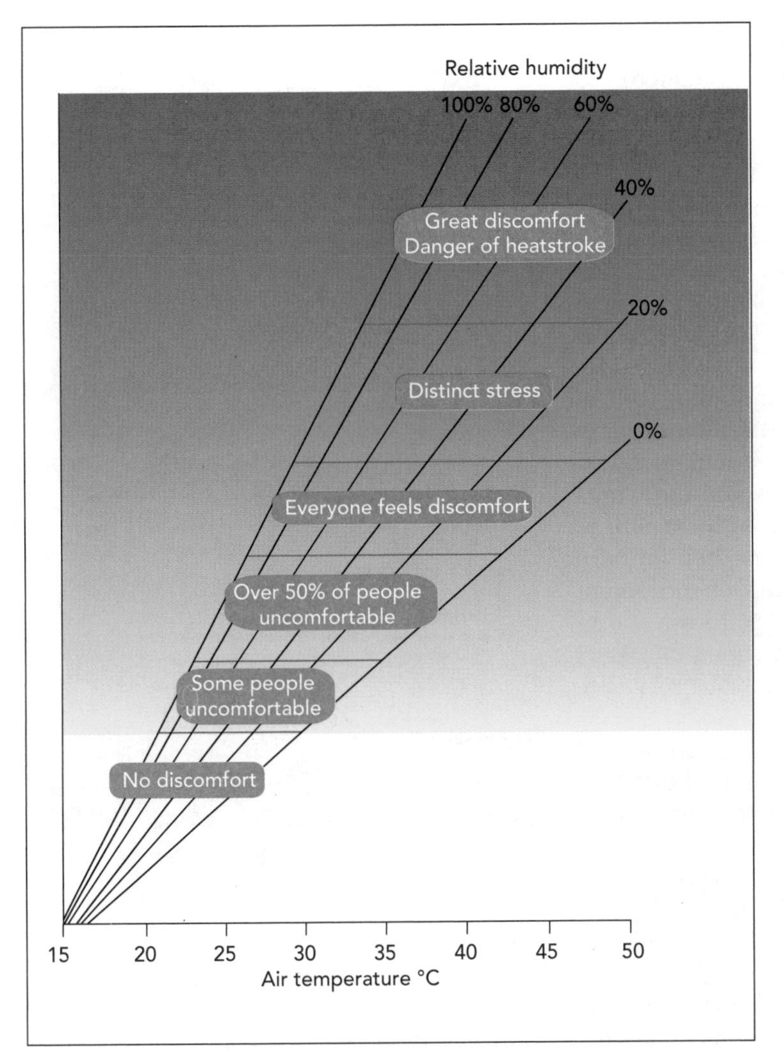

Figure 9.4
Comfort Index

tropical air. Most weather systems affecting mid-latitude areas develop due to the interaction of these two different air masses. These weather systems, zones of low pressure called cyclones, and zones of high pressure called anticyclones, move across our globe resulting in day-to-day changes in the weather we experience.

During World War I, a Norwegian meteorologist named Bjerknes put forth a theory that helped to explain the weather experienced throughout the mid- and high latitude areas of the world. He referred to his theory as the **frontal wave theory of cyclonic development**. Bjerknes likened the meeting of cold, polar air and warm, tropical air to the front lines in World War I that separated the two opposing armies. At times, one army would invade the space of the other, resulting in a displacement of the opposing forces. Bjerknes noted that, when cold and warm air meet, the warm air, being less dense, is forced to rise over the colder air mass. Along the front, this rising air would create a low pressure zone called either a **depression**, a **cyclone**, or a **cyclonic disturbance**, and a wave would form as this developing depression moved westward across the globe. As explained in Chapter 8, winds blow towards a low pressure area and, due to the Coriolis force, circulate counterclockwise in the northern hemisphere. By contrast, winds blow out of a high pressure area or anticyclone and, in the northern hemisphere, circulate clockwise. The Coriolis force causes the winds to be deflected to the right of their path of movement in the northern hemisphere and to the left of their path of movement in the southern hemisphere. In the southern hemisphere, therefore, the clockwise, counterclockwise directions are reversed. (See Figure 9.5.)

Figure 9.6a represents an idealized drawing of the development and death of a mid-latitude cyclone according to Bjerknes' wave theory. Figure 9.6a also illustrates the weather that would be experienced at the earth's surface at each stage of the wave development. Note the associated cloud types in Figure 9.6b. Clouds, which are concentrations of atmospheric moisture, or water vapour, are good indicators of the weather conditions which we can expect to experience.

The exact mechanism that triggers the development of the waves and thus the cyclones is not fully understood. In the late 1940s, Jules Charney put forth the **baroclinic wave theory of cyclones**. Charney's mathematical models pointed out that temperature differences from the poles to the equator combined with increasing wind speeds with altitude would generate cyclones even in the absence of a well-defined polar front.

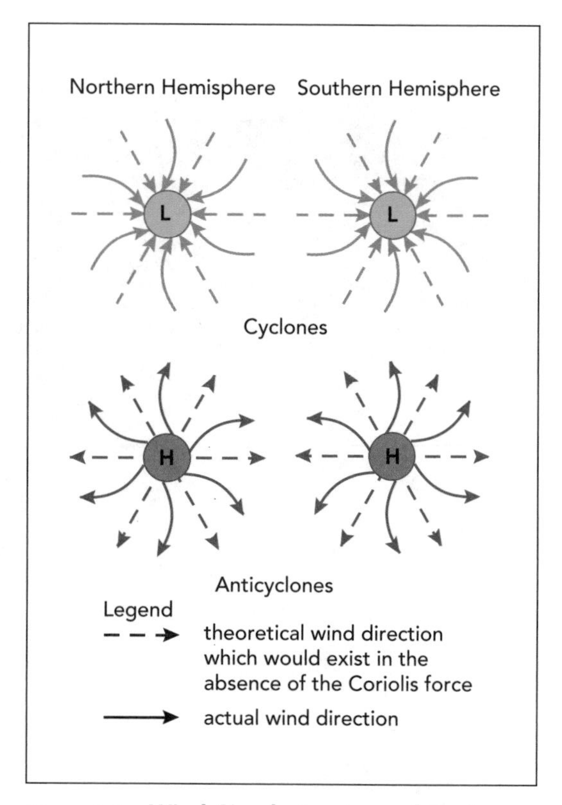

Figure 9.5 **Wind Circulation Around Cyclones and Anticyclones in Both Hemispheres**

Wind speeds increase with altitude because the effect of friction with the earth's surface is reduced. In addition, as altitude increases, the effect of the Coriolis force decreases and thus the wind adjusts itself to blow at right angles to the **isobars**, the lines joining points of equal pressure at the earth's surface. There exists, therefore, a wind shear or tension as we move upwards from the earth's surface. Both wind speed and direction change with altitude. This wind shear contributes to the "spinning motion" of the cyclone.

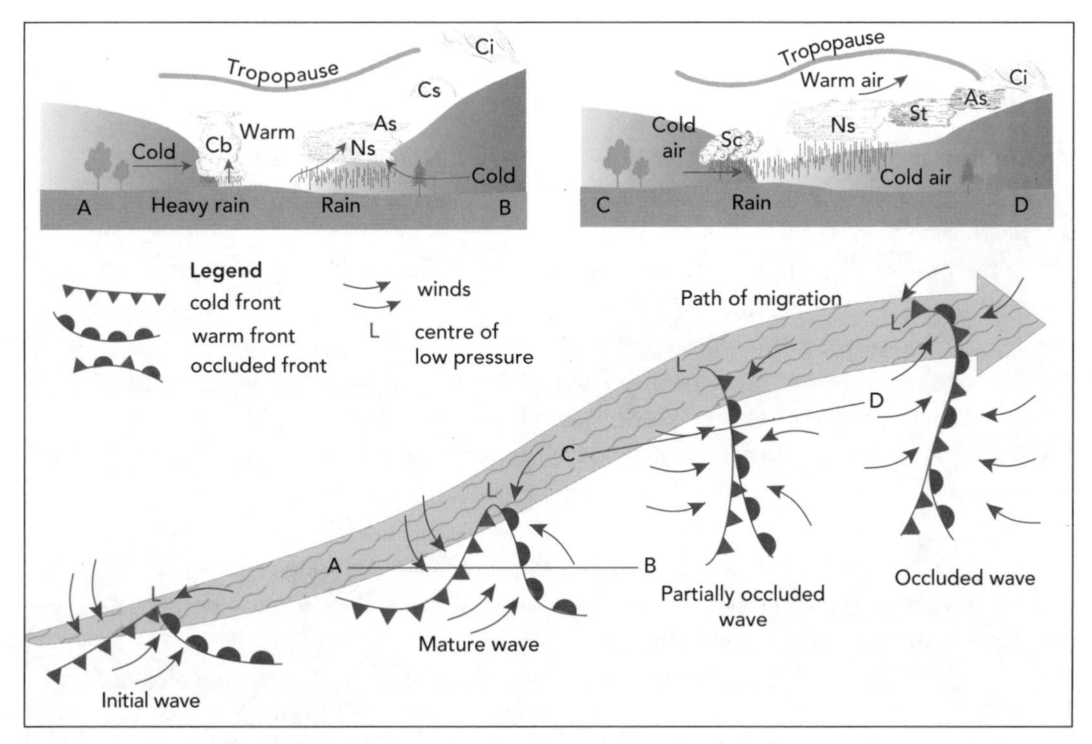

Figure 9.6a The Stages of a Mid-Latitude Cyclone The cyclone begins as a wave along the contact between tropical and polar air. As the cold front "catches up" to the warm front, an occluded front is formed. The cross-sections show the corresponding weather conditions which exist at two of the stages. A cloud classification is included as the types of clouds indicate the stages of the cyclone.

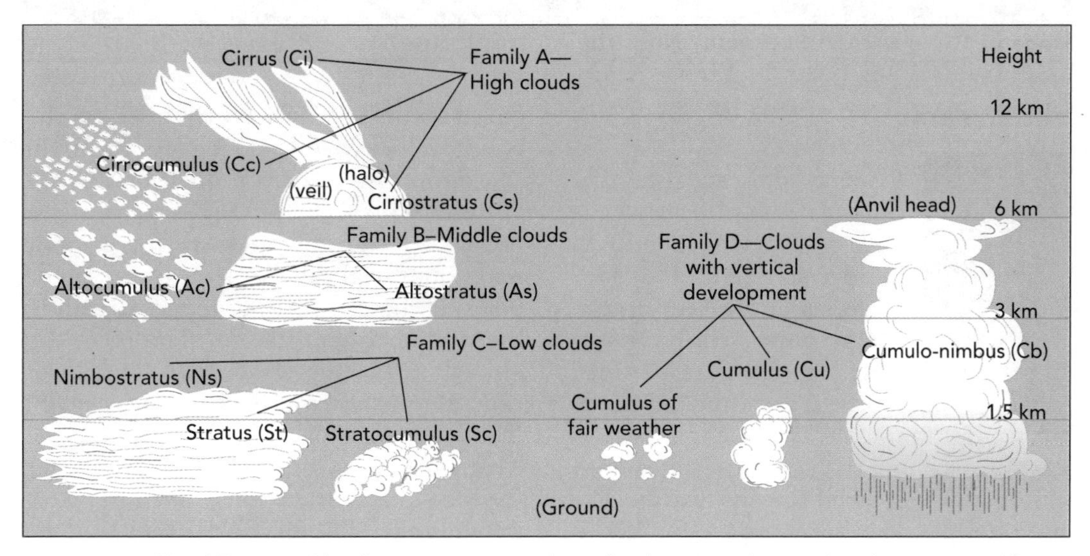

Figure 9.6b Cloud Types Cloud types are grouped into families according to altitude, range, and form.

Imagine

- Imagine your desk top as a section of the earth's surface in the northern hemisphere.
- To the top of your desk, imagine dry, cold air. The wind in this area is blowing from east to west, or from right to left.
- To the bottom of your desk, there is warm, moist air. The wind in this section is blowing from west to east, or left to right.
- Between the two different types of air is a front.
- The wind speed in both sections of your desk is increasing with altitude, up to perhaps a metre above the surface. It increases because there is less friction the higher up you go.
- The warm air, to the bottom of your desk, is more buoyant than the cold air to the top of your desk. The warm air will tend to rise from the surface, while the cooler air sinks and moves down, or south, to replace it. Due to the Coriolis effect at the surface, the winds will begin to "spin" in a counterclockwise direction around the rising warm air. Remember, if your desk is in the northern hemisphere, winds near the desk top will be deflected to the right of their path of movement.
- You have now created in your mind a depression or cyclonic disturbance, the area of rising air. In addition, you have air flowing into the depression and being deflected counter-clockwise by the Coriolis force. The upper air movement is contributing to the "spinning" of your depression.
- The imagining of this complex model is not easy. Try it a number of times and discuss it with the teacher and other members of the class. If it makes it easier to understand, attempt to sketch the model in stages as you add more and more to it.

Jet Streams

Evidence now suggests that upper atmospheric winds have a direct influence on the development of surface cyclones and anticyclones. **Jet streams** are bands of upper air movement at altitudes between 9000 and 12 000 m. They flow at speeds up to 450 km/h. As these winds speed eastward around the globe, they meander, creating waves called **Rossby waves**. The meanderings appear to trigger the development of low pressure cells on the earth's surface and contribute to the counterclockwise surface flow of air in the depressions in the northern hemisphere. The Rossby waves also allow polar air to advance towards the equator and equatorial air to move towards the poles. If you are north of the jet stream in the northern hemisphere, then cold, dry air will affect the weather in your location. If you are south of the jet stream, then warmer, moist air will influence the weather. If your location is directly under the jet stream, then you will most likely experience storms and unsettled or changeable weather.

QUESTIONS

5. Think of the last time that a major change in weather occurred in your area. Was the change associated with the arrival of a warm front or cold front? Explain.
6. Warm fronts are those that precede warm air masses, while cold fronts are the leading edges of cold air masses. Which type of front would usually be moving the fastest across the surface of the earth? Explain. (*Hint:* Consider the density and buoyancy of warm versus cold air and the effect of friction on air masses.)
7. Would each of the following weather occurrences be more likely associated with the arrival and/or passage of a warm or cold front? Explain.
   a) thunderstorms
   b) clear skies
   c) a drop in air pressure
   d) a drop in temperature
   e) cumulo-nimbus clouds
   f) an increase in humidity
8. Examine Figure 9.7 which shows a satellite picture of a mid-latitude cyclone travelling from west to east during July. Using Figures 9.5 and 9.6 as your guide, summarize the changes in weather that you would expect to experience as this low pressure system approached, passed over, and then left the area in which you are living. In your answer, refer to temperature, wind, humidity, air pressure, and precipitation changes.

Figure 9.7 **Satellite Photo of a Mid-Latitude Cyclone**

9·3 Equatorial Weather

Equatorial regions generally have more predictable weather patterns than do the mid-latitudes. In equatorial areas, we do not often have the meeting of radically different air masses and, therefore, the creation of well-defined fronts as are known in the mid-latitudes are rare. The easterly flow of the trade winds dominates air movement in the equatorial zones. (See Figure 8.13, page 144.) The trade winds blow towards an area of low pressure at the equator and often are associated with weak **troughs** or fronts of low pressure that result in convectional rainfall as warm, moist air rises in these troughs. As these fronts pass a location in the tropics, clear skies give way to a day or two of thunderstorm activity. The fronts seldom develop through the stages associated with the mid-latitude cyclones illustrated in Figure 9.6.

Near the equator itself is found the **intertropical convergence zone**, the area where northeast and southeast trade winds meet. The air in the tropics is usually unstable to begin with due to intense surface heating and evaporation. At the intertropical convergence zone, this instability is enhanced by the meeting of the trade winds. The rising warm, moist air results in convectional rainfall, thunderstorms, and squalls. In some locations, these storms occur on a daily basis. In drier areas, it is not uncommon for all the rainfall associated with the storm to be evaporated before it hits the ground! Figure 9.8 illustrates the development of convectional rainfall and contrasts the formation of this type of rainfall with orographic rainfall and cyclonic rainfall. It should be noted that Figure 9.8 illustrates three different ways to achieve the same result. Precipitation occurs when warmer, moist air is forced to rise and cool. When the condensation point is reached, precipitation occurs. Only the mechanism by which the warmer, moist air is forced to rise changes.

Hurricanes and Typhoons

Due to the great amount of water vapour usually associated with equatorial air masses, a special type of cyclone called a **hurricane** or **typhoon** sometimes develops. These storms, the most dramatic of all the earth's weather systems, develop over warm oceans whose surface water temperatures are at least 27°C, usually between 8° and 15° north and south of the equator. The absence of the Coriolis force at the equator itself seems to prevent the circular pattern of winds in a depression from developing and this in turn prevents the storms from originating there.

A hurricane or typhoon starts as a small, low pressure system moving westward over the warm, subtropical ocean. As more water is evaporated by the depression, it gains in strength and becomes extremely unstable. The warm, moist air blowing into the centre of the depression carries a great amount of stored energy, having brought water vapour into the depression from a large ocean area. As condensation occurs, heat is released which further warms the unstable rising air. As long as there is a warm, moist ocean to provide energy into

It's a Fact...

Raindrops have a radius from 0.5 mm to 2.5 mm and usually fall at a rate of between 4 and 9 m/s. Drizzle drops have a radius from 0.05 to 0.5 mm and are just heavy enough to fall out of the clouds. Drops with a radius greater than 25 mm break up into smaller drops due to aerodynamic considerations. Snowflakes usually have a diameter of between 10 and 40 mm and fall at a rate of between 1 and 2 m/s.

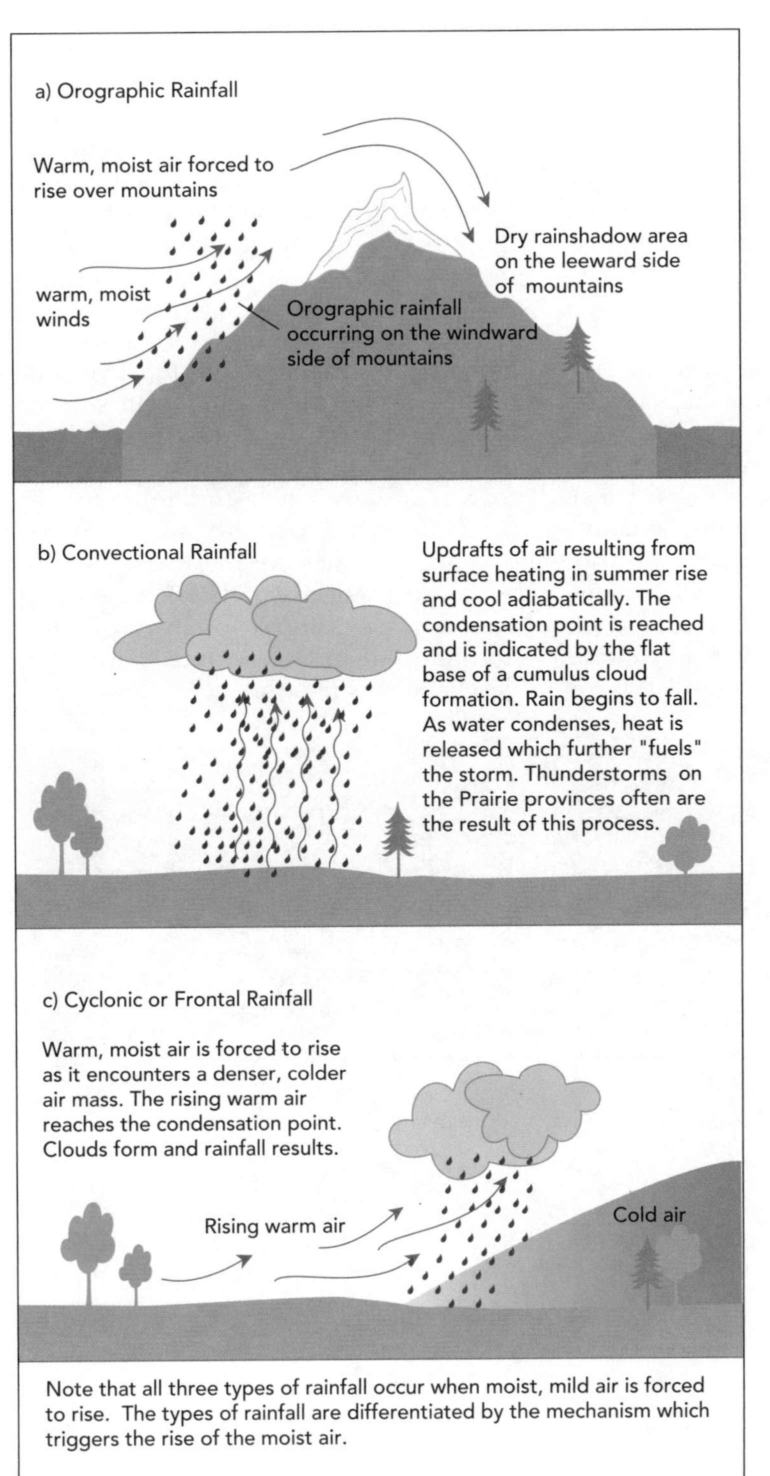

Figure 9.8
Types of Rainfall

the system, the tropical cyclone will continue to develop and intensify. When the source of moisture is taken away (for example, when the storm moves inland or moves to cooler locations), the hurricane "dies". Figure 9.9 categorizes these tropical low pressure systems or cyclones into four categories based on their wind speeds. Figure 9.10 shows a cross-sectional view of a typical hurricane.

Figure 9.11 illustrates the paths which hurricanes and typhoons typically take. With an average life span of six to twelve days, these storms can cause incredible damage if they collide with populated coastal areas. Direct damage from their winds, which by definition exceed 117 km/h, is augmented by storm surges as the winds push ocean levels far above their normal height. In addition, rainfall amounts of 200 to 300 mm/h have often been experienced as the storm passes an area, and rainfall amounts as high as 1000 mm in 24 h have been recorded. Widespread flood damage is, therefore, a common characteristic of these tropical cyclones, especially in low-lying coastal areas.

The occurrence of hurricanes and typhoons corresponds with those periods of the year in which the ocean surface water is the warmest. In the northern hemisphere, this period runs from June to November, while in the southern hemisphere, the period from December to March is when we find the most violent storms. In areas that have consistently warm waters, for example, the northern Indian Ocean, these storms can develop in any month.

It's a Fact...

"Hurricane" is derived from an Indian word used in the Caribbean. It means "big wind". One explanation of the word "typhoon" is that it is derived from the Chinese expression "tai fung", which means "wind which strikes". Some believe, however, that it refers to the Greek mythological monster Typhon or Typhoeus who was the father of all storms.

a) A Tropical Disturbance:
No true low pressure system has developed at the surface. There are no closed isobars and no strong winds.

b) A Tropical Depression:
Rotary circulation of winds exists at the surface. Closed isobars define an identifiable low pressure system. Winds do not exceed 62 km/h.

c) A Tropical Storm:
Wind speeds are between 62 and 117 km/h as the low pressure system intensifies.

d) Hurricane:
Wind speeds in excess of 117 km/h. An intense low pressure system now exists and a pronounced circular motion of winds is present.

Figure 9.9
The Classification of Tropical Cyclones

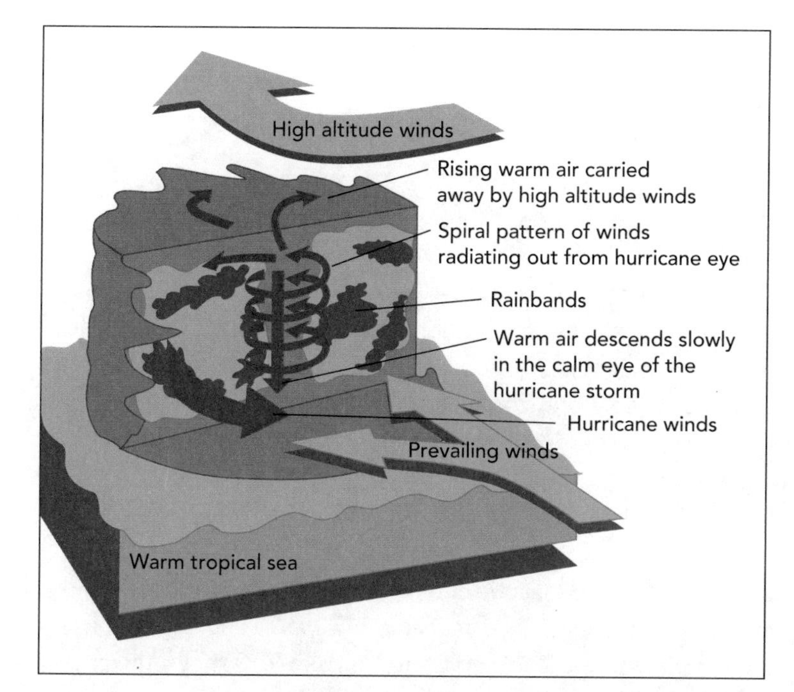

Figure 9.10

A Cross-Sectional View of a Hurricane The warm ocean supplies the energy that creates and continues to supply hurricanes. Water vapour is sucked up around a tropical storm over seas with temperatures at or above 30°C. When it condenses to form water droplets (rain), the water vapour releases heat at high altitudes, warming the air around it. As the warmed air continues to rise ever more quickly, more moist winds are pulled up from below to fuel the storm.

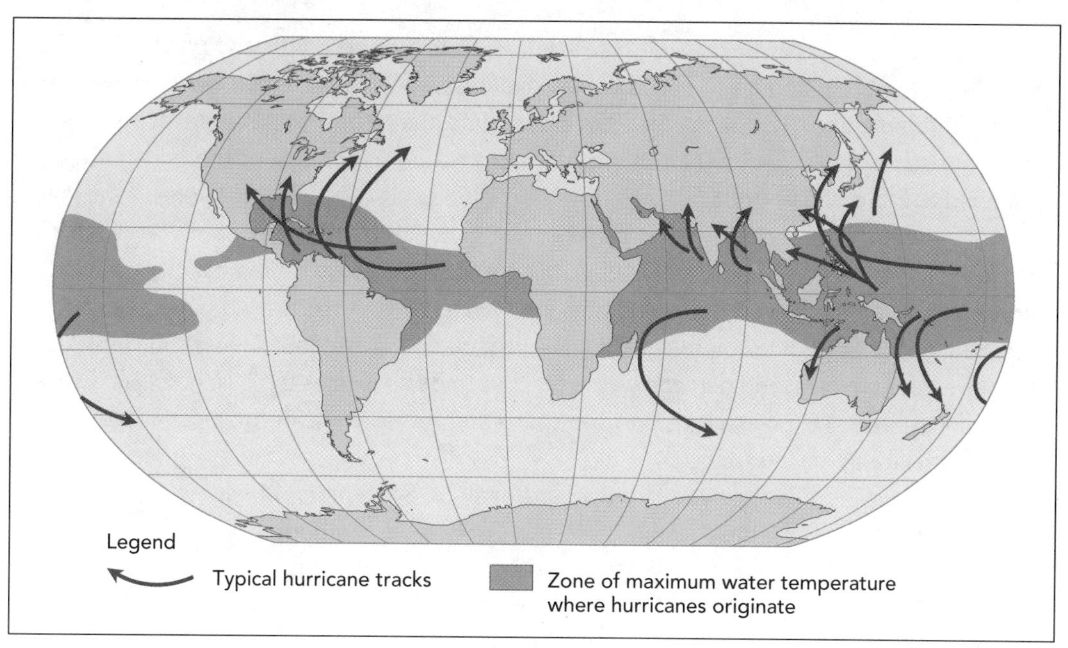

Figure 9.11 **Paths of Hurricanes and Typhoons**

Case Study: Hurricane Gilbert

In the last days of August, 1988, a tropical disturbance, one of hundreds that form each year in the equatorial waters of the Atlantic, was recorded off the African coast of Guinea by weather observers. Pushed by the northeast trade winds, this disturbance began a deadly voyage westward across the Atlantic. By the first week of September, the weather system had intensified into a tropical depression. Unstable, warm, moist air continued to flow into the depression. Winds and precipitation associated with the depression intensified as the air pressure in the centre of the system continued to drop. Feeding on the warm waters of the equatorial Atlantic, the tropical depression developed into a tropical storm by the end of the first week in September.

In its path lay the islands of the Caribbean and the coasts of Mexico and Texas. Warnings were issued as the storm grew and became more threatening. The tropical storm swept over the islands of St. Vincent, Dominica, and Guadeloupe causing widespread floods and crop damage. On September 10, wind speeds had reached 117 km/h, and this tropical cyclone had officially become a hurricane. It was named Gilbert. Figure 9.12 shows the path of Gilbert as it moved through the Caribbean.

On September 11, the "eye" of Hurricane Gilbert slipped south of Puerto Rico, the Dominican Republic, and Haiti. These islands experienced wind and flood damage and loss of life. By September 12, Gilbert had developed into the largest hurricane ever recorded. Wind speeds had reached 300 km/h and the air pressure in the eye of the hurricane had dropped to 885 millibars (= 88 500 pascals = 88.5 kilopascals = 26.13 inches), one of the lowest air pressures ever recorded. Compared to most hurricanes, Gilbert's path was relatively straight and predictable. Jamaicans had warning that Gilbert would pass directly over their island. However, the storm's force and damage stunned even experienced hurricane watchers. Four out of five homes on the island were damaged or destroyed. At least 500 000 people were left homeless, and damage was estimated to be as high as ten billion dollars. Crops were devastated and power lines and transportation routes were destroyed. Miraculously, the loss of life was limited to 25.

Gilbert travelled directly over Jamaica, and early on September 14, struck the Mexican resort area of Cancun on the Yucatan Peninsula. Thirty thousand people were left homeless, and an additional 250 deaths were attributed to Gilbert and the flooding that it brought. Over 100 000 people in Texas, Louisiana, and Mississippi left their homes and fled inland to avoid the storm. However, Gilbert did not move north to the coastal areas of the United States. The hurricane continued in its westward direction inland where it lost its source of moisture and thus its energy. Before dying out, however, the storm system spawned upwards of 50 tornadoes, which caused widespread damage in the southern United States.

Figure 9.12 Path of Hurricane Gilbert

Some meteorologists suggest that storms such as Gilbert will become more common in the future as the greenhouse effect warms our atmosphere and increases the chances for the development of hurricanes.

Tornadoes

A third type of cyclone, much smaller in size than the mid-latitude or tropical cyclones, is the **tornado**. It has a central low pressure area around which rising unstable air is found. Tornadoes are almost always formed in association with thunderstorms along the cold front of a mid-latitude or tropical cyclone. Figure 9.13 shows the areas that experience the most tornadoes in North America.

They are triggered when a thick layer of warm, humid air is forced to rise over cooler, drier air. When a strong upper atmospheric wind is present, a tornado is born and grows as more and more air is sucked into the vortex to replace the rising air. As the low pressure zone in the centre intensifies, often to the point where it is 100 mb less than the surrounding air, winds increase and the risks to humans and property become considerable. Figure 9.14a shows a cross-sectional view of a typical tornado and Figure 9.14b summarizes the characteristics of a tornado.

Tornadoes vary greatly in their intensity, size, and path. The most destructive tornadoes have winds that reach speeds as high as 500 km/h and travel for hundreds of kilometres across the earth's surface. They can be hundreds of metres in width. Damage from tornadoes occurs in many ways. First, the wind speed itself can amount to pressures of 1000 kg/m². Wind speeds of 500 km/h exert a direct

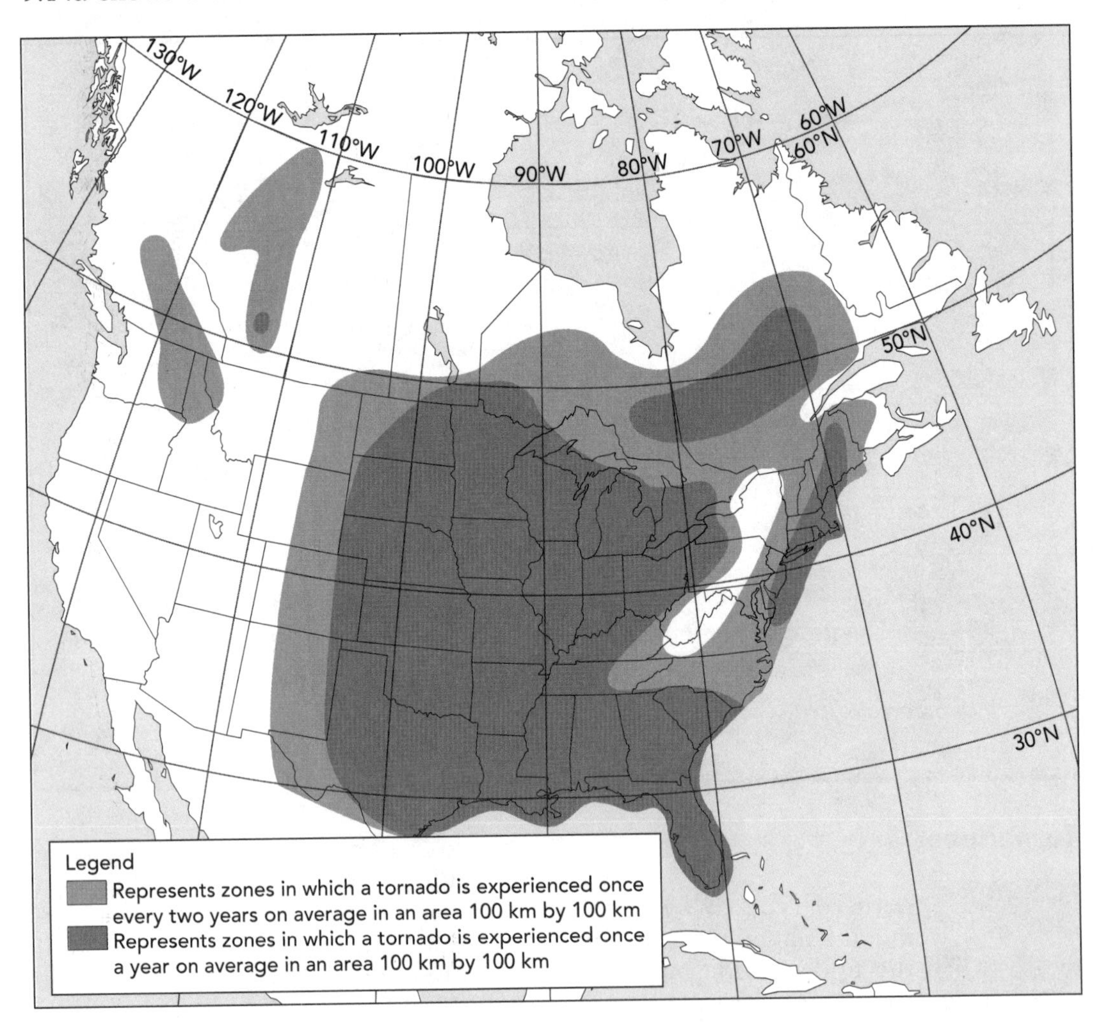

Figure 9.13 Tornado Frequencies in North America

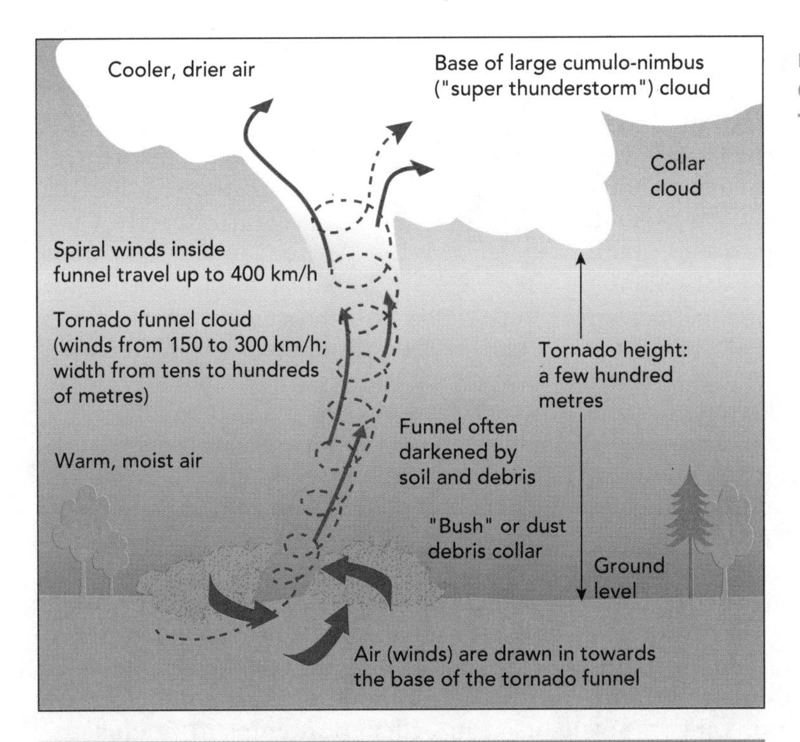

Figure 9.14a
Cross-Section of Typical Tornado

Time of Day during which tornadoes are most likely to occur is mid-afternoon, generally between 3 and 7 p.m., but they have occurred at all times of day.

Direction of Movement is usually from southwest to northeast. (*Note:* Tornadoes associated with hurricanes may move from an easterly direction.)

Length of Path averages 6 km, but may reach 500 km. A tornado travelled 498 km across Illinois and Indiana on May 26, 1917, and lasted 7 hours and 20 minutes.

Width of Path averages about 100 to 130 m, but tornadoes have cut swaths a kilometre and more in width.

Speed of Travel averages from 45 to 80 km/h, but speeds ranging from stationary to 100 km/h have been reported.

The Cloud directly associated with a tornado is a dark, heavy cumulo-nimbus (the familiar thunderstorm cloud) from which a whirling funnel-shaped pendant extends to the ground.

Precipitation associated with the tornado usually occurs first as rain just preceding the storm, frequently with hail, and as a heavy downpour immediately to the left of the tornado's path.

Sound occurring during a tornado has been described as a roaring, rushing noise, closely approximating that made by a train speeding through a tunnel or over a trestle, or the roar of many airplanes.

Figure 9.14b
Some General Characteristics of a Tornado

force 100 times that of winds blowing at 50 km/h. The force of the wind is proportional to the square of the wind speed. Secondly, wind speeds within the narrow vortex of the tornado are often so unequal that the tornado literally twists objects around. Thirdly, it is thought that pressure gradients of 6 kPa and more can exist within a tornado. If buildings are subjected to such a sudden difference in air pressure between inside and out, they can explode outwards. Finally, upward wind currents of 300 km/h can exist within the tornado. Objects such as cars and trees, and even human beings, can, therefore, be lifted into the air and dropped. A tornado in Mississippi in 1931 lifted an 83 t rail coach with 117 passengers 25 m into the air. The damage from a tornado can be severe, but it is also random as the tornado "touches down" at certain points destroying property and life, while locations just a few metres away are left undamaged.

QUESTIONS

9. Write a description of the worst storm you have ever experienced. Include in your description reference to precipitation patterns, wind conditions (strength and direction), damage done, and the behaviour of yourself and those around you.
10. Compare mid-latitude cyclones, hurricanes, and tornadoes. Develop a chart to organize your answer. Include such characteristics as size, life span, location, trigger mechanism, appearance.
11. If you wished to avoid any possibility of experiencing a hurricane, during which months should you avoid travelling to a) the Caribbean, b) Madagascar, and c) Hong Kong? Explain your answers.
12. What precautions can be taken to minimize the damage caused by a) hurricanes and b) tornadoes?

9·4 Predicting the Weather

In theory, the old cliché, "Nothing is as unpredictable as the weather", is incorrect. To predict the weather, you simply need to know what type of air masses are influencing or approaching your location. In the real world, however, despite the use of detailed measurements, weather radar, and weather satellites, the old cliché contains a great deal of truth. The atmosphere is very complex, and the interplay of high and low pressure systems, the effect of local lakes and landforms, and the influence of factors that we still do not fully understand combine to make weather forecasting a much maligned and seldom praised profession.

Four methods of data collection are used in order to attempt to predict the weather. The first and oldest method is the ground observations and measurements made at locations across the globe. In Canada alone, there are more than 2700 weather observation sites, over 2000 of which are managed by volunteers. The simplest of these stations record maximum and minimum temperatures and rainfall and snowfall amounts, while professional weather stations continuously record temperature, wind, air pressure, cloud cover, humidity, and precipitation values. These ground stations produce data that allow weather forecasters to evaluate the nature and direction of movement of the air masses that are likely to influence a location.

Weather or radiosonde balloons are still in widespread use today. These balloons carry intruments up to 30 000 m above the earth's surface. The balloons have sensing elements that measure the pressure, temperature, and humidity of the air as they ascend. The measurements are sent back to ground stations through the use of radio transmitters. The flight and path of the balloons also give information as to wind speed and direction at various heights in the atmosphere.

Television weather reports frequently display the third method of collecting data, weather radar. Weather radar is used to detect, locate, and measure the amount of precipitation in the atmosphere. Microwave energy is emitted into the atmosphere to scan for precipitation: rain, snow, sleet, or hail. On encountering precipitation, some of the microwave energy is reflected back to the radar antenna. This permits an accurate mapping of the location, extent, intensity, and path of atmospheric precipitation. (See Figure 9.15.) Each radar station is able to map an area of the earth's surface approximately 650 km in diameter. New radar techniques, such as the Doppler weather radar, have the potential to measure the speed of water droplets within a weather system and thus hold the promise of being able to predict more accurately such weather phenomena as tornadoes and thunderstorms.

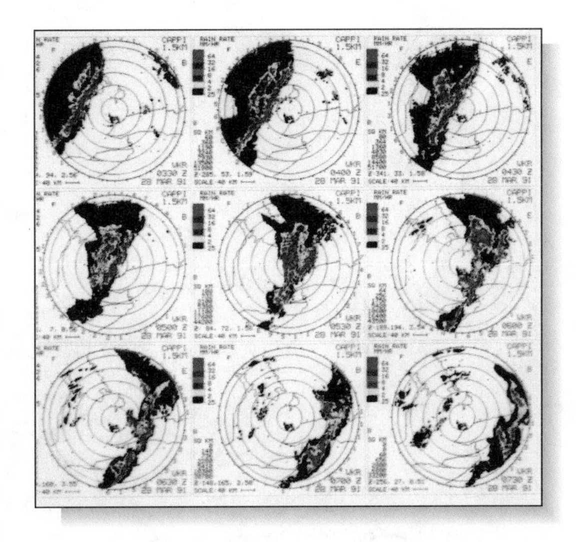

Figure 9.15 Radar Weather Map Showing Precipitation

In the early 1960s, weather forecasting took a giant step forward as the fourth type of data collection became available. For the first time, we began to receive images of our atmosphere taken by satellites orbiting far above the earth's surface. The existence of weather fronts, jet streams, and storm systems could now be observed and tracked by studying successive satellite photographs. A group of such successive images is shown in Figure 9.16. Satellites allow meteorologists to make long-term forecasts concerning the movement of weather systems. These long-term

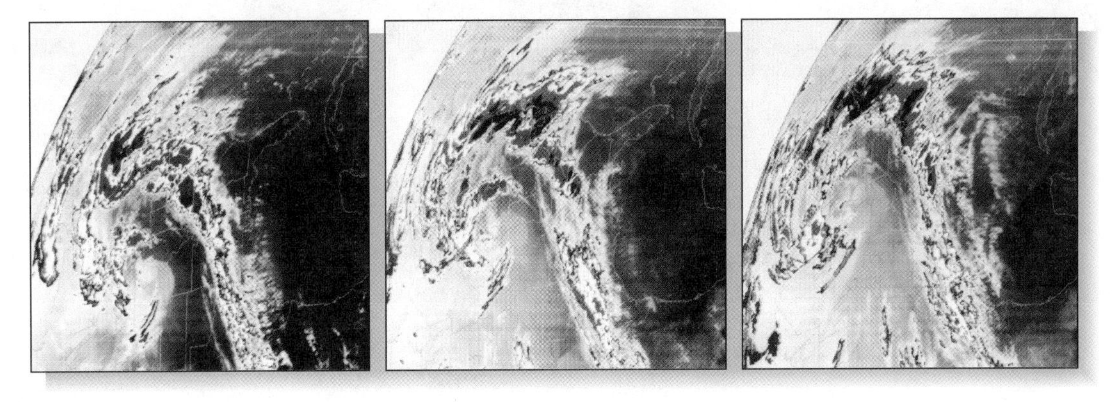

Figure 9.16 Successive Satellite Photos Showing Movements of Weather Systems

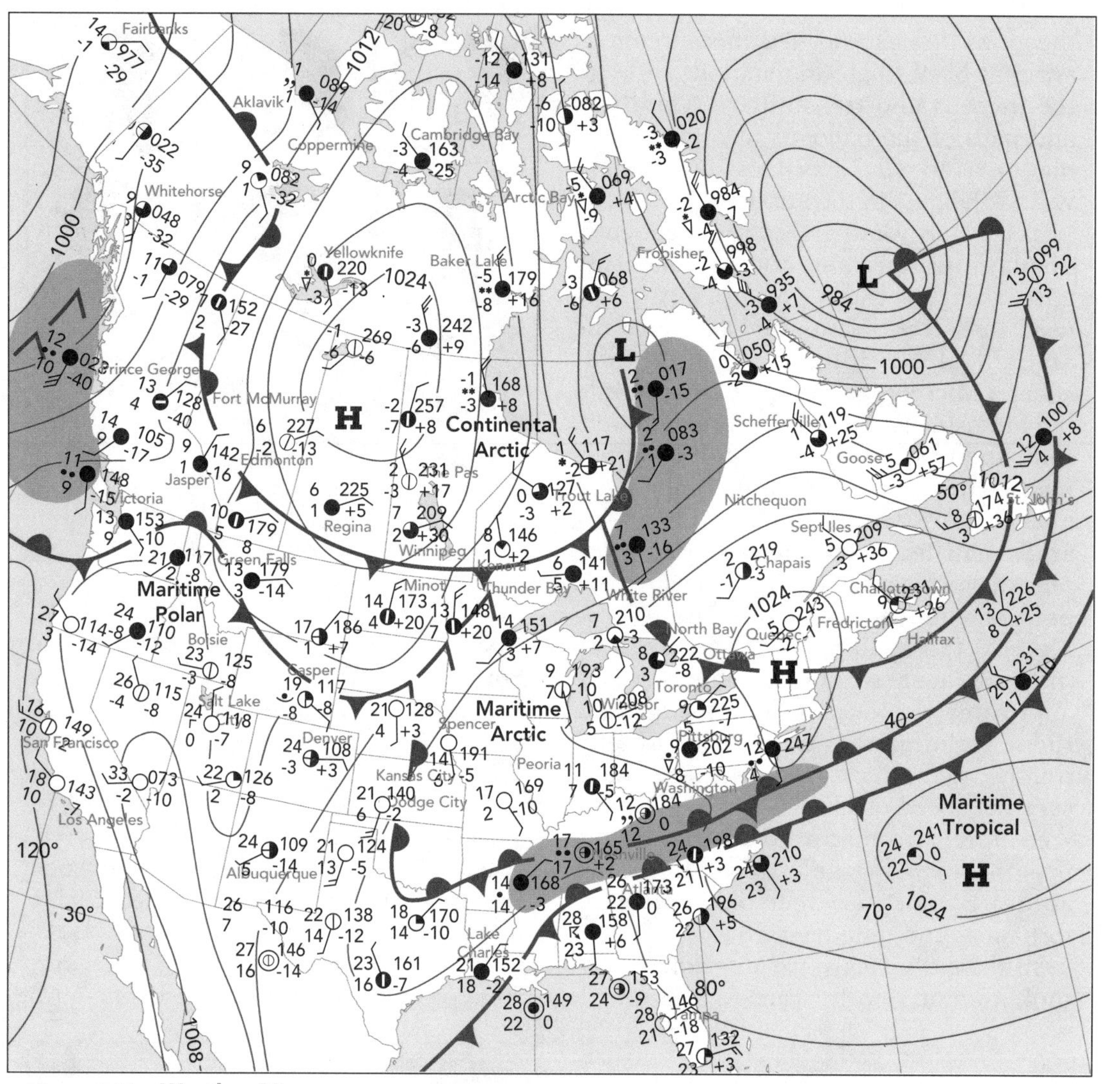

Figure 9.17 Weather Map

forecasts become particularly important when we are dealing with hurricanes and other storm systems. In addition, weather satellites have allowed forecasters to view atmospheric conditions in areas which were previously difficult to obtain data for, such as over the oceans and in uninhabited areas of the earth's surface.

At the moment, Canada receives information from four weather satellites. Two satellites are located over the equator at altitudes of 36 000 km. They are called geostationary because they rotate with the earth and always maintain the same position relative to the earth's surface. The Goes-East satellite is positioned at longitude 75°W, while Goes-West is positioned at 136°W. The images from these satellites are often now part of the daily weather forecasts seen on television.

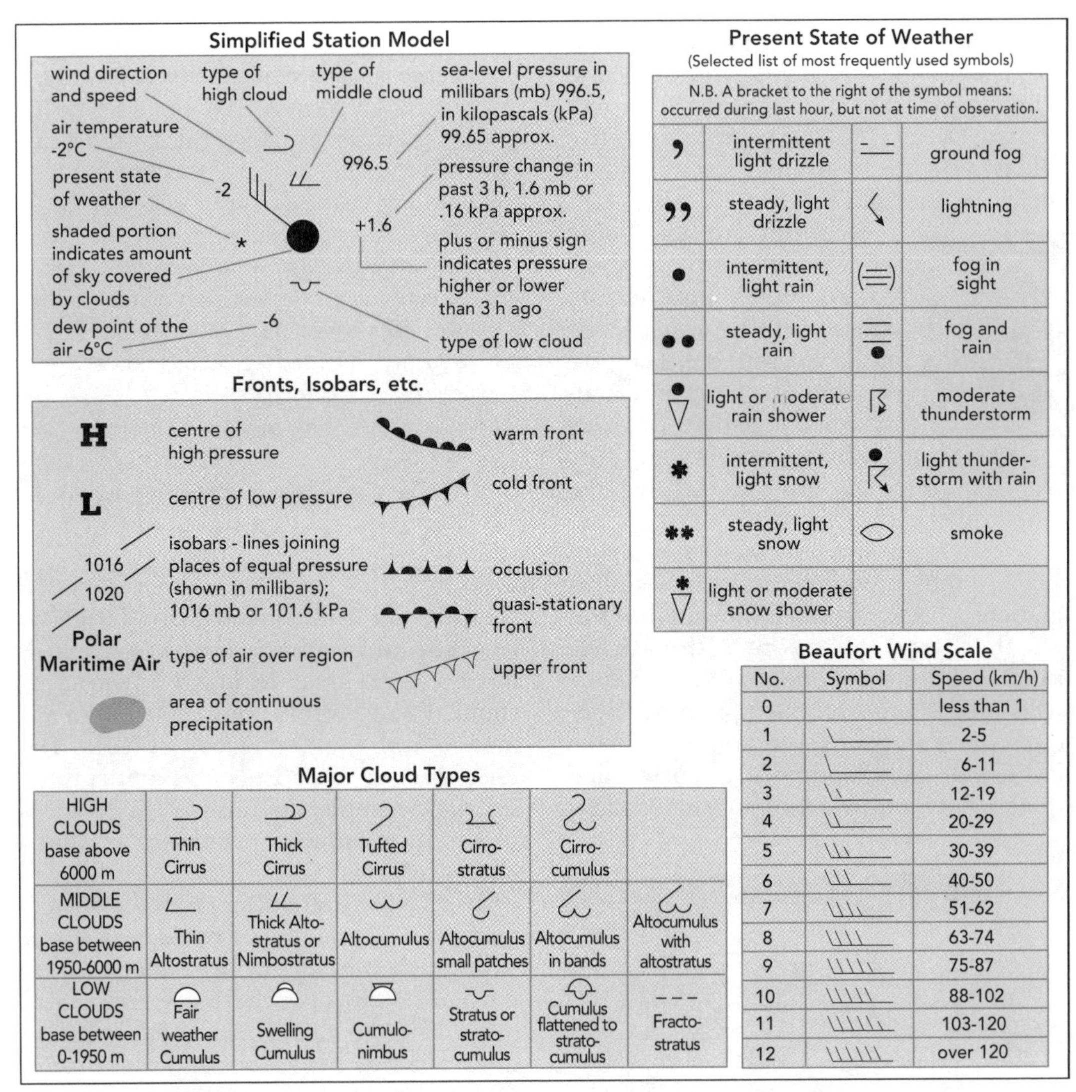

Figure 9.17 **Weather Map: Legend**

When time-lapse sequences are made from a series of satellite images, the result is a dramatic view of the dynamics of the earth's atmosphere, which allows for a much better understanding of our weather systems. The other two satellites are over the polar region at 860 km above the earth. They orbit the earth approximately 14 times a day and are thus not "geostationary". These satellites give accurate information concerning the atmospheric conditions in the high northern latitudes as well as information about ice conditions in the Arctic areas.

The World Meteorological Organization shares satellite images from many different satellites owned by a number of countries. In Canada, the Atmospheric Environment Service (AES) operates satellite receiving stations in Halifax, Toronto, Edmonton, and Vancouver. Canada also jointly operates a station in Greenland with Denmark, which provides coverage of the North Atlantic Ocean. The images from these five receiving stations are distributed to 16 major weather offices across Canada.

16•2/16•3

Satellites are now being used to measure greenhouse gases and ozone amounts in the upper levels of the atmosphere. They are capable of giving us accurate temperature profiles through the atmosphere and in the future will be used to measure wave heights and amplitudes in the oceans.

The modern weather map (see Figure 9.17) is thus a compilation of the information collected from all four methods of weather data collection. Although we tend to blame the weather forecaster whenever it rains on our picnic, weather forecasting has improved greatly over the last few decades, both in its accuracy and in its ability to predict long-term trends.

QUESTIONS

13. Watch the weather report on television tonight. List all the information that the weather report gives (for example, the report might tell you the wind speed, or that a low pressure system is approaching). For each piece of information, indicate what would have been the most likely source — ground observations, weather balloons, weather radar, or satellites.
14. Devise and clearly outline a procedure you could use to test the accuracy of the weather forecasts given in your area.
15. How far into the future do most weather reports attempt to forecast? Why are "long-range" forecasts less reliable than forecasts concerning tomorrow's weather?
16. Examine Figure 9.17. Choose three locations on the map, one location in front of a warm front, one location between that warm front and the cold front following it, and one location behind the cold front. Use the information given on the weather map to describe the weather at each location.
17. Compare Figure 9.17 with Figure 9.6. Locate the cyclone near Hudson's Bay on the weather map. Name the stage of development of this cyclone. In your own words, describe what is likely to happen to this cyclone.

9•5 Our Complex Atmosphere as Illustrated by an El Niño Event

Despite our increased knowledge of weather and climatic systems, we still have a great deal to learn and understand. To stress this point, consider a climatological event known as **El Niño**. The phenomenon of an El Niño event emphasizes the complexity and interrelatedness of the atmospheric processes we have been studying.

An El Niño event refers to the displacement of an upwelling of cold ocean water with much warmer surface waters. Under normal conditions, there are five major upwellings of deep ocean water found near coastal areas around the globe. These are areas where deep, cold ocean water replaces warmer surface waters that have been propelled away from the coast due to the combined effects of the prevailing wind system and the rotation of the earth. These coastal upwelling areas are rich in nutrients brought to the surface from the ocean depths, and thus they support rich and diverse populations of marine life.

The five major coastal areas where this upwelling phenomenon is found are off the coasts of California, Peru, Nambia, Mauritania, and Somalia. In certain years when the prevailing winds are weak or reverse themselves, warm ocean water flows into these coastal areas and disrupts the supply of the cold upwelling water from the ocean depths. If this disruption continues for a year and exists over a large area of ocean, surface water temperatures can rise by up to 7°C above normal. It appears that major El Niño events are associated with extreme and disruptive weather. As early as 1921, observers noted that droughts in the central equatorial Pacific areas were associated with weak trade winds and abnormal ocean currents. During the El Niño events of 1940-41 and 1957-58, climatologists took note of disruptions in world wind and pressure patterns, and of unusual and extreme weather in many parts of the world. Bjerknes hypothesized the following explanation to explain the link between El Niño and world weather.

Bjerknes hypothesized that a cell of air movement, which he called the Walker Cell, exists over the equatorial regions of the Pacific Ocean. The cell involves the movement of air in a westerly direction from over the colder waters of the eastern Pacific to the warmer water region of the western Pacific. When this movement of air, which is being pushed by the southeast trade winds, reaches the western Pacific, it rises and moves at high altitudes back to the eastern Pacific where it sinks, cools, and begins the flow all over again. The measured temperature and air pressure readings from the eastern and western Pacific areas tend to support the existence of this cell. When an El Niño event occurs, the eastern Pacific becomes warmer and the Walker Cell is disrupted. Wind, ocean currents, precipitation, and air pressure patterns change and a corresponding impact on agricultural and biological systems occurs. The suggestion is that a major El Niño event can disrupt global weather patterns, influencing even the jet streams at high latitudes and the thermocline layer in the ocean depths.

A major El Niño event occurred off the Peruvian coast in 1972-73. During those years, the USSR suffered major setbacks in agricultural production due to a severe drought. Severe droughts were also experienced in Central America, parts of Australia, and northern and western Africa. Floods wrecked havoc in the Philippines, Kenya, and parts of Australia. World food production declined in 1972 for the first time since 1948, and the Peruvian economy came close to collapse as its rich fishing industry suffered major setbacks. In the aftermath of the El Niño event of 1972-73, many scientists proposed a direct cause and effect link between El Niño and the disruption of world weather patterns.

In 1982-83, a very large El Niño event occurred. Unlike "typical" El Niño events, this particular one seemed to have its origin in the central and western Pacific. The events that followed this El Niño event shocked even those scientists who had issued warnings of its possible impact. Figure 9.18 summarizes some of the occurrences that have been attributed to the 1982-83 El Niño event.

Many scientists caution us that weather is always changing and is highly variable from one year to another. They point out that we still do not know enough about our atmosphere to understand the causes and effects of occurrences such as an El Niño event. No one, however, disputes the influence that our climate and weather systems exert on agricultural, biological, economic, and, ultimately, political systems. Whether we are discussing rainfall amounts, hurricanes, or El Niño events, continued research into climate and weather systems will allow us to better understand and plan for the variable nature of our atmosphere.

QUESTIONS

18. Attempt to explain an El Niño event in language that a ten-year-old child would understand.
19. Locate the five major locations for cold water upwelling on an atlas map that shows ocean features. What do all these locations have in common? In what ways are they different?
20. At some point, we will understand enough about El Niño events to be able to predict them with accuracy. What do you suppose could be done to try to limit the negative impacts of the events if we could predict them?

CONTINENT	UNUSUAL OCCURRENCES
Australia and Oceania	Australia suffered the worst drought in two hundred years. Six tropical cyclones devastated Tahiti and Polynesia in less than six months.
Asia	The monsoon wind system failed to develop in Indonesia and the Philippines, resulting in crop failure and food shortages.
Africa	Widespread droughts in Africa were labelled the worst of this century and caused crop failures from South Africa to Zimbabwe to the West African Sahel.
South America	Torrential rains fell in normally dry areas of Peru, Ecuador, Brazil, Bolivia, and Paraguay, causing widespread flood damage. Commerical fishing off the west coast of South America was disrupted, causing severe economic hardship.
North America	The worst flooding in decades was experienced in the Midwest and South due to abnormally heavy winter and spring rains. Drought in the north central region of the continent reduced corn and soybean harvests. Winter temperatures were the warmest in 25 years. Coastal storms in California caused mudslides and flooding, resulting in over one billion dollars worth of damage.

Figure 9.18
Occurrences Attributed to the 1982-83 El Niño Event

Review

- Wind systems move air masses across the earth's surface and influence our weather patterns.
- Mid-latitude areas are places where differing air masses meet. The interaction between the different air masses leads to the development of mid-latitude cyclones.
- The jet streams are high altitude air movements that influence the development of the mid-latitude cyclones.
- Hurricanes and tornadoes are destructive types of cyclones that can cause serious property damage and loss of life.
- Weather data collection techniques are improving all the time. This improves the accuracy of weather forecasting.
- The complexity of the atmosphere means that we still have a great deal to learn about weather and climate.

Geographical Terms

polar front
frontal wave theory of cyclonic development
depression
cyclone
cyclonic disturbance
baroclinic wave theory of cyclones
isobar
jet stream
Rossby wave
trough
intertropical convergence zone
hurricane
typhoon
tornado
El Niño

Explorations

1. Keep your own weather diary for the next two weeks. On each day at 8:00 a.m. and at 8:00 p.m., record the wind direction, the temperature, the cloud cover, and any precipitation which is falling. Graph your results in a manner that allows you to compare wind direction with temperature, cloud cover, and precipitation. State conclusions from your graphs. Explain your conclusions by referring back to the material in this chapter.
2. Take a poll of at least ten local residents who have lived in your area over the last few decades. Ask them if the weather in your area has changed very much over that period of time. Summarize your results in a visual way. Do you think your findings are valid? Explain.
3. In this chapter, we studied modern methods of weather forecasting. Many people still rely on sayings or clichés to forecast the weather. Based on what you have learned in this chapter, comment on the degree of validity or truth which the following sayings contain.

   "Red sky at night, sailors' delight.
   Red sky in the morning, sailors take warning."

   "When the wind is in the east, it's good for neither man nor beast."

   "A rainbow in the morning
   Is the shepherd's warning.
   A rainbow at night
   Is the shepherd's delight."

   "Rainbow to windward
   Foul fall the day.
   Rainbow to leeward
   Damp runs away."

   "If Candlemas Day [February 2] be fair and bright,
   Winter will have another flight;
   But if Candelmas Day brings clouds and rain,
   Winter is gone and won't come again."

CHAPTER

10

THE HYDROSPHERE AND THE HYDROLOGIC CYCLE

OBJECTIVES:

By the end of this chapter, you will be able to:

- recognize that water moves through a cycle, with places where it is stored and paths which it moves along between storage places;
- understand that the oceans are very important influences on global climate patterns;
- appreciate that the world's water resources are limited and must be protected from misuse.

CHAPTER 1: The Nature of Physical Geography
CHAPTER 2: Earth: Its Place in the Universe
CHAPTER 3: The Earth in Motion
CHAPTER 4: The Earth's Interior
CHAPTER 5: The Earth's Crust
CHAPTER 6: The Lithosphere in Motion: Plate Tectonics
CHAPTER 7: Solar Radiation
CHAPTER 8: Climate
CHAPTER 9: Weather
CHAPTER 10: The Hydrosphere and the Hydrologic Cycle
CHAPTER 11: Natural Vegetation and Soil Systems
CHAPTER 12: Denudation: Weathering and Mass Wasting
CHAPTER 13: Distinctive Landscapes: Humid and Arid Environments
CHAPTER 14: Distinctive Landscapes: Glacial, Periglacial, and Coastal Environments
CHAPTER 15: Natural Hazards: Disrupting Human Systems
CHAPTER 16: The Disruption of Natural Systems
CHAPTER 17: Fragile Environments

Introduction

The **hydrosphere** is that part of our planet where water, in its varied forms, is found. Water makes our planet distinct from all the others in the solar system. It shapes and reshapes the earth's surface, wearing down mountains, transporting sediments, and building new rock layers under the oceans. We use water as an energy source, for transportation, for recreation, and to irrigate our crops. In fact, water is the basis of all life as we know it. Increasingly, we are beginning to appreciate that water is among the most valuable of our natural resources, a resource that needs to be understood, protected, and conserved.

10•1 The Hydrologic Cycle

Figure 10.1 illustrates the location of water on the earth and gives average values for the amount of water found in each location at any time. While these values are relatively constant, the water itself frequently changes location, state, and chemical makeup.

The **hydrologic cycle** refers to the movement of water through the hydrosphere. The hydrologic cycle is a closed cascading system because water is neither added nor taken away from the system. (See Figure 10.2.)

It's a Fact...

The water in the Marianas Trench in the Pacific Ocean reaches a depth of 11 000 m, which is over 2000 m greater than the height of Mount Everest.

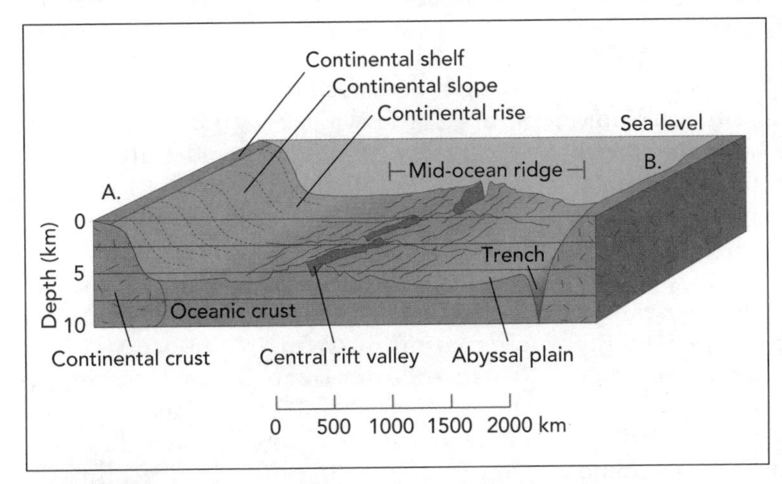

Figure 10.5
Simplified Diagram Showing the Major Features of the Ocean Floor The ocean bottom is not the flat, uninterrupted plain we once believed it to be. Flat areas of the abyssal plain are interrupted by mountain ranges, trenches, and by mid-oceanic ridge systems that formed when two of the earth's plates moved away from one another.

The world's oceans contain salt water — a complex mixture of water and at least 80 other chemical elements. In addition to the elements of hydrogen and oxygen which make up the water itself, eight other elements (chlorine, sodium, magnesium, sulphur, calcium, potassium, bromine, and carbon) are of prime importance. Many of these elements combine to form chemical compounds called **salts**, the most common of which are sodium chloride, magnesium chloride, and sodium sulphate. (See Figure 10.6.) These elements and compounds are carried into the oceans by rivers. Over time, they accumulate and increase the salinity of the water. **Salinity** refers to the total weight of dissolved salts. The average salinity of sea water is 3.5 percent, or 35 000 parts per million; however, the salinity can vary due to temperature and location. In areas of high evaporation and low rainfall, such as the Red Sea, salinity values exceed 4 percent. In seas which have colder water and a greater input of fresh water, such as the Baltic Sea, the salinity can be less than 1 percent.

Name of Salt	Chemical Formula	Grams of Salt per 1000 g of Water
Sodium chloride	$NaCl$	23.0
Magnesium chloride	$MgCl$	5.0
Sodium sulphate	Na_2SO_4	4.0
Calcium chloride	$CaCl_2$	1.0
Potassium chloride	KCl	0.7
With other minor ingredients to total		34.5

Figure 10.6 **Composition of Seawater**

Variations in salinity combined with the variations in water temperatures give rise to two distinctive zones in the oceans. The upper zone, heated by sunlight, is distinct from the lower zone, in which the water is much colder and more dense. The boundary between the two zones — the **thermocline** — usually begins at a depth of about one kilometre. The characteristics of these two zones are so different that there is only a very limited exchange of water between them. Some recent estimates by oceanographers suggest that a molecule of water, on average, remains for a thousand years in the deeper ocean zone before it moves into the warmer, more active surface zone.

The study of the oceans is made more complex by the constant movement of water through them. Ocean currents are relatively shallow flows of water usually found in the upper 1000 m of ocean water. They are of prime importance in distributing heat from equatorial zones to mid-latitude and polar areas. But this surface flow cannot take place without a counterbalancing replacement flow of water; this deep water circulation occurs in oceans at depths below 1000 m, carrying cold, dense water towards the equator. Figure 10.7 illustrates the deep water circulation that occurs in the world's oceans. Figure 8.18 (page 149) showed the warmer surface flows. In Antarctica, the dense, cold water flows down the continental shelf to form a great deep-sea current at a depth of four kilometres. This current is called the **Antarctic Bottom Water** and is believed to be one of the triggers to the worldwide circulation of water in the oceans. The **North Atlantic Deep Water Current**, in the northern hemisphere, has a similar effect. It carries a flow of water 20 times that of all the freshwater rivers in the world combined. Deep water flows in the

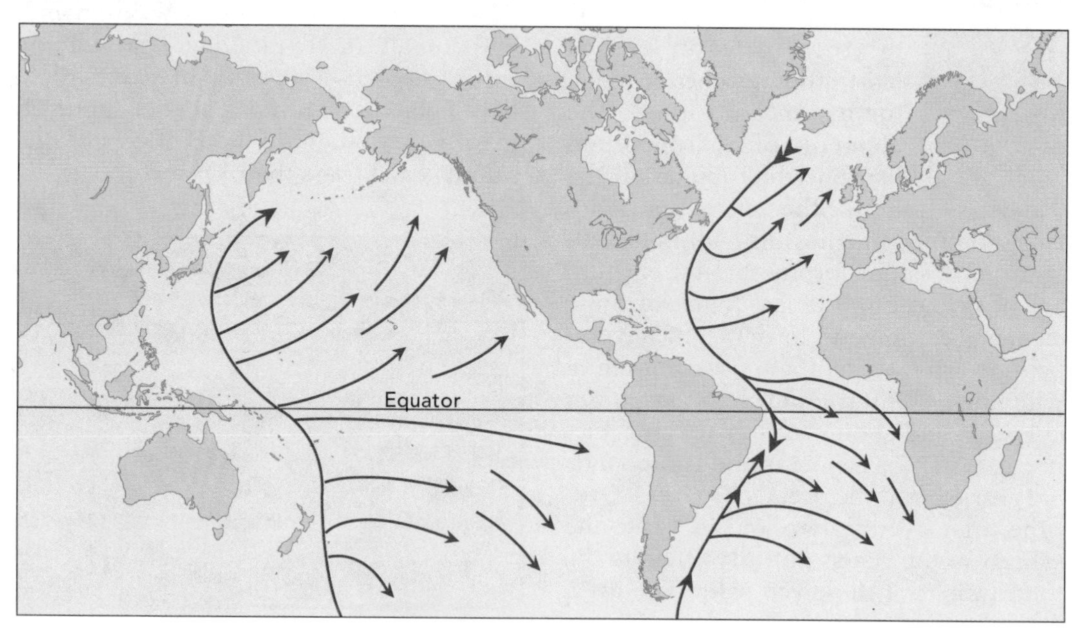

Figure 10.7 **Deep Water Circulation**

ocean combine with the surface flows and with the global wind systems to distribute heat over the globe.

The abundant life of the ocean's ecosystems is concentrated in the continental shelf and slope areas along the edges of land masses. In these areas, minerals and nutrients are fed into the water by run-off from the adjacent land masses. Sunlight penetrates most of the way to the bottom and provides energy for photosynthesis. Microscopic plants and animals thrive here forming the basis for rich food chains. However, we are just beginning to explore the ocean's depths and a great deal is yet to be learned concerning the deeper areas of the oceans. For example, in 1960 when scientists explored the dark, cold waters of the Marianas Trench in the bathyscaphe Trieste, their spotlights revealed a fish at a depth of almost 11 000 m. Until this voyage it was thought that no life could exist at such depths. Recently, more evidence has emerged of rich ecosystems existing around, and depending upon, ocean vents, which emit hot gases and heat, providing the energy and nutrients for life forms. These, and other explorations, are leading to new understandings of the ocean storehouses of water.

QUESTIONS

5. Viewed from a satellite the earth is really not earth at all, as its surface is primarily water. Suggest two other names for our planet other than "earth".
6. Use an atlas to list the eight largest saltwater bodies found on the earth's surface. Record your list in order from largest to smallest.
7. Ocean levels have fluctuated a great deal throughout geologic history. Use a physical map of the world from an atlas to identify those areas of the coasts that would be flooded if sea level were to rise by 100 m. Shade this area on an outline map of the world.
8. Examine Figures 10.7 and 8.18. What patterns, similarities, or differences can you observe between these two maps of ocean currents?

10•3 Freshwater Storehouses

Glaciers and Ice Sheets

Seventy-five to seventy-seven percent of all the fresh water on earth is frozen in glaciers and ice sheets. Most of this water, approximately 96 percent, is found in the earth's two major ice sheets, one located on the island of Greenland and the other located in Antarctica. The Antarctic ice sheet reaches depths of over three kilometres and covers 13 million km^2. Attached to the Antarctic ice sheet are a number of ice shelves. An ice shelf is a large mass of ice, part of which is anchored to land and part of which is floating. The largest is the Ross ice shelf, with a surface area of over 500 000 km^2. It is feared that a warming of the earth's climate might cause the

Ross ice shelf to slip off the land mass of Antarctica into the ocean, causing a global rise in sea level.

In comparison with Antarctica, Greenland's ice sheet appears small. However, the Greenland ice sheet contains over 8 percent of the earth's fresh water and covers an area twice the size of Ontario. As with the Antarctic sheet, the depth of the ice on Greenland exceeds three kilometres in places.

The remainder of earth's frozen water is locked up in alpine glaciers found in many of the world's mountainous areas. Glaciers and glaciation are discussed in Chapter 13.

Groundwater

The second most important freshwater storehouse in terms of volume is **groundwater** — water found beneath the earth's surface. This means of storage contains 22 to 24 percent of the earth's fresh water. Some of this groundwater is found at very shallow depths and remains in the soil or rock for only very short periods of time. This water often seeps into rivers and lakes to become part of the surface flow or is used by vegetation. Most of the groundwater, however, is located at depths below 600 m and can remain in the earth's crust for thousands of years.

The **water table** is the level beneath the earth's surface below which the soil and rock are saturated with groundwater. At this level, all the pores and air spaces are full of water. The depth of the water table varies with the climate of an area and the properties of the bedrock. In areas that have humid climates and where impermeable rock layers are found near the earth's surface, the water table can be as shallow as one metre, while in areas of permeable rock and dry climates, the distance below the surface to the water table can exceed several hundred metres. This combination makes it difficult and expensive to access groundwater for surface use. Sometimes the geological structure assists rather than hinders the extraction of groundwater. Alternate layers of permeable and impermeable rock can give rise to artesian springs. (See Figure 10.8.)

Aquifers are important storehouses for fresh water deep in the earth's crust. These are underground layers of porous and permeable rock through which water moves easily. The United States obtains one quarter of its freshwater supply from a network of underground aquifers, taking out of the groundwater which has been held in the earth for thousands or even millions of years. One such aquifer is

It's a Fact...

On a per capita basis, Canada has twice as much fresh water available than any other country in the world. In absolute amounts, Brazil has the most available fresh water in the world.

It's a Fact...

The words "permeability" and "porosity" are often misunderstood. Porosity refers to the percentage of the total volume of rock that is occupied by openings, whereas permeability refers to the ability of the rock to transmit water. The two terms have no direct relationship to one another. A rock might be porous, but the pores might be small and not connected; therefore, the permeability might be low. If the pores are large and are interconnected, then the permeability will be high.

the Ogallala, which supplies New Mexico, Texas, Wyoming, Colorado, Oklahoma, and Kansas with much of their freshwater needs. Underlying an area of over 550 000 km^2 and at depths ranging from 400 to 1800 m, the water contained in the Ogallala is being consumed 50 times faster than nature can replace it. In some areas, the ground has sunk and wide trenches have formed as the water was removed from underground. Some estimates suggest that 40 percent of the water stored in the Ogallala has already been consumed and that the remaining amount will be reduced by a further 40 percent over the next 30 years. Tens of millions of

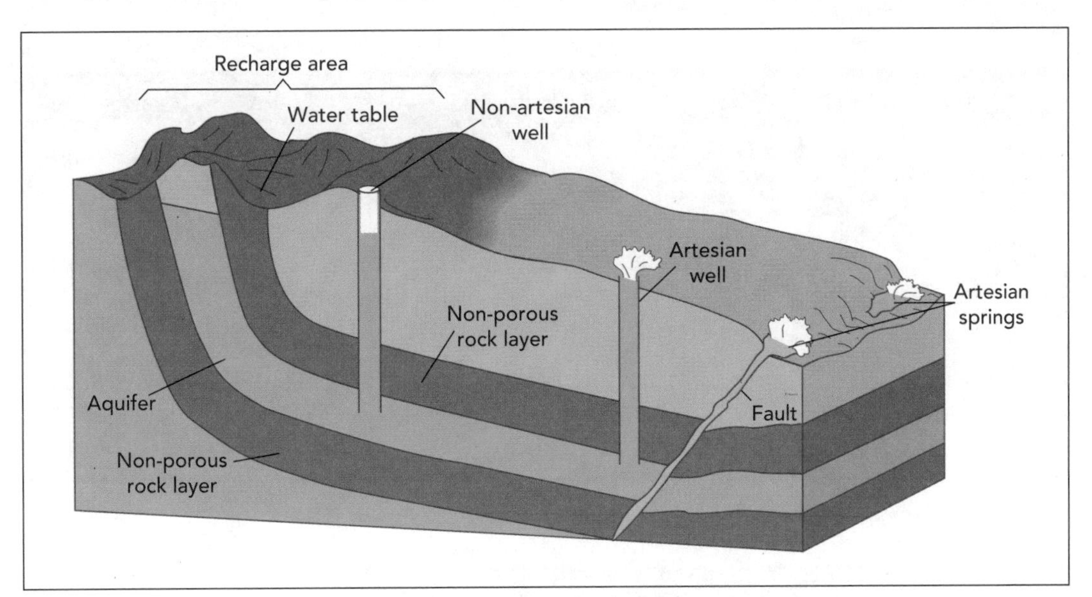

Figure 10.8 **Artesian Wells, Artesian Springs, and Aquifers** Two conditions are necessary for an artesian system: a confined aquifer and water pressure sufficient to make the water in a well or spring rise above the aquifer. The water in a non-artesian well rises to the same height as the water table in the recharge area. In the artesian well, water flows out at the surface without pumping due to the water pressure built up in the aquifer.

people and a very significant percentage of America's agriculture is dependent on an aquifer which is rapidly disappearing.

Rivers and Lakes

Although the earth's rivers and lakes are the most visible of the freshwater storehouses, and the ones which most often directly affect our lives, they represent less than one half of one percent of the planet's fresh water. In addition to being perhaps the most important of the earth's sculptors, rivers provide us with hydro-electricity, transportation routes, recreational waterways and, in some cultures, with religious symbols. All the rivers in Canada discharge approximately 68 000 m^3 of water a second into the world's oceans, with two river systems, the St. Lawrence and the Mackenzie, accounting for a third of this total.

Lakes occupy natural and human-made depressions in the earth's crust. They can range from small ponds to huge water bodies which rival the earth's seas in size and importance.

The Great Lakes of North America hold close to 20 percent of the world's freshwater supply found in lakes and rivers. Most of the lakes in Canada were filled with glacial meltwater some 10 000 years ago. The input of ground, atmospheric, and river water into these lakes today has established an equilibrium with the amount flowing and evaporating out of the lakes. Careful management of our fresh waters is vital.

Ranked by Length		Ranked by Drainage Basin		Ranked by Discharge	
River	Length (km)	River	Area of Drainage Basin (1000 km^2)	River	Discharge (m^3/s)
Nile	6 690	Amazon	6 150	Amazon	175 000
Amazon	6 570	Congo	3 822	Congo	39 000
Mississippi	6 020	Mississippi	3 230	Negro	35 000
Yangtze	5 980	Plata	3 100	Yangtze	32 190
Yenisey	5 870	Ob-Irtysh	2 990	Orinoco	25 200

Figure 10.9a Some of the Earth's Major River Systems

River	Length (km)	Area of Drainage Basin (1000 km^2)	Discharge (m^3/s)
St. Lawrence	3 460	1 463	14 160
Mackenzie	4 063	1 766	7 930
Yukon	3 380	855	5 098

Figure 10.9b Major Canadian River Systems

Ranked by Surface Area		Ranked by Volume of Water		Ranked by Depth	
Lake	Area (km^2)	Lake	Volume (km^3)	Lake	Depth (m)
Superior	82 261	Baikal	22 995	Baikal	1 620
Victoria	62 940	Tanganyika	17 827	Tanganyika	1 471
Huron	59 580	Superior	12 258	Malawi	706
Michigan	58 020	Malawi	6 140	Great Slave	614
Tanganyika	32 000	Michigan	4 940	Crater	589

Figure 10.10 The Earth's Major Freshwater Lakes

It's a Fact...

In terms of water use per capita, the United States leads the world with a usage rate of 1986 m^3 annually. Canada is in second place with an annual usage rate of 1172 m^3.

Atmosphere

Although miniscule in terms of the amount of water which it holds, the earth's atmosphere plays a fundamental role in distributing and circulating water through the hydrologic cycle. Without the process of evaporation, virtually all of the earth's water would eventually end up in the oceans. The earth's rivers and lakes would dry up and soil moisture in the important upper levels of the crust would disappear. It is the atmosphere that picks up water from the oceans, lakes, and rivers and transports it to the land. The energy for this process is provided by solar radiation.

Evaporation is the change of state from liquid to vapour, while **condensation** is the change of state from vapour to liquid. Evaporation can be thought of as the starting point for the hydrologic cycle. The amount of evaporation is primarily a function of air temperature; hot air can hold more water vapour than cold air. Figure 10.11 illustrates the effect that the temperature of air has on its ability to hold water vapour. Wind strength, the nature of the surface, and air turbulence are other factors that can influence the evaporation rate.

It's a Fact...

Every year, approximately 453 000 km^3 of water are evaporated from the surfaces of the world's oceans. (1 km^3 = 1 billion m^3 = 1 trillion litres)

The term "humidity" refers to water vapour in the atmosphere. If you collected and weighed all the water vapour contained in a given volume of air, then you would have the **absolute humidity**, which is usually expressed in g/m^3. The amount of water vapour in a given volume of air compared to the maximum amount of water vapour which that air could hold is the **relative humidity** of the air. The relative humidity is a ratio usually expressed as a percentage. A relative humidity reading of 100 percent means that the air is saturated, that is, it is holding all the water vapour that it possibly can at that temperature. Cooler air has less ability to hold water than warm air; thus, if the temperature of an air mass decreases but the amount of water vapour in the air remains the same, the relative humidity will increase. The temperature at which air is saturated is called the condensation or dew point; at this point the air cannot hold any more water vapour. If the temperature of saturated air is lowered, condensation occurs.

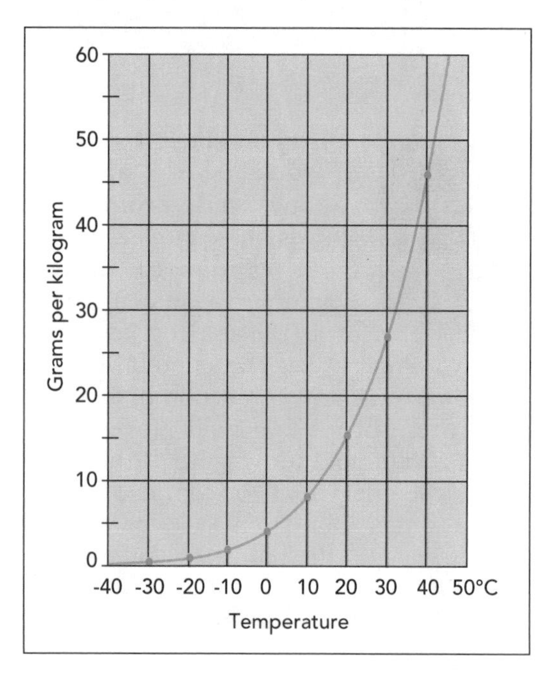

Figure 10.11 Maximum Amount of Water Which a Mass of Air at Various Temperatures Can Hold

Condensation can take many forms. Dew commonly occurs at night as the air temperature falls and water vapour slowly condenses on exposed surfaces. People wearing glasses who move from the cold outdoors to a warm house also experience dew as the cold glass causes the air in contact with it to cool, leading to condensation on the glass. Fog is also a form of condensation and occurs most often when warm, moist air is cooled by coming into contact with a cold surface. Fogs occur regularly in coastal regions where air which originates over a warm current of water contacts colder air found over a cold ocean current and condensation takes place. For example, the dense fogs which form over the Grand Banks off the coast of Newfoundland are the result of the mixing of the air masses associated with the cold Labrador Current and the warmer Gulf Stream. Fogs can also occur due to the cooling that results from the radiation of longwave energy from the earth's surface at night. As the earth cools, the temperature of the air above it drops and condensation occurs.

In order to condense in the atmosphere, water vapour needs **hygroscopic particles** in the air. Hygroscopic particles are microscopic in size and can be dust, pollen, pollution, salt crystals, or a number of other solid particles that can attract water vapour. Often ice particles form the nuclei around which water vapour condenses. In the case of fogs and clouds, the water vapour droplets do not reach sufficient size to be precipitated out of the atmosphere. When the droplets of water or ice in the clouds become too large and heavy for the air to support, precipitation occurs. The most common forms of precipitation are rain and snow. Hail, sleet, and freezing rain are the other types of precipitation. Raindrops can be as small as 0.5 mm; at this size, precipitation is described as "drizzle". The normal size of raindrops is between 1 and 2 mm. When larger than 5 mm, raindrops become unstable and usually break up.

Snow occurs when atmospheric temperatures are below the freezing point, while sleet, a mixture of rain and snow, often occurs when the air temperature is close to the freezing point. Freezing rain can occur when rain falls through a layer of sub-zero air and is cooled to the point where it can freeze upon contact with a cold surface. Hail is the most damaging form of precipitation. It occurs when strong updrafts of air in thunderstorms transport water droplets upwards where they freeze, grow in size, and fall. This process can occur over and over with the hailstones growing in size each time they are carried upwards. When the hail eventually reaches the earth, it can inflict severe damage on crops and buildings and has been known to injure people.

In recent years, scientists have come to realize that the hydrosphere, and especially the oceans, are of fundamental importance in influencing global systems which involve the energy and chemical balances of the lithosphere, biosphere, and atmosphere. The concept of the oceans as merely large basins or containers of water has been shattered as we gain more knowledge of the crucial role they play in regulating the globe's climate and the atmosphere's chemical composition.

QUESTIONS

9. Fresh, clean water is becoming a scarce resource. List all the ways in which fresh water is used in your home. Identify five conservation methods that could be instituted today in your home in order to reduce the amount of fresh water that your family consumes.
10. Suppose you are the Water Commissioner for the State of Oklahoma. Recent studies have shown you that the Ogallala aquifer is fast being depleted. Suggest five actions that the state could take to reduce the consumption of fresh water from the Ogallala aquifer.
11. One example of the process of condensation is when eye glasses "fog up" when you come into a warm room from the cold outside. Identify three other forms of condensation that you have witnessed or know about. What conditions triggered the condensation?
12. As the temperature of an air mass drops, its ability to retain moisture decreases. Similarly, as the temperature rises, more moisture can be held. Show this relationship in a diagram.

Review

- The earth's water moves from one storage place to another in the hydrologic cycle.
- Evaporation can be thought of as the starting point for the hydrologic cycle.
- Oceans contain most of the water on earth. They contain dissolved salts which remain as the water is evaporated.
- Fresh water is stored in glaciers, in lakes and rivers, under the ground, and in the atmosphere.
- Human use of underground water supplies has depleted aquifers to the point where the ground above is subsiding.
- The earth's water resources are finite and must be protected from human mismanagement.

Geographic Terms

hydrosphere
hydrologic cycle
abyssal plain
salt
salinity
thermocline
Antarctic Bottom Water
North Atlantic Deep Water Current
groundwater
water table
aquifer
evaporation
condensation
absolute humidity
relative humidity
hygroscopic particle

Explorations

1. Explain why even the largest lakes have fresh water, while all oceans have salt water.
2. The issue of water quality is becoming more and more important today. Poll 25 people in order to find out how many of their households use a water filter or purification system and how many use bottled water. Graph your results and state two conclusions arising from your study.
3. Find out the source of your household's tap water. Explain the purification process which this water goes through before coming out of your tap. You might need to use a library for this research.
4. Some parts of the world do not have enough fresh water to meet human needs. At the same time, great volumes of water are stored in the ocean and as ice. Identify one area of the world that could benefit from greater water supplies and describe a system to supply the needed fresh water. Make a poster to illustrate your water supply scheme.
5. Examine Figure 10.12. Summarize in your own words the way in which the hydrologic cycle is changed by urbanization.

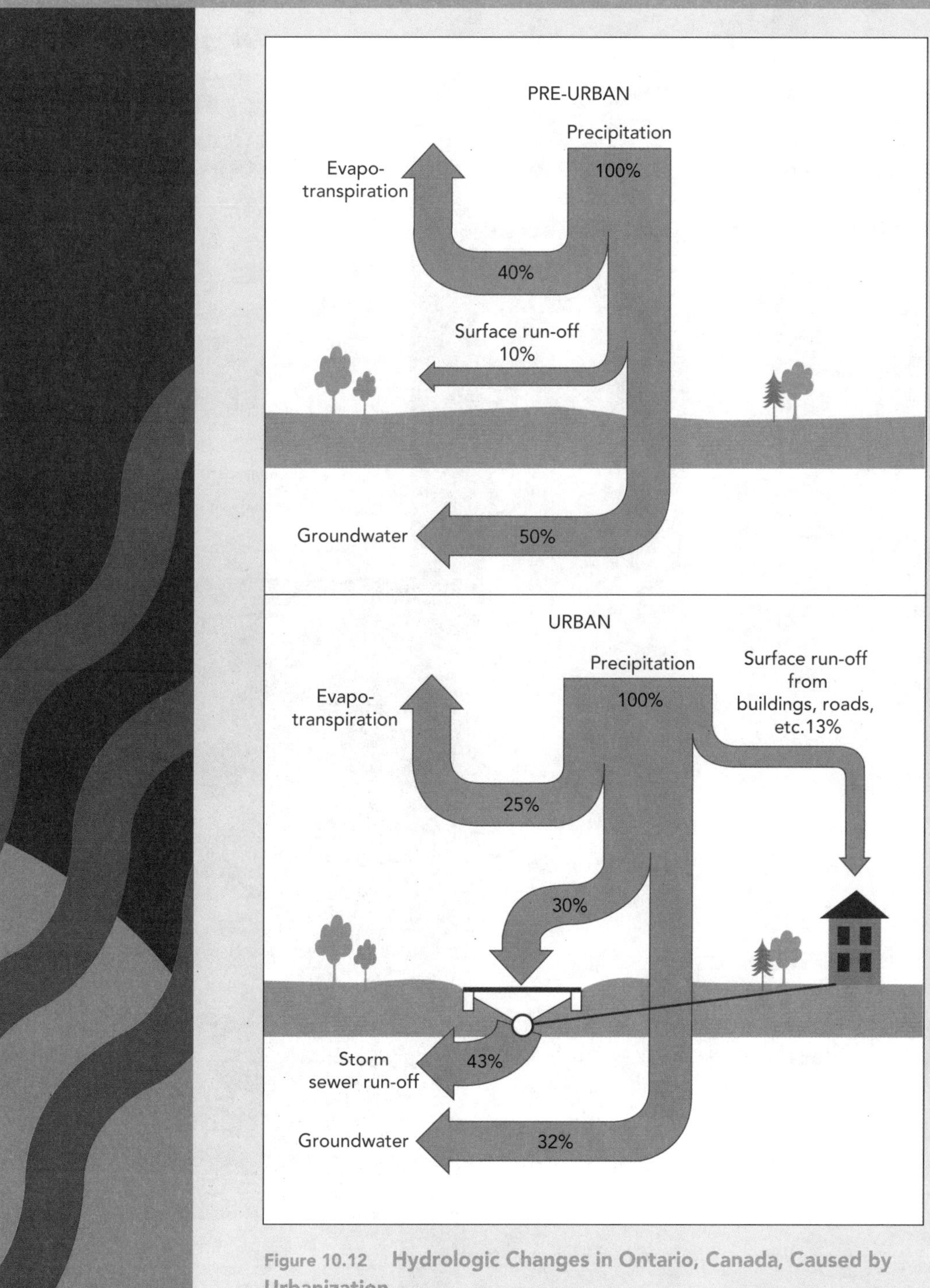

Figure 10.12 **Hydrologic Changes in Ontario, Canada, Caused by Urbanization**

CHAPTER

11

NATURAL VEGETATION AND SOIL SYSTEMS

CHAPTER 1: The Nature of Physical Geography

CHAPTER 2: Earth: Its Place in the Universe

CHAPTER 3: The Earth in Motion

CHAPTER 4: The Earth's Interior

CHAPTER 5: The Earth's Crust

CHAPTER 6: The Lithosphere in Motion: Plate Tectonics

CHAPTER 7: Solar Radiation

CHAPTER 8: Climate

CHAPTER 9: Weather

CHAPTER 10: The Hydrosphere and the Hydrologic Cycle

CHAPTER 11: Natural Vegetation and Soil Systems

CHAPTER 12: Denudation: Weathering and Mass Wasting

CHAPTER 13: Distinctive Landscapes: Humid and Arid Environments

CHAPTER 14: Distinctive Landscapes: Glacial, Periglacial, and Coastal Environments

CHAPTER 15: Natural Hazards: Disrupting Human Systems

CHAPTER 16: The Disruption of Natural Systems

CHAPTER 17: Fragile Environments

OBJECTIVES:

By the end of this chapter, you will be able to:

- understand the meaning of the term "ecosystem" and appreciate the complexity and fragility of ecosystems;
- recognize how solar energy is converted into food by plants through the process of photosynthesis, making possible all other life forms;
- describe and classify the global patterns of natural vegetation and soils;
- show the close relationships between climate, natural vegetation, and soils on a global scale;
- understand the dynamic processes responsible for the formation of soils;
- develop an awareness of the vulnerability of vegetation and soil systems and a concern about their wise use;
- increase your understanding of, and respect for, all life forms on the planet and the vital role they play in maintaining the planetary life support system.

Introduction

Plants are the most visible and vital life forms on earth. They harness the power of the sun and make it available for all other forms of life. Because of plants, webs of life developed and spread through the oceans and across the continents. Plants, and the animals dependent on them, developed a wide variety of forms and specializations as they adapted to the many climatic environments across the face of the earth. They also played a major role in creating the different types of organized dirt that we call soils. All of these amazing aspects of the plants and soils of the world will be investigated in this chapter.

11•1 The Ecosphere: The Home of Earth's Life Layer

The first life on earth was microscopic in size and of only minor influence on the surface of the planet. By the beginning of the Cambrian Period, about half a billion years ago, life had spread throughout the oceans. About 425 million years ago, in the Silurian Period, life first invaded land. By the Mesozoic Era, 245 million years ago, plants and animals formed an almost continuous layer over the earth's surface, a layer now called the **biosphere**. The biosphere has two divisions: life forms based on land make up the **terrestrial biosphere** and those living in the seas and oceans are part of the **aquatic biosphere**. It is estimated that between two and four million different kinds of organisms exist within the biosphere.

1•1/1•3

The **ecosphere** includes both the biosphere and hydrosphere and extends into the atmosphere and lithosphere. It is the zone in

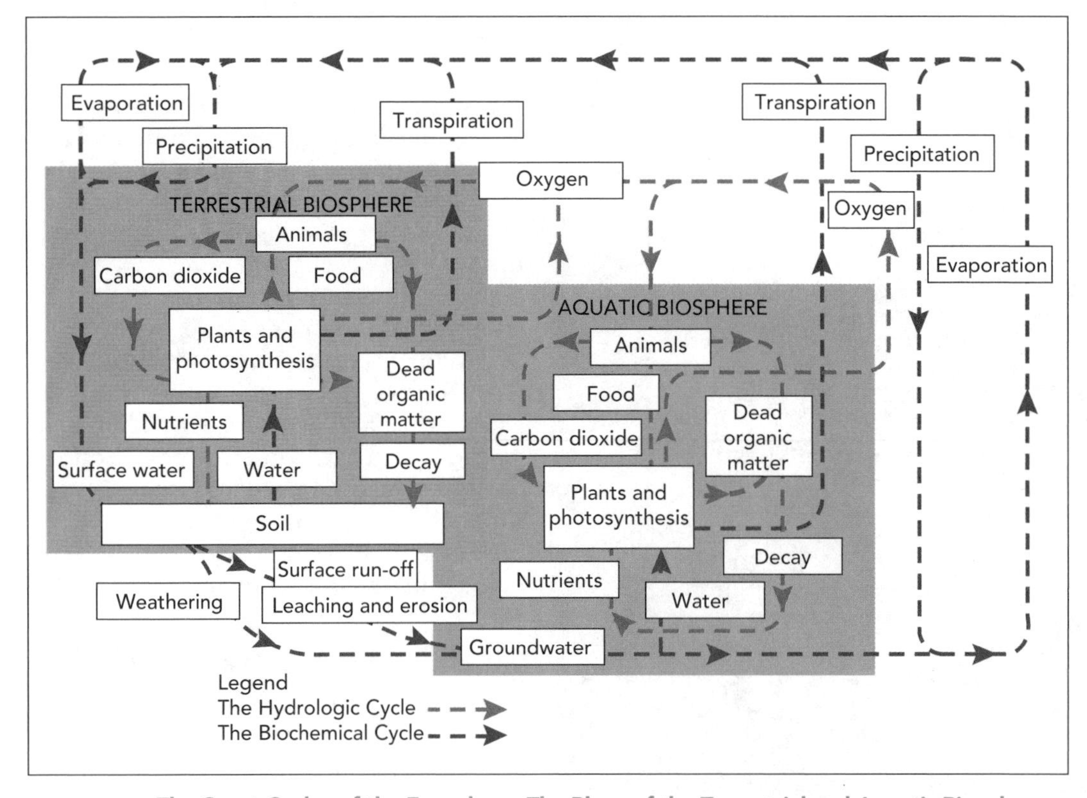

Figure 11.1 **The Great Cycles of the Ecosphere: The Place of the Terrestrial and Aquatic Biospheres**

which all the major processes that affect life operate.

Within the ecosphere, life forms exist in astounding diversity and complexity from the bottom of deep ocean trenches to the top of the troposphere. This great variety of life makes an interacting complex of living organisms, not just a collection of isolated groups or communities.

Interacting Systems in the Ecosphere

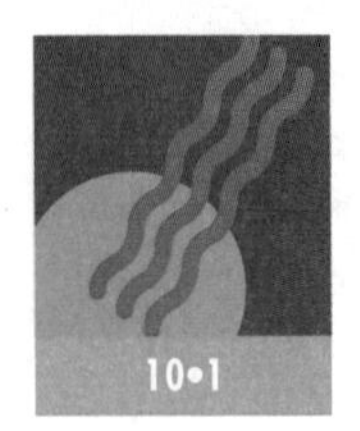

The ecosphere is influenced by three cascading systems:

- The hydrologic cycle provides water, the essential ingredient of life. This cycle is powered by solar energy.
- The rock cycle includes the weathering of rocks. Combined with the activities of plants and animals, weathered materials form the vital few centimetres of soil needed to support life. The rock cycle is largely powered by energy from the earth's interior.
- **Biochemical cycles** move the different chemical elements necessary for the survival of life. They consist of many interacting subcycles, each with its own pathways through the ecosphere and powered by energy from both above and below the earth's surface.

Carbon, oxygen, nitrogen, and phosphorus are four biochemical cycles, which together make up 99 percent of all the atoms contained in the life forms of the biosphere. These elements constantly circulate into, through, and out of the ecosphere, making them available for the growth of plants, as shown in Figure 11.1. As plants and animals die, their organic remains are broken down. The chemicals of which they are made are returned to the soil or ocean waters to be recycled into life forms or incorporated into the larger hydrologic or rock cycles.

The key life-supporting chemicals move at varying speeds through the biochemical cycles. When the cycles are interrupted by natural or human actions, there can be important effects on life in the fragile ecosphere.

Creating the Earth's Atmosphere

Some scientists believe plants help create the special conditions needed for their own survival. The high free oxygen content of the atmosphere is different from the atmosphere that might have developed on a lifeless planet. This oxygen was built up over millions of years by the process of photosynthesis. Plants found within the oceans are estimated to account for 70 percent of earth's annual production of oxygen. It has been suggested that, without life, earth's atmosphere would be similar to those of Mars and Venus, with their high carbon dioxide and low oxygen and nitrogen content.

QUESTIONS

1. What are the differences between the ecosphere and the biosphere?
2. a) Describe, in your own words, the various cycles shown in Figure 11.1.
   b) Summarize three important relationships between the biosphere and the cycles that are shown in this diagram.
   c) What sorts of disruptions could upset the flow of energy or materials in one of the systems shown in Figure 11.1?
3. a) Of all the cycles shown in Figure 11.1, which is the slowest moving?
   b) Why is this the case?
   c) What implications might this have for the other cycles that operate in the ecosphere?
4. a) Point out how human activities have altered the natural system of food production shown in Figure 11.1.
   b) How might human actions, which are increasing the carbon dioxide levels of the atmosphere, affect the biochemical cycles shown in Figure 11.1?

11·2 Nutrient Cycles

Energy is a vital ingredient in the maintenance of life, but is not able to support life on its own. Also needed are certain key chemical elements, known as **nutrients**. The three basic nutrients needed for plant growth are carbon, oxygen, and nitrogen. The processes that circulate these important ingredients of life are **nutrient cycles**. These nutrient cycles are part of the broader category of biochemical cycles discussed on pages 214-15. These cycles are powered by the hydrologic and rock cycles which operate throughout the biosphere.

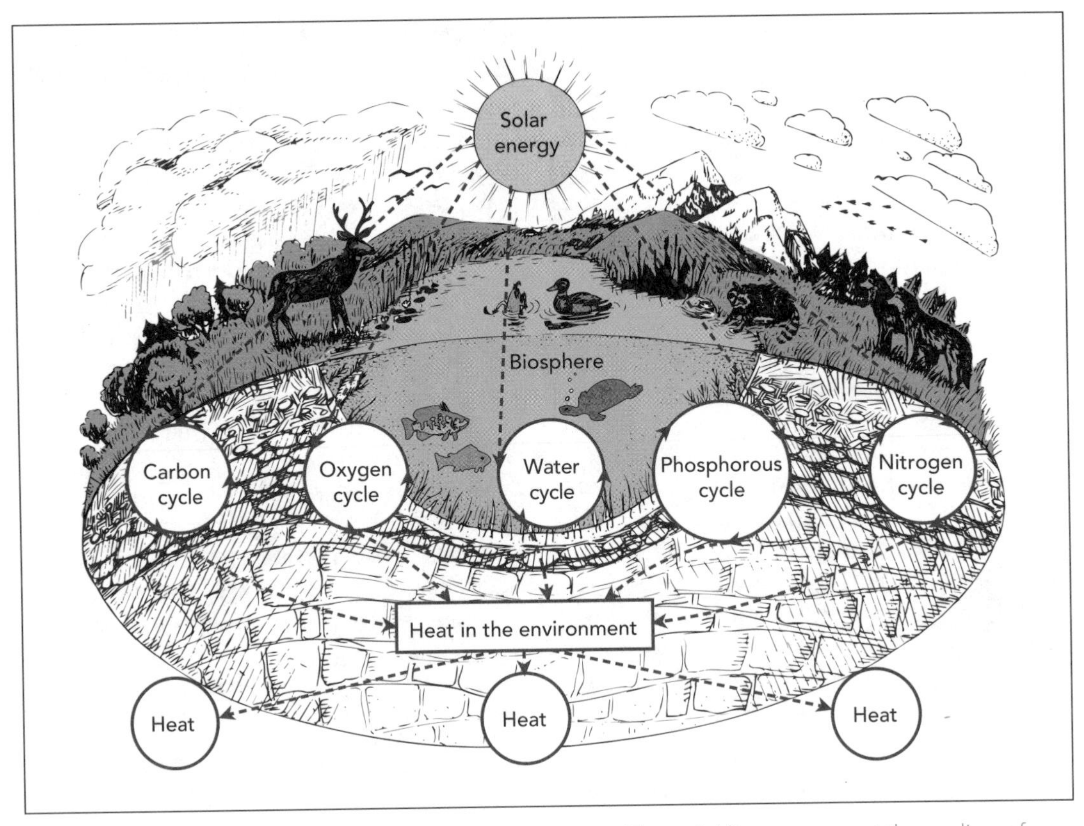

Figure 11.2 Nutrient Cycles and the Biosphere *Note:* The solid lines represent the cycling of critical chemicals. The dashed lines represent the one-way flow of energy through the biosphere.

The Carbon Cycle

Carbon is a key element because it can form long chains of organic molecules and combine with other elements in many ways. For example, carbon dioxide is used in photosynthesis, while calcium carbonate builds skeletal structures in animals. Although carbon by itself is rare in the lithosphere, atmosphere, and hydrosphere, it is the third most abundant element in the biosphere.

Many scientists believe that the amount of carbon dioxide in the atmosphere regulates the earth's temperatures through the "greenhouse effect". Studies of air bubbles in ice cores dating back over 100 000 years show that warm interglacial periods occurred during periods when large amounts of carbon dioxide were present in the atmosphere. This evidence suggests that the higher the carbon dioxide level, the higher the world's average temperatures will be.

The Oxygen Cycle

Oxygen's importance is shown by the fact that it makes up 29.5 percent of the earth and 45.2 percent of the earth's crust by weight. Oxygen is present as a gas

Symbol	Name	Atomic Mass
Basic Building Blocks		
H	Hydrogen	1
C	Carbon	12
N	Nitrogen	14
O	Oxygen	16
Macronutrients		
Na	Sodium	23
Mg	Magnesium	24
P	Phosphorus	31
S	Sulphur	32
Cl	Chlorine	35
K	Potassium	39
Ca	Calcium	40
Trace Elements		
F	Fluorine	19
Si	Silicon	28
V	Vanadium	51
Cr	Chromium	52
Mn	Manganese	55
Fe	Iron	56
Co	Cobalt	59
Cu	Copper	64
Zn	Zinc	65
Se	Selenium	79
Mo	Molybdenum	96
Sn	Tin	119
I	Iodine	127

Figure 11.3 The Chemical Elements of Life

molecule in the atmosphere (O_2), as a gas combined with carbon (CO_2), in liquid form as water (H_2O), in many organic compounds (sugars, starches, and proteins), and in many kinds of rocks (for example, limestone and quartz). Much of the "free" oxygen in the atmosphere was produced by the process of photosynthesis. Small amounts of additional oxygen were released during volcanic eruptions. On the other hand, oxygen is lost through the process of respiration. This occurs when the energy in foods such as glucose is released through oxidation, producing water and carbon dioxide.

The Nitrogen Cycle

Nitrogen is the most abundant gas in the atmosphere (78 percent of its total volume), but is an inert gas that is not usable by plants because it is not soluble in water. This element is made available for plant growth by microscopic nitrogen-fixing organisms, such as blue-green algae in the oceans and bacteria that live in soils and on the roots of legumes, for example, soybeans, peas, clover, and alfalfa. Nitrogen is used to create amino acids, the building blocks of proteins. Nitrogen, in the form of soluble nitrates and ammonia, is needed in large quantities by plants and is a common ingredient in fertilizers.

Certain bacteria return nitrogen to the atmosphere as a gas. Without this process, the atmosphere would have been depleted of nitrogen long ago.

QUESTIONS

5. In what ways do the hydrologic and rock cycles play an important part in nutrient cycles?
6. Why is carbon such a critical element, despite being a minor component of the earth's crust?
7. Based on information already presented in this chapter, create a drawing or diagram pointing out the importance of oxygen as an element making up the earth.
8. List and explain the key processes in the nitrogen cycle.

11•3 Photosynthesis: A Vital Process

Photosynthesis is the process that uses sunlight to convert carbon dioxide and water into glucose and oxygen. Glucose is made up of carbon, hydrogen, and oxygen atoms that "fix" sunlight as "bound" energy. Glucose is the only way that the sun's energy is made available to animals living on land and in the oceans. Only about 0.1 percent of the solar energy that reaches earth's surface is captured by photosynthesis.

The reverse of photosynthesis is **respiration**. In this process, glucose is burned, or oxidized, releasing the energy that was bound within the plant food during photosynthesis. An example of oxidation is the burning of wood in a fireplace. The wood is quickly oxidized, releasing its bound energy as heat, which warms the room.

Photosynthesis occurs only in green plant cells that contain chlorophyll. Its efficiency is related to the amount of direct sunlight received by the plant, as well as to the supply of nutrients (water, phosphorus, nitrogen, potassium, calcium, and trace elements) available to the plant. Glucose production is greatest under high temperature and moisture conditions, and least under cool, dry conditions. Since plants are able to make their own food, they are classified as **autotrophic** life forms.

Animals Depend on Plant Foods

Plants usually produce a surplus of glucose, which is stored in tissues or roots. This surplus glucose is the food source for all animals. Because they are unable to manufacture their own food, animals are classified as **heterotrophic** life forms.

Animals that consume plants directly are **herbivores**; animals that prey on herbivores as their primary source of food are **carnivores**. Carnivores are indirectly dependent on the foods produced by plants. An animal whose diet includes both animal flesh and plant foods is classed as an **omnivore**.

QUESTIONS

9. Into which category of food consumer are you classified: herbivore, carnivore, or omnivore? Explain.
10. A bumper sticker asks: "Have you thanked a green plant today?" Give two reasons why you should.
11. Draw a diagram to illustrate the process of photosynthesis.
12. In establishing a human colony on the moon, or on another planet, what conditions and materials would be needed to set up a sustainable food production system? Give an answer in either diagrammatic or written form.

It's a Fact...

It is estimated that there are over 400 000 species of different plants within the earth's biosphere.

11•4 Ecosystems: Webs of Life

Ecosystems are communities of plants and animals that interact with one another and with the environment in which they live. Ecosystems vary in size and complexity from the life in a small pond to all organisms in a continental or oceanic region. A typical **food chain**, the general flow of energy and nutrients in an ecosystem, is shown in Figure 11.4. Most ecosystems contain both autotrophic and heterotrophic organisms.

The largest numbers of species and biomass are found at the bottom levels of food chains. The number of species drops off with each level up the chain. The pyramid structure of a typical ecosystem reflects the fact that only 10 to 50 percent of the energy makes it from one level to the next. Thus, the number and mass of organisms decreases up the system. Four levels of consumers is the norm for most ecosystems. In mature ecosystems, the organisms within the system are in balance or equilibrium with the other members of the system. When a natural event or human interference upsets the system — for example, by killing off or reducing the numbers of certain species — this balance is upset. A healthy ecosystem should re-establish this equilibrium if the interference is not too severe. A useful illustration occurred in the western national parks of North America when the natural predators of elk and deer were almost eliminated. The elk and deer population increased dramatically and damaged the balance between many plant species in the ecosystem. By restoring wolves and other predators, park managers made it possible for the ecosystem to re-establish its natural balance.

QUESTIONS

13. a) Study Figure 11.4 and suggest what parts of an ecosystem are included under each of the four headings: environment, producers, consumers, and reducers.
    b) Why can this example be called a "system"?
14. Give examples of how disruptions in an ecosystem might have an impact on your life at some point, or on the lives of people in general.

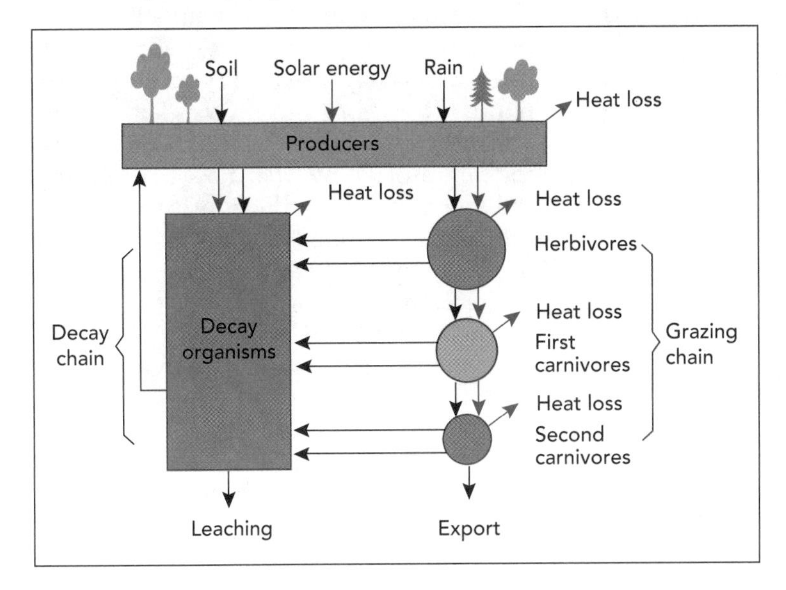

Figure 11.4
The General Flow of Energy and Nutrients in an Ecosystem
Note: The red lines represent the flow of energy. The black lines represent the flow of materials. Although most of the energy obtained by photosynthesis is lost as heat energy through respiration of all organisms, many materials are eventually recycled.

15. a) In a group, make a list of the components of the "ecosystem" in which you live and survive. To simplify your task, consider just the energy and nutrient flows that would support your group for a day.
    b) Draw a diagram to summarize the interrelationships in this "human ecosystem".
    c) How many of the steps in your group "ecosystem" do you directly observe on a regular basis?

11•5 Vegetation Systems: The Major Biomes

Ecosystems are communities of related organisms. A **biome** is a group of ecosystems with similar plant and animal species that exist under a similar climate. Terrestrial biomes are based on the dominant plant species, while oceanic biomes are defined using the predominant animal species. This section will examine the biomes that dominate the continental or land areas of the globe.

The Major Biomes

The major world biomes, based on the natural vegetation characteristics, are:

Tundra
Boreal or taiga forests
Temperate deciduous forests
Temperate grasslands
Schlerophyll forests
Deserts
Tropical rainforests
Tropical deciduous forests
Tropical savanna and grasslands
Complex highland vegetation

The distribution of the major biomes is shown in Figure 11.5, while the close relationship between the type of natural vegetation and annual temperatures and precipitation is summarized in Figure 11.7.

The boundaries between vegetation biomes are wide zones of transition, rather than distinct boundaries, as illustrated by the vegetation "cross-sections" shown in Figure 11.6. For example, mixed forests of coniferous and deciduous trees form wide transition zones between the taiga and temperate deciduous forests in North America and Eurasia. This holds true for almost all vegetation boundaries, except where steep mountain ranges create sharper, more rapid changes over short distances.

Tundra

The **tundra** biome develops under a climate with very low annual temperatures, long, bitterly cold and windy winters, extended periods of winter darkness, and short, cool summers. Annual precipitation is low, although the land remains snowbound for up to eight months of the year.

Common plants are lichens, mosses, grasses, and annual flowering plants, species adapted to little moisture and extreme temperatures (Figure 11.7). During the short summers, the upper layer of permafrost melts, creating a soggy landscape of shallow lakes, marshes, bogs, and ponds. Hordes of mosquitoes, blackflies, and other insects provide a food source for migratory birds that nest here during the summer. Along the southern edge of the tundra, stunted coniferous trees, no higher than one or two metres, survive in sheltered hollows or river valleys.

The low rate of organic decomposition, the shallow, waterlogged soils, and slow plant growth make this one of the world's most fragile biomes. It takes a long time for vegetation destroyed by human interference to grow back to its undisturbed state.

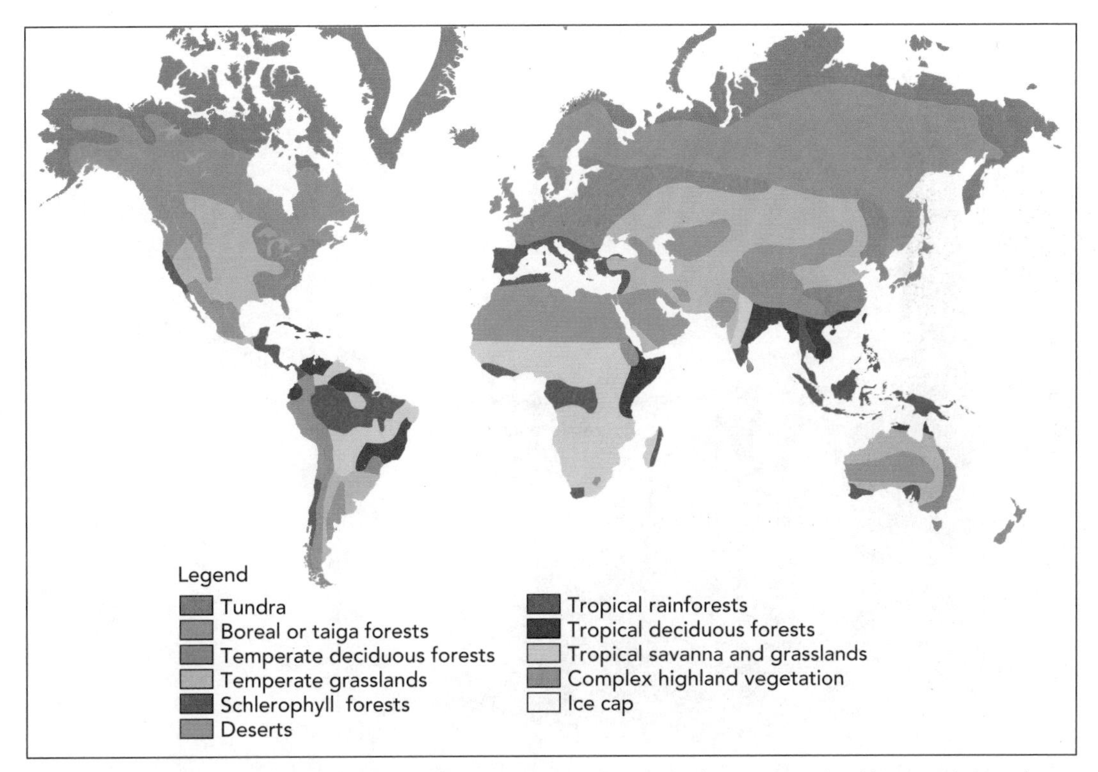

Figure 11.5 World Pattern of Vegetation Biomes There are few undisturbed biomes left in the world today. Only in isolated regions of the continents, and in the oceans, is the impact of human activities minimal.

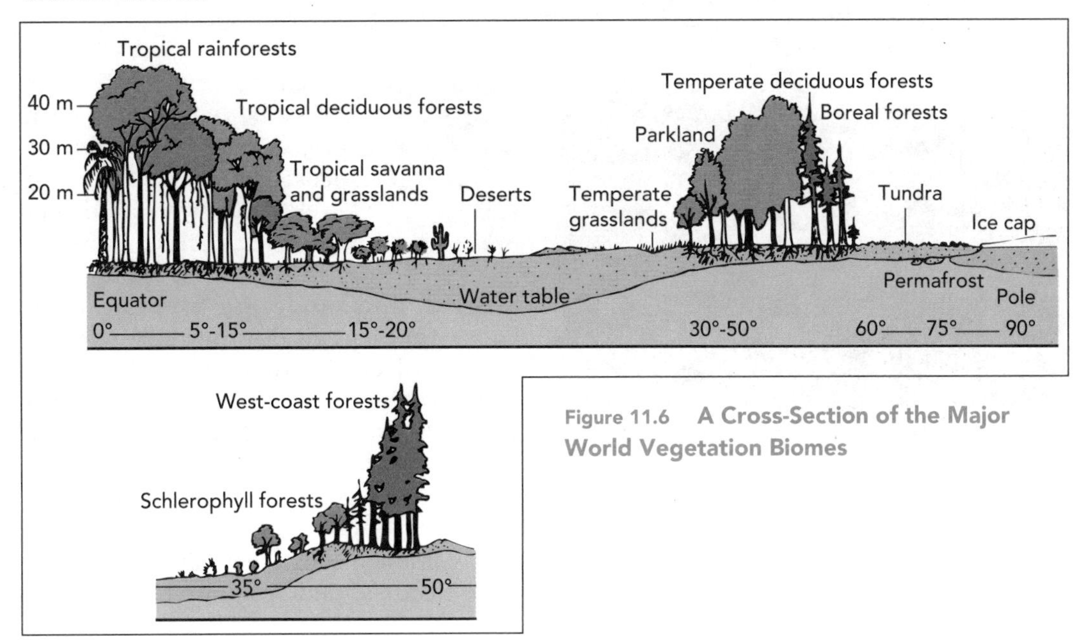

Figure 11.6 A Cross-Section of the Major World Vegetation Biomes

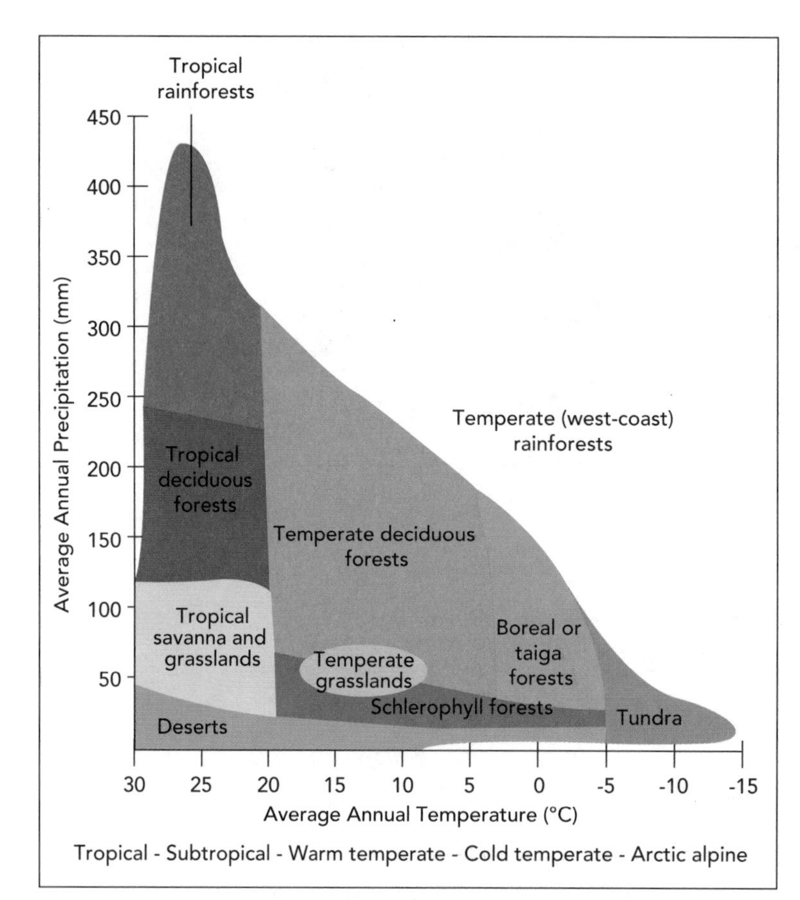

Figure 11.7 **Natural Vegetation and Climate: How They Are Related**

Figure 11.8 **A Typical Tundra Biome Landscape in Summer Along the Labrador Coast**

Boreal

Boreal forests, sometimes known by the Russian term "**taiga**", refer to **coniferous** forests (Figure 11.9) that develop under subarctic climates with short, summer growing seasons, long, cold winters, and moderate precipitation. These forests are dominated by a few species of coniferous evergreen trees, including pine, spruce, fir, cedar, and larch. Because of the harsh climate, the diversity of plant species is low.

Coniferous trees are adapted to survive the cold climates and infertile, acidic, poorly drained soils. Their short branches, waxy needles, and resinous bark reduce

Figure 11.9 **Boreal Forest in Northeastern British Columbia**

Figure 11.10 **Autumn Colours, Temperate Deciduous Forest, Niagara Escarpment, Southern Ontario**

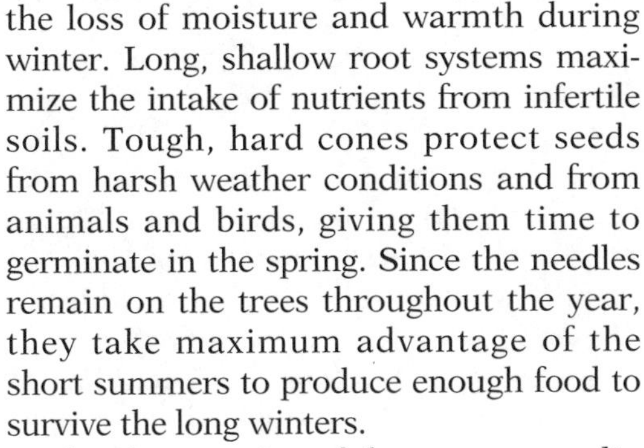

the loss of moisture and warmth during winter. Long, shallow root systems maximize the intake of nutrients from infertile soils. Tough, hard cones protect seeds from harsh weather conditions and from animals and birds, giving them time to germinate in the spring. Since the needles remain on the trees throughout the year, they take maximum advantage of the short summers to produce enough food to survive the long winters.

The close spacing of the trees means little sunlight penetrates to ground level and lower levels of vegetation are sparse or nonexistent. The waxy, acidic nature of coniferous needles, combined with slow rates of decay, results in a thick mat of needles and forest litter covering the ground.

Temperate Deciduous Forests

Temperate deciduous forests are found where summers are relatively long, winters are mild (minimum winter temperatures not below -12°C for any length of time), and there is abundant moisture year round (annual precipitation generally between 750 and 2000 mm). The longer growing season allows time for the growth and development of leaves each spring. The long summers give ample time for the production of enough food to survive the cold, but relatively short, winters. Deciduous forests are dominated by a few species of tall, broad-leafed trees, including maple, oak, beech, walnut, hickory, poplar, and chestnut (Figure 11.10).

The ground under deciduous forests is generally covered with leaf litter, which breaks down into humus, releasing basic nutrients for plant growth.

Temperate Grasslands

In the interior and rainshadow areas of the continents, lower precipitation and higher evaporation rates result in the dominance of grasslands, rather than forests.

Grasses are adapted to the semi-arid and sub-humid climates, growing quickly, and producing seeds in the short, rainy periods. The seeds germinate and grow when spring moisture is available. In dry years, the seeds lie dormant and the upper part of the grass plant withers, surviving on the foods stored in its roots, until moisture arrives. Long roots reach deep into the soil to find moisture. The ability of grasses

Figure 11.11 Typical Temperate Grasslands Biome in Coromandel, New Zealand

to lie dormant during the cold winters is an important adaptation for survival.

Growing conditions vary widely within the temperate grasslands biome. As the trees thin on the margins of humid forests, the open spaces are entirely covered by tall grasses, up to one metre in height. As the climate becomes drier, the grasses become shorter and shorter and eventually are replaced by bunch grasses, with areas of bare soil between. Along the margins of deserts, the spacing between the bunch grasses becomes greater. This transition is illustrated in Figure 11.6.

Schlerophyll Forests

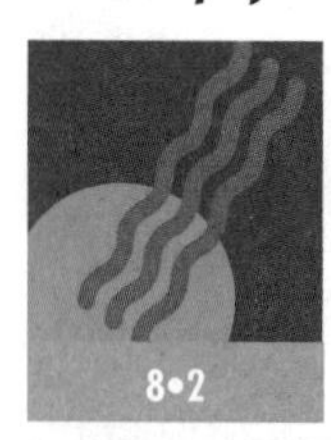

Schlerophyll forests develop under the Mediterranean climate of the Köppen Climate Classification System. The areas fringing the Mediterranean Sea contain the largest areas of oak-shrub-grassland vegetation known as the **schlerophyll** forests biome. This biome is also found in California, South and West Australia, South Africa, and Chile. Northern hemisphere schlerophyll forests include trees such as the cork oak, live oak, white oak, olive, and various species of pine. In Australia and Africa, eucalyptus and acacia trees are common. Shrubs include the wild lilac, poison oak, and mountain mahogany. A typical schlerophyll forest in southern California is shown in Figure 11.12

Figure 11.12 A Schlerophyll Forest Biome Near San Diego in Southern California

The vegetation has adapted to a climate where precipitation occurs mainly during the cooler winter months and summers are hot, sunny, and rainless. The adaptations to dry summers include long root systems, thick bark on gnarled branches and trunks, leaves that are light coloured, small, and leathery, and a low to medium tree height. The shrubs and grasses survive by remaining dormant in summer and growing during the cooler, moister winters. The trees cover anywhere from 25 to 60 percent of the ground surface, with the remainder covered by shrubs and grasses.

Deserts

Desert vegetation must adapt to dry periods that can last several years. The name "**xerophyte**" is applied to plants, such as the cactus, that have adapted to extremely dry climates. Adaptations include thick, waxy skins protecting large, pulpy interiors that store moisture; long, thin, protective needles for leaves; and wide spacing of plants competing for scarce moisture. Common shrubs, such as sagebrush and creosote bushes, and grasses have deep root systems and an ability to become dormant during long dry periods. When desert rainshowers do occur, the plants must make the most of the moisture. Annuals adapt by quickly sprouting from seeds, flowering, and going to seed in the short period after infrequent showers. The tough seeds survive for decades until sufficient moisture becomes available for germination and growth.

A typical desert biome is shown in Figure 11.13. The plants cover only a small percentage of the ground surface, less than any other biome except for the tundra.

Figure 11.13 **Widely Spaced Grasses, Shrubs, and Bushes in a Desert Biome in Southern Utah**

Tropical Rainforests

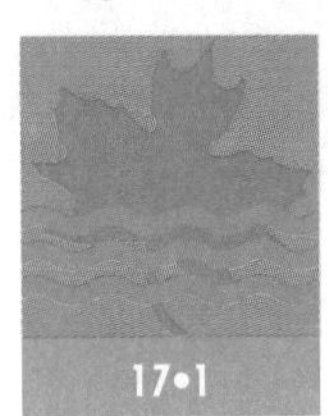

In tropical rainforests, precipitation is almost a daily event. The air is constantly humid and the hot temperatures scarcely change from day to night or from month to month. Growing conditions are so favourable that tropical rainforests contain by far the largest number of plant, insect, and animal species of any biome.

The dominant species are tall, broad-leaf evergreen trees. The tops of these trees form a continuous "canopy", usually about 30 m above the ground, that blocks most of the sunshine from reaching the ground below. Shorter trees are able to survive in breaks in the canopy. Ferns grow in the dim light and abundant moisture at ground level, but the canopy blocks the sun so effectively that few species grow on the forest floor. Climbing vines use the trunks of tall trees as supports for growing upward into the sunlit areas of the upper canopy.

Up to 300 different species of trees may be found in one hectare of the forest. The most valuable trees are the tropical hardwoods, including ebony, ironwood, and mahogany.

The leaves of the trees are large, with leathery surfaces that protect them from the intense solar radiation and high temperatures. The trees of the rainforest are never dormant because of the great competition among trees for sunlight and nutrients and the constant high temperatures and precipitation that enable growth to occur all year round.

While the trees are evergreen, they do lose their leaves in a steady stream throughout the year. Almost no leaves

Figure 11.14 **A Typical Rainforest Biome in the Central Philippines**

accumulate on the forest floor because decomposers quickly break them down into nutrients that are absorbed by the shallow tree roots or lost through rapid leaching.

Tropical Deciduous Forests

In tropical climates where rainfall is concentrated in a distinct wet season, tropical deciduous or monsoon forests develop (Figure 11.15). Such forests must survive dry seasons where temperatures are high and evaporation greatly exceeds rainfall. These conditions occur in the monsoon climates of India, Myanmar, Thailand, Brazil, and northern Australia.

Figure 11.15 Typical Tropical Deciduous Forest Biome in the Calcutta Region, India

High evaporation and low rainfall make it impossible for trees to be evergreen, so monsoon forests are dominated by tropical deciduous trees that drop their leaves during the dry season. The trees are shorter (12 to 35 m in height), more widely spaced, and less diverse than in tropical rainforests. This reduces the competition for light and nutrients and allows dense thickets of shrubs or bamboo to grow at ground level.

Tropical Savanna and Grasslands

Figure 11.16 Typical Tropical Savanna and Grassland Biome in Reunion, off the African Coast

Africa has the greatest extent of the tropical savanna and grassland biome which develops under climates with one or two wet seasons separated by long, dry periods. Temperatures and evaporation rates are high and the pattern of precipitation is similar to tropical monsoon climates, but the annual totals are much lower. As a result, there is only enough moisture to support widely spaced trees, with tall tropical grasses between them. The result is an open, parklike landscape, as shown in Figure 11.16. The trees generally have flattened tops and thick, rough bark. Some are xerophytic with small leaves and thorns, such as the acacia, while others are deciduous and shed their leaves during the dry season. Closer to the deserts, the trees are smaller and more widely scattered, and the grasses either become shorter or are replaced by thorn scrub. Eventually, the trees disappear and are replaced by short thorn bushes and bunch grasses.

Complex Highland Vegetation

Vegetation within mountain regions varies widely depending on climate, altitude, and latitude. Changes in temperature and exposure give rise to a **vertical zonation** of vegetation biomes over short horizontal distances.

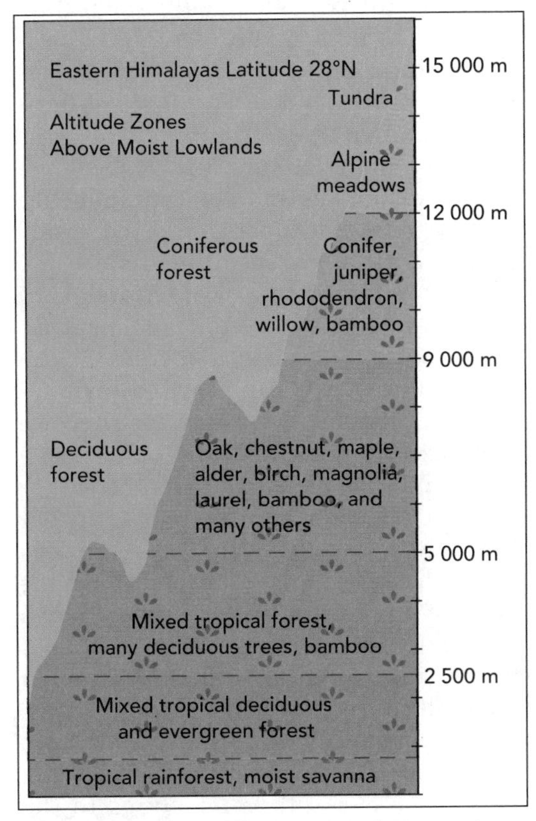

Figure 11.17 Vertical Zonation of Biomes in the Himalayas of Asia

In temperate regions, lower, windward slopes of mountains may be clothed in deciduous or coniferous forests. As elevations increase and temperatures decrease, the tree sizes decrease and species may change from deciduous to coniferous. If the mountains are high enough, the climate may become too cold or windy for the survival of trees, and mountain meadows with short, flowering plants, mosses, lichens, and bushes take over. These meadows are similar to the tundra biomes of higher latitudes. High tropical mountains, such as the eastern Himalayas of Asia illustrated in Figure 11.17, have vegetation zones that include many of the world's biomes.

QUESTIONS

16. a) Name the type of natural vegetation that is native to your local region, based on the world map shown in Figure 11.5.
    b) If there are any areas of relatively undisturbed natural vegetation left in your local region, suggest reasons for their survival into the present day. If not, why has the natural vegetation been modified in your region?
    c) Examine your own values and indicate if you feel that the natural vegetation in at least some part of your local region should be preserved or restored. Explain your position.
17. Complete a comparison chart to summarize the characteristics of the major vegetation biomes, using headings such as Location, Precipitation Patterns and Amounts, Temperatures, Dominant Vegetation Types, and Major Adaptations.
18. a) Draw a representative climograph for any three natural vegetation biomes. Climate statistics appear in Figure 8.36, page 167, of this book, in atlases, or in books on physical geography in your resource centre. Explain how you know each of your climographs is representative of the chosen biome.
    b) Prepare a sketch or diagram to show how the vegetation found in each of these biomes has adapted to the climatic conditions shown by the representative climograph.
19. a) Using Figure 11.7, draw one vegetation profile of South America along the 10°S line of latitude and a second of Africa along the 20°E line of longitude. Label the vegetation biomes and any important physical features on the profiles.
    b) Explain the variations in the vegetation patterns across each of these profiles.

11·6 Soil Systems: Between Living and Non-Living Matter

Soils are complex mixtures of animal, mineral, and organic materials, capable of supporting plant life. Since plants form the base of the food chain for all life forms, soils are, therefore, indispensable to all terrestrial animal life.

For soil to form, two basic processes must occur. First, water must be active in the soil, bringing about physical and chemical changes in the original weathered rock material. Second, the activities of soil organisms must cause further changes. The result is the development of different layers within the soil.

The Four Components of Soils

Soil is composed of four main materials: minerals, organic matter, air, and water. The **mineral matter** consists of weathered rock, known as **parent material**. Parent material may be either decomposed rock or sediment that has been deposited by running water, wind, or glacial ice. The size of the parent material particles determines the **texture** of the soil. **Organic matter** is the remains of plants and animals living on, or in, the soil. It may form a distinct layer on the soil surface or be mixed with the mineral matter. **Air** provides oxygen used in the decay of organic matter and the growth of plants. In addition to dissolving soluble chemical compounds needed for plant growth, **water** transports mineral particles and soluble chemical compounds through the soil.

The Properties of Soils

Soil Texture

Soil texture is determined by the sizes of the rock particles of the parent materials. Sands are the coarsest, while clays are the finest, soil particles. Most soils, however, are composed of many sizes of soil particles. The classification system illustrated in Figure 11.18 is used to identify soil texture classes. The term "**loam**" describes a mixture of sand, silt, and clay that combines many of the individual properties of these three textures into a more favourable structure for agriculture.

Soil texture has an important influence on soil properties. Coarse-textured, sandy soils are usually well-drained since the sand particles do not hold water very well. Sandy soils dry out and heat up more quickly than other soils in the spring and early summer, so farmers refer to them as "warm" soils. They are often used to grow vegetable crops, since they can be ploughed and planted earlier than other soils, and farmers can get their crops to market first, when prices are higher. Because sandy-textured soils are easy to plough, they are often known as "light" soils. Sandy soils are also less fertile since water moves easily through them, flushing away plant nutrients.

Clay soils, on the other hand, hold water very well because the particles are close together and water clings to the clay particles. Since clay soils take longer to warm up in the spring, farmers refer to them as "cold" soils. The compact nature of clay soils means they are often poorly drained and hard to plough, giving them a reputation as "heavy" soils. Because water does not move through them easily, they hold nutrients effectively. As a result, clay soils are best used for pasture land and for growing alfalfa or hay crops.

Loamy-textured soils combine the properties of sand and clay-textured soils. The clay portion gives them the capacity to hold water and nutrients, while the sand component makes the soil lighter, easier to plough, and better drained.

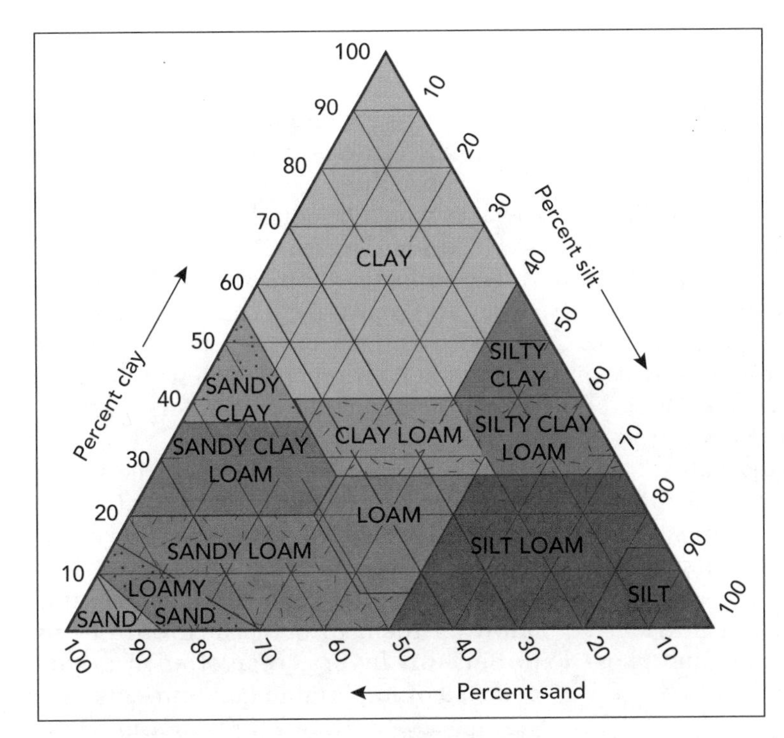

Figure 11.18
Soil Texture Classes
The three sides represent base lines for sand, silt, and clay with the apices opposite representing 100 percent of each soil texture. Percentages can be read off to give the textural name for any soil sample.

Soil Structure

Soil structure refers to the tendency of individual grains of the soil to group together to form larger units called "**peds**". Peds increase soil porosity, making more space for air and water in the soil, as well as providing places for worms and other microfauna to live. Peds also help a soil resist erosion by holding individual soil particles in larger clusters.

Organic matter and decayed humus encourage the formation of peds, as they provide the cement that binds the soil particles together. In organic-rich soils, the spaces between the peds can make up to 60 percent of the volume of the upper layers of the soil.

Soil Colour

Soil colours vary greatly from place to place and within a soil profile. Soils of humid tropical regions are usually red to reddish brown; those of the temperate grasslands are often black to dark brown; coniferous forest soils are normally grey in their upper levels and brown to yellow at lower levels.

Reasons for differences in soil colour include the amount of iron and organic matter and the chemical state of the iron in the soil. Organic matter and humus give soils a dark brown to black colouring, such as those under grassland biomes. Where the bedrock has a high iron content, soils may have a reddish to reddish brown colour. The reddish brown colour of tropical soils results from their high oxidized iron content. In poorly drained soils where oxygen is lacking, reduced iron produces greenish and grey-blue colours in the soil. The ash-grey colour of coniferous forest soils results from a lack of organic matter and iron and a high silica content. In arid regions, concentrations of calcium carbonate give soils a white colour.

Organic Matter

Organic matter may lie upon the soil surface, where it has accumulated as partly decayed plant and animal remains, or be mixed with the mineral matter of the surface layers of the soil. The rate of organic decay is related to temperature, the amount of water and volume of air in the soil, and the activity of soil bacteria. The higher the temperatures and soil moisture, the greater the activity of soil bacteria and rate of decay. However, if too much water fills the pores in the soil, bacteria will not have the oxygen needed to break down organic matter.

Partly decomposed organic matter within a soil is known as **humus**. Large quantities of humus give soils a dark brown to black colour. The decay of humus releases soluble plant nutrients, including nitrogen, potassium, and phosphorus, into the soil.

Air and Water

The proportion of air in soils is controlled by the soil texture and structure. Air penetrates into soils through the pores between the soil peds. The coarser the soil texture, the greater its porosity and the greater the amount of air it contains. Sandy soils contain more air than clayey soils.

The amounts of air and water in soils are complementary, in that the higher the content of one, the lower the content of the other. Fine-textured soils, such as clays, hold water more readily than coarse-textured soils, such as sands. Organic matter also holds water in the soil, acting like a sponge. Where soil pores are saturated with water, the plant growth is reduced.

Soil Chemistry

Soil fertility is closely related to clay and humus content. Microscopic particles of clay and humus are bound together into clusters and held in suspension in the soil water. These particles have negative electrical charges that attract and hold positively charged ions, or **cations**, such as those of calcium, magnesium, potassium, and sodium. Since these plant nutrients are soluble, they would be quickly leached from the soil unless held by the clay-humus complex.

The strength of attraction of the cations varies and weakly held cations can be displaced by others that are more strongly attracted. A hydrogen cation will displace any of the other cations from its place on the clay-humus complex.

In some parts of the world, the water contains an abundance of hydrogen cations because of acid rain. Since the hydrogen cations are more active, they quickly displace the plant nutrients, allowing them to be washed out of the upper soil layers, resulting in a soil depleted of important plant nutrients.

Soil acidity is a measure of the concentration of hydrogen ions found within it, as measured on the **pH scale**. The lower the pH value of a soil, the higher its hydrogen ion content and acidity. The higher the pH value, the more alkaline the soil. Alkaline soils have high concentrations of bases, such as calcium carbonate, or salts. Many important food crops can only be grown successfully within a certain range of pH values. Where the soil is too acidic, lime can be added to "sweeten" or raise the pH value and raise yields of acid sensitive crops.

Soil Horizons

Soils have distinct horizontal layers called "**horizons**" created by the movement of water, upward and downward, through the soil. As it moves, the water transports soluble compounds and fine clay particles from one soil layer to another. Over thousands of years, this process creates horizontal layers — horizons — within the soil that have distinct textures, colours, chemical compositions, and structures.

The horizons of a typical soil profile are shown in Figure 11.19. The A-horizon, also known as **topsoil**, normally contains the highest proportion of organic matter. The B-horizon receives material removed from the A-horizon and is usually lower in organic matter than the A-horizon. The C-horizon is largely made up of parent materials, little altered by soil-forming processes. Acidic soils, developed under coniferous forests, often have an E-horizon that is lighter in colour because it contains less organic matter and clay than neighbouring horizons. Generally speaking, the moister and warmer the climate, the deeper the soil and each of its horizons will be.

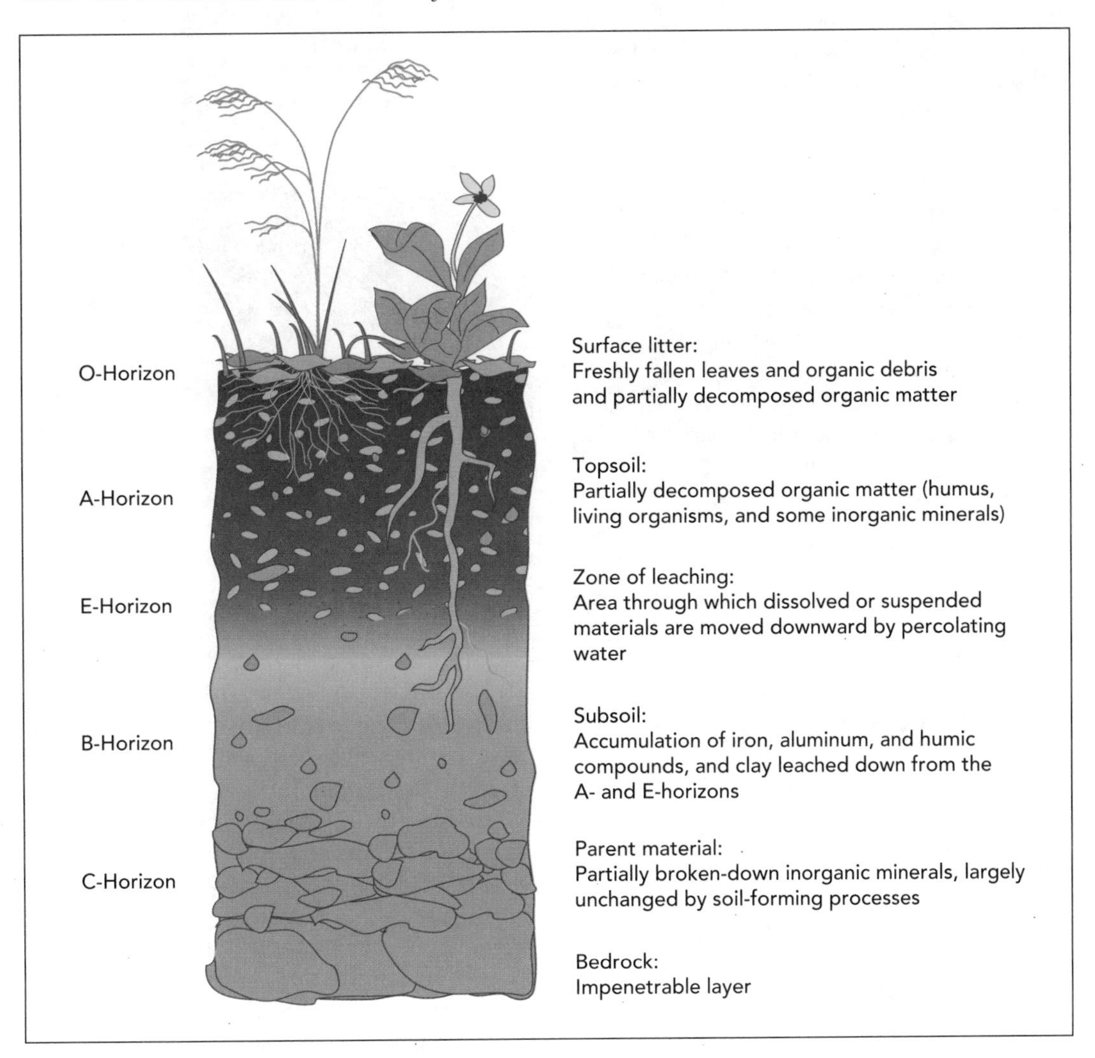

Figure 11.19 The Soil Horizons of a Typical Soil

QUESTIONS

20. What is your attitude towards soils and their importance? For example, do you agree with the attitude expressed in detergent commercials that soil is only a nuisance that gets clothes dirty?
21. a) What are soils and how are they formed?
    b) Identify the characteristics that distinguish a real soil from a mound of dirt piled up from an excavation or construction project.
22. Get a soil sample from your backyard or schoolyard. Describe the properties of the soil sample using the information in this section of the chapter.
23. a) Identify some detrimental effects of acid rain on soils.
    b) Explain how acid rain can affect soils in these ways.
    c) Explain one way in which such effects can be counteracted.
24. a) Why is it important to put only topsoil on a garden or lawn?
    b) What effect would the removal or erosion of the topsoil have on soil development?

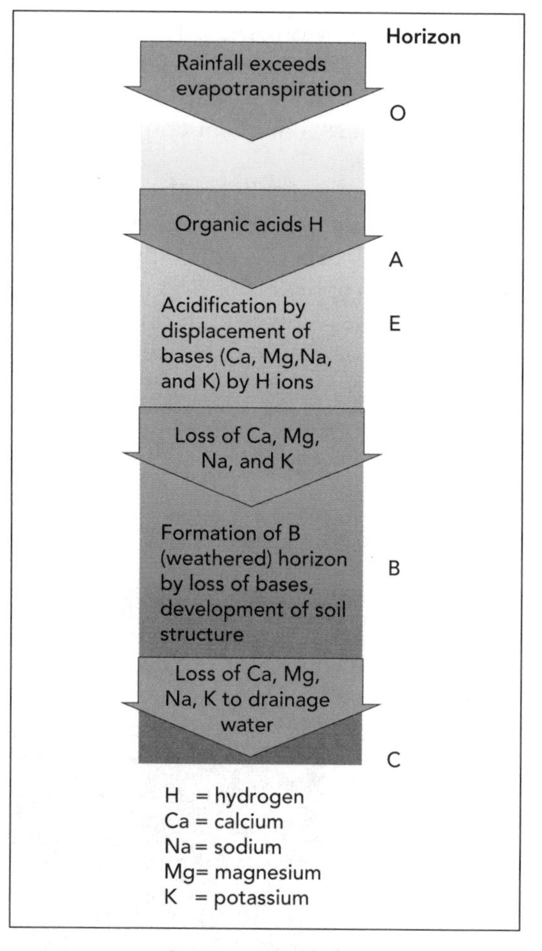

Figure 11.20 The Process of Leaching

11·7 Soil Formation

A number of dynamic systems and factors produce the soils found on the earth's surface. The term "dynamic" indicates they are ongoing and respond to changes in the environment within which the soil is forming.

Leaching: Water in Action in the Soil

The direction and volume of water movement is a critical factor in soil development. **Leaching**, the process by which water removes materials in solution from one horizon to another, is illustrated in Figure 11.20. In general, the greater the amount of water moving through the soil, the greater the amount of leaching. Where the water contains weak acids, such as carbonic, hydrochloric, sulphurous, or nitrous acids, leaching is increased. The greater the acidity of the water, the greater the leaching effect.

Under moist, tropical climates, leaching occurs all year round and is extremely effective in removing soluble minerals from the soil. In cooler climates, winter freeze-up stops leaching action.

In soils of arid and semi-arid climates, high evaporation rates lead to the formation of a layer of calcium carbonate or other soluble salts near the surface of the soil. This is much like the salt crust left in a glass when salty water is completely evaporated. When this deposit develops into a hard layer, it is known by the Spanish word "**caliche**".

Factors Affecting Soil Formation

Climate

Humidity, evapotranspiration, precipitation, and the duration of sunshine are climatic factors that influence soil development, directly or indirectly. **Evapotranspiration** is the combined processes of evaporation of water to the atmosphere and the transpiration of water by plants. It affects the amount of water available for soil-forming processes and the growth of organisms that influence soil formation.

Temperature is the other major aspect of climate controlling soil formation. In general, higher temperatures are more effective in breaking down organic and inorganic materials. Weathering in tropical climates is almost ten times more effective than in polar regions (Figure 11.21). Deeper weathering of rocks and soils is characteristic of tropical regions; many tropical areas are covered with weathered materials up to 50 m in depth.

Organisms and Plants

Living organisms affecting soil development are bacteria and fungi, along with earthworms and burrowing insects and animals. Bacteria and fungi cause the initial breakdown of plant and animal tissues. Earthworms and termites help to integrate partially decayed organic materials with the mineral particles in the soil.

Plants also have a major influence on soil development. Plant roots help to open up the soil, allowing water and air to penetrate below the surface. Plant tissues act as storage reservoirs for soluble nutrients that would otherwise be leached from the soil in humid areas. When plants die, their organic remains decay and release nutrients into the soil for new plants to take up in their root systems. In this way, vital plant nutrients are continually recycled by the vegetation.

The importance of plants can be seen when the vegetation of an area is altered and the equilibrium between the soil and the natural vegetation is upset. For example, where forests are cleared for farming, the natural supply of organic matter to the soil is largely cut off and the plant nutrients originally recycled by the soil-plant system are removed in the form of crops. They are also lost by increased soil erosion and leaching. Farmers must compensate for this by adding manure (organic matter) and fertilizers to maintain soil structure and fertility.

Relief

The configuration of the land surface is an important variable in soil formation. Where slopes are steep, erosion often prevents

	Average Soil Temperature (°C)	Days of Weathering	Relative Weathering Factor
Arctic	10	100	1
Temperate	18	200	2.8
Tropical	34	360	9.5

Figure 11.21
Weathering Effectiveness Related to Temperatures

the development of a deep soil profile, whereas level areas may suffer from drainage problems. The orientation of slopes is also important. In the northern hemisphere, south-facing slopes are often warmer than north-facing slopes. Because of this, north-facing slopes may be covered by forests since temperatures are cooler, evaporation less, and the soil more moist. South-facing slopes may have tall grasses and scattered trees as a vegetation cover since the soils are drier and evaporation rates higher. Such differences in vegetation affect the soils that develop beneath them.

Parent Material

Parent material is the weathered mineral matter from which soils originally develop. The size of fragments largely determines soil texture. Texture influences the speed with which soil-forming processes operate, all other factors being equal. For example, in areas of clay texture, soil profiles develop very slowly since the small pore spaces of clays mean water moves slowly through the soil, slowing down the soil-forming process. In sandy soils, large soil particles and pore spaces mean water can move materials in solution and suspension easily, so the soil profile develops quickly.

A second influence of parent materials is their pH reaction. Parent materials developed on acidic rocks, such as granite, have a low pH. Soils developed from rocks such as limestones often have a basic or neutral pH.

Time

Soils require time to form. In general, young soils closely reflect the nature of the parent materials; as time passes, the influences of natural vegetation, climate, slope, and other factors become dominant. These influences show up as increased organic matter, more clearly defined horizons, and a difference in colour between the parent material and the upper horizon of the soil. In general, the soil develops an equilibrium with all of the environmental factors influencing it, not just the parent materials. The large number of factors makes it difficult to say how long it takes a soil to reach this equilibrium, but it often is hundreds to thousands of years.

The Soil as a System

The factors of soil formation are parts of a system that operate to create soil. It is difficult to separate the effects of individual factors since they are so closely related to one another. Figure 11.22 shows an open system in which soil results from additions (or inputs), such as solar energy and water, and losses (or outputs) of such things as water, nutrients, and soil particles through drainage, leaching, and erosion.

QUESTIONS

25. Describe the ways in which water is involved in the development of soil horizons.
26. a) Devise a demonstration to show how water can move upward against the force of gravity.
    b) What name is given to water that moves through the soil in this way?
    c) Under what conditions is this type of water movement important in soil formation?
27. Draw a diagram to summarize the factors that increase the effectiveness of leaching in a soil.
28. a) Using Figure 11.22, write an account that summarizes the various systems and processes that operate over time to form a soil.

b) Describe how a change in any one of a number of factors could bring about changes in an already mature soil profile.

29. Having completed this section, review the answer you gave to Question 20 in this chapter. Has your attitude towards soils and their importance changed as a result of these studies? Explain.

11·8 Soil Classification and Distribution

The classification of soils is hotly debated by soil scientists and physical geographers. Soils can be grouped using many variables, including texture, structure, organic content, fertility, horizon characteristics, colour, and human modification. Agreement on the important variables is made more difficult because soil classification is often tied to local and national ways of looking at soils. For example, different soil classification systems are used in Britain, the United States, France, Russia, Australia, Canada, and many other nations. As a result, there are many different approaches to naming and grouping soils, making it difficult to reach agreement on an acceptable, worldwide classification system.

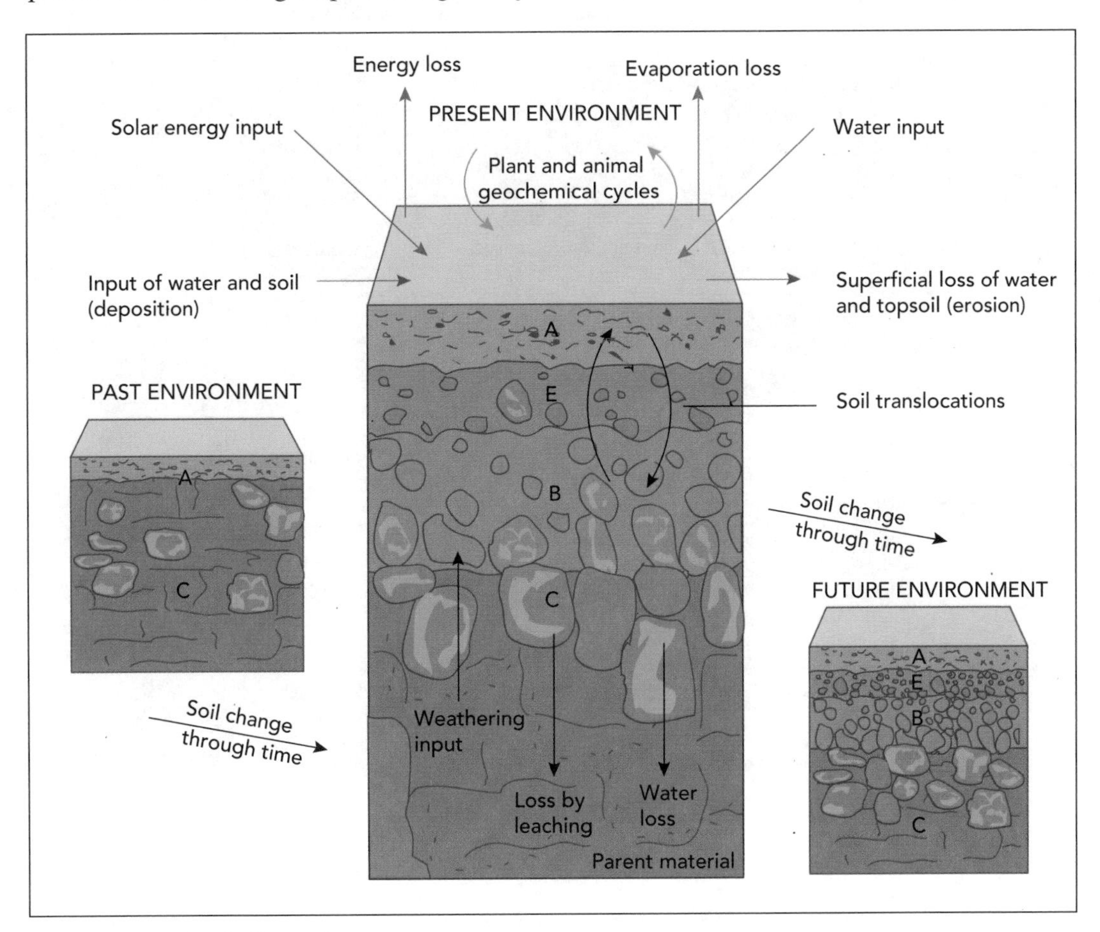

Figure 11.22 The Soil as a System

The 1938 USDA Soil Classification System will be used in this book. This system recognizes the importance of climate, with its related vegetation types, as a fundamental influence on soil development. It divides the world's soils into three major orders: zonal, intrazonal, and azonal. These major orders are divided into nine suborders, which are then subdivided into great soil groups. The suborders and important great soil groups are shown in the table in Figure 11.23.

ORDER	SUBORDER	GREAT SOIL GROUP
Zonal	Soils of cold regions	Tundra
	Light-coloured soils of arid regions	Desert soils Red desert soils Sierozems Brown soils Reddish-brown soils
	Dark-coloured soils of semi-arid to humid temperate and tropical grasslands	Chestnut soils Reddish-chestnut soils Chernozem soils Prairie soils Reddish prairie soils
	Soils of the temperate forest-grassland transition	Degraded chernozem Non-calcic brown soils
	Light-coloured podzolized soils of temperate forest regions	Podzol soils
		Brown podzols Grey-brown podzols
	Ferrallitic soils of the humid forested tropical and subtropical regions	Red and yellow ferrallitic soils Red and yellow latosols Red and yellow podzols
Intrazonal	Saline and alkali soils	Saline soils Alkali soils – solonetz and soloth soils
	Waterlogged soils of poorly drained areas	Bog soils Alpine meadow soils
Azonal	No suborders	Alluvial soils Regosols (including dry sands)

Figure 11.23 **The 1938 USDA Soil Classification System**

Soil Orders

Where soil-forming factors have the time to develop distinctive horizons and soil characteristics different from the parent material, soils are placed in the **zonal** soil order. Zonal soils show the direct influence of the climate and vegetation biomes under which they have developed.

Soils that have not developed distinct horizons because of local factors, such as distinctive parent materials or poor

MAJOR SOIL CHARACTERISTICS	MAJOR LAND USES
Shallow, unleached, organic-rich, acidic, waterlogged soils formed in the active layer in permafrost areas; dark A-horizon of undecayed humus (peat)	Largely unused, except for isolated short-season vegetable gardens, grazing
Thin, unleached, humus-poor, nutrient-rich, grey to red soils; moderate to heavy salt accumulation in upper soil; may form a caliche layer; red and brown colours due to iron and magnesium oxides staining soil particles	Grazing, intense irrigation agriculture where water is available; crops dependent on local climate and markets
Thick, mildly leached, neutral to slightly acidic, humus-rich, highly fertile soils; black to dark brown A-horizon rich in organic materials; B-horizon rich in calcium carbonate; the darker the colour, the higher the organic content	Grains (especially wheat in semi-arid to sub-humid, corn in humid areas), grazing, sugar beets, field crops
Similar to above soils, except stronger leaching results in weak or non-existent calcium carbonate layer in B-horizon	Grains, field crops, grazing, legumes, mixed livestock farming
Shallow, highly leached, very acidic, infertile soils of coniferous forests; thick forest litter and surface humus layer; ash-grey, leached, silica-rich A-horizon; yellow-brown B-horizon rich in oxides of aluminum and iron	Dairy farming, potatoes, general livestock farming, grazing, pastures
Average depth, moderately leached, mildly acidic, humus-rich, relatively fertile soils of deciduous forest; grey-brown A-horizon with moderate humus content; yellow-brown B-horizon rich in clays	General livestock farming, grains, field crops, pasture, fruits, and vegetables
Deep, intensively leached, mildly alkaline to neutral, red and yellow soils of tropical rainforests and savannas; clay soils rich in iron and aluminum oxides; poor in silica, humus, and plant nutrients; podzols are less leached, neutral, relatively infertile soils of broadleaf evergreen forests of humid subtropics; richer than ferrallitic soils in silica, humus, and plant nutrients	Ferrallitic – plantation crops (sugar, bananas, pineapples, coffee), cattle grazing, subsistence shifting agriculture; podzols – cotton, citrus, subsistence, grazing
Very alkaline soils, hard clay layer, salt crusts, and horizons	Little used due to high salt content
Similar to tundra soils in characteristics, except conditions due to poor drainage (bog soils) or cold climate (alpine meadow soils)	Little used bog soils, except for cranberries; high mountain pastures
Little or no time to develop soil horizons or alter original parent materials; soils constantly being eroded and redeposited by rivers, ice, etc.	Alluvial soils often rich in nutrients and suitable for intensive farming (flood plains, deltas, etc.)

drainage, are known as **intrazonal** soils. **Azonal** soils are immature soils that have not had enough time to develop horizons. They are found where erosion and deposition constantly rework the soils, as in alluvial flood plains and deltas and in areas where landslides, mudflows, and avalanches are common.

QUESTIONS

30. Point out some reasons for the great number of soil classification systems in use in the world today.
31. a) What great soil group is found in your local area, as shown by the world soil map in Figure 11.24?
    b) Point out other countries or regions of the world that have a similar soil type.
    c) Are there places in your local area where azonal or intrazonal soils might be found? Explain.
32. a) List the various great soil groups found in Canada.
    b) Using the information in this chapter on the great soil groups, draw up a chart that classifies each of these soils into one of three categories: Low Natural Fertility, Moderate Natural Fertility, and High Natural Fertility.

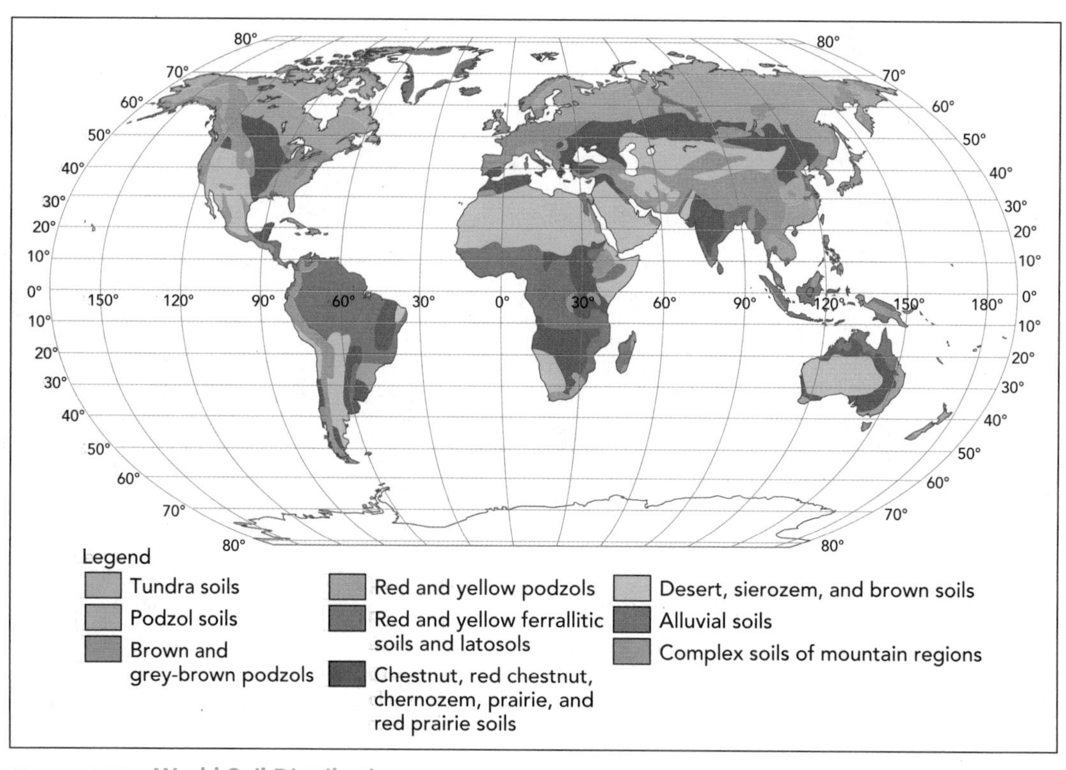

Figure 11.24 **World Soil Distribution**

Review

- The ecosphere is the zone of the earth within which all the major processes that affect life operate.
- Carbon, oxygen, and nitrogen are three key nutrients that cycle through the ecosphere and are used, along with sunlight and water, in the production of foods through the process of photosynthesis.
- The productivity and species of plants vary over the earth as a result of variations in climate and the availability of nutrients.
- Ecosystems are interactive communities of plants and animals.
- The major vegetation biomes of the world are groups of ecosystems that have developed under different climatic conditions.
- Soil is composed of air, water, organic material, and minerals.
- The varying characteristics of the four components of soil lead them to have different textures, structures, colours, organic compositions, air and water volumes, and soil chemistry.
- Mature soil profiles have layers or horizons created by the upward and downward movements of soil water.
- Leaching is the process by which soluble minerals are moved downward through soils by water.
- Soils are formed through the influence of climate, organisms, and plants, relief of the land, nature of the parent materials, and time.
- One classification system, used to analyse the distribution of soils around the world, has orders, suborders, and great soil groups.

Geographic Terms

biosphere
terrestrial biosphere
aquatic biosphere
ecosphere
biochemical cycle
nutrient
nutrient cycle
photosynthesis
respiration
autotrophic
heterotrophic
herbivore
carnivore
omnivore
ecosystem
food chain
biome
tundra
boreal
taiga
coniferous
schlerophyll
xerophyte
vertical zonation
soil
mineral matter
parent material
texture
organic matter
air
water
loam
ped
humus
cation
pH scale
horizon
topsoil
leaching
caliche
evapotranspiration
zonal
intrazonal
azonal

Related Careers

- biologist
- pedologist
- ecologist
- botanist
- geochemist
- agronomist
- laboratory technician
- conservationist
- silviculturalist
- land-use planner
- hydrologist
- landscaper
- horticulturalist
- farmer
- civil engineer
- naturalist
- park ranger
- biotechnician

Explorations

1. a) Find five examples of species of plants or animals that have become extinct.
   b) How might extinction of species affect ecosystems or affect you personally?
2. a) Find a detailed natural vegetation map of your local region to determine what type of natural vegetation is found there. You might use a national or regional atlas or a textbook dealing with the physical geography of your region or province. Use this source to draw up a chart summarizing the similarities and differences between the two descriptions.
   b) Explain why there are differences between the general world description and the one for your local region or province.
   c) What special conditions in the local region might account for these differences from the general world description?
3. a) Draw up a natural vegetation map of Canada and the United States, based on the world map in Figure 11.5.
   b) Using an atlas, road atlas, or encyclopedia, try to find at least three national parks in each of the natural vegetation regions for each of the countries.
   c) Based on the information on the map, how representative are the national parks of the major vegetation biomes of the two countries? Explain your answer.
   d) Based on your research, are there any locations where you would suggest further national parks be developed to more effectively preserve important vegetation biomes of Canada or the United States? Explain.

Natural Landscapes

UNIT

Tectonic processes create the major features of the earth's crust, powered by energy from below. At the same time, the processes of denudation, powered by energy from above, work to level the earth's surface by breaking down, eroding, transporting, and depositing the materials of the crust. As they carry out this purpose, the agents of running water, wind, ice, and waves use different processes to create distinct landscapes, each with its own unique set of surface forms.

As you work through this unit, consider these questions:

- What processes are slowly at work to break down the solid rocks of the crust?
- What evidence is there that "the eternal hills" are not eternal at all?
- What major factors determine the types of landforms found within a given landscape?
- How can the landscapes of the world be classified and understood?
- What source of energy powers the denudational processes operating on the earth's surface?

CHAPTER

12

DENUDATION: WEATHERING AND MASS WASTING

CHAPTER 1: The Nature of Physical Geography
CHAPTER 2: Earth: Its Place in the Universe
CHAPTER 3: The Earth in Motion
CHAPTER 4: The Earth's Interior
CHAPTER 5: The Earth's Crust
CHAPTER 6: The Lithosphere in Motion: Plate Tectonics
CHAPTER 7: Solar Radiation
CHAPTER 8: Climate
CHAPTER 9: Weather
CHAPTER 10: The Hydrosphere and the Hydrologic Cycle
CHAPTER 11: Natural Vegetation and Soil Systems
CHAPTER 12: Denudation: Weathering and Mass Wasting
CHAPTER 13: Distinctive Landscapes: Humid and Arid Environments
CHAPTER 14: Distinctive Landscapes: Glacial, Periglacial, and Coastal Environments
CHAPTER 15: Natural Hazards: Disrupting Human Systems
CHAPTER 16: The Disruption of Natural Systems
CHAPTER 17: Fragile Environments

OBJECTIVES:

By the end of this chapter, you will be able to:

- determine the meaning of the terms associated with the processes that work to lower the surface of the earth;
- understand that the end product of denudational processes is the general levelling of the earth's surface to sea level;
- apply a classification system for the two types of weathering, physical and chemical, and the various processes that operate under each type;
- understand the conditions that affect the rate and distribution of weathering processes;
- determine the nature of mass wasting and the influence gravity has on the movement of weathered materials.

Introduction

Energy from below pushes up high mountain ranges, transforming what were once ocean sediments into snow-covered peaks. The Rocky Mountains of North America stand in testimony to the constructive forces of such energy. But, there are other forces on our planet that slowly and surely work to reduce these mighty mountains to mere featureless plains. Two of these forces — weathering and mass wasting — are the subjects of this chapter.

12·1 Denudational Processes and Weathering

Landforms result from the uplift caused by tectonic forces and the processes of **denudation**, or levelling, that are constantly at work to lower the earth's surface to a common level. Uplift is powered by energy from below, while denudation is powered largely by energy from above. The processes of denudation work to reduce the land surface to **base level**, the lowest level to which a land surface can be eroded. Sea level is the ultimate base level, although there are desert depressions, such as Death Valley in California and the Dead Sea in Israel and Jordan, that lie below sea level.

At the present rate of denudation of the earth's surface, all the continents would be worn down almost to sea level in 10 to 12 million years. Only the constant uplifting of the rocks of the lithosphere by plate tectonics keeps the denudational processes from completing this task.

The processes of denudation can be divided into two categories:

1. **Degradation** includes the processes of weathering, erosion, and transportation. **Weathering**, the first step in denudation, is defined as the disintegration or decomposition of rocks in places on or near the earth's surface. **Erosion** is the removal and movement of rock debris and associated organic matter. **Transportation**, whether carried out by running water, ice, wind, or wave action, is an integral part of erosion.
2. **Aggradation** involves building up the land surface by the deposition of rock materials. **Deposition** occurs when a drop in energy slows the transporting agent to the point where it deposits some of its rock materials. The end result of degradation and aggradation is the levelling out of the land surface at, or close to, base level.

Figure 12.1 **The Effects of Weathering on Statues on the Facade of the Louvre in Paris, France** *Note:* The statue on the left has been restored.

We are aware of the most spectacular forms of denudation that occur. Events such as floods or major landslides often grab the headlines in newspapers or TV news broadcasts because they cause great death and destruction in very short periods of time. But, most denudational processes are so commonplace that they often go unnoticed.

Denudation occurs very slowly in terms of human experience. Such changes are revealed in subtle ways, such as the letters on a tombstone slowly disappearing over the decades, the fine detail on stone carvings gradually being eaten away by rainwater, the rusting of the family car, peeling paint on the woodwork of a building, and soil-laden streams and rivers during spring flood season. These reminders illustrate the constant changing of the face of the earth by weathering and erosion.

The processes of gradation and aggradation often operate so that long periods of little action are broken by short, very destructive episodes, such as major floods, wind storms, coastal storm waves, or the rapid advance of a glacier. Weathering, erosion, and deposition rates vary from place to place, and from time to time.

Weathering does not end when the rock fragments are carried away by agents of erosion. It continues while the rock particles are being transported until they are finally deposited in oceans, seas, or lakes.

The remainder of this chapter studies the processes of weathering and mass wasting in detail. Chapters 13 and 14 look at how erosion, transportation, and deposition have created the many types of landscapes we see over the earth's surface.

QUESTIONS

1. Create an illustration, model, or series of photographs to demonstrate the differences between weathering, erosion, and deposition.
2. Point out how the processes of weathering, erosion, and deposition are part of the hydrologic cycle. You might choose to answer this in the form of a labelled diagram.
3. a) Speculate on what the world would look like if no further denudation were to occur for the next:
   i) decade or so;
   ii) 25 million years or so.
   b) Identify the positive and negative effects of these changes over both the short and longer terms.
4. What evidence of weathering have you personally witnessed? What examples of erosion and deposition have you seen?

12·2 Weathering: A Key Process

Weathering reduces solid rocks to smaller and smaller particles. This breakdown is important because without weathering, water, wind, waves, and ice would be far less effective erosive agents. The weathering of rock material is also an essential step in providing the parent materials needed for soil formation.

The two major sources of energy for the breakdown of rock materials are:

1. solar energy, which powers chemical activity, the hydrologic cycle, and winds; and
2. gravitational energy, which brings down moisture as rain and causes stream flow and other downhill movements.

The intensity of weathering decreases with depth below the surface.

Physical Weathering

Weathering can be divided into two major divisions: physical (or mechanical) and chemical. **Physical weathering** involves the disintegration or fragmentation of rocks into smaller particles without chemical alteration of the minerals that make up the rock.

Frost Shattering

As water seeps into fractures and joints within rocks, it freezes into ice crystals and expands in volume by 9 percent, splitting the rocks apart and forming jagged and pointed boulders (Figure 12.2). This process, known as **frost shattering**, is most effective where temperatures fluctuate frequently about the freezing point. Some geographers refer to this as **ice wedging**.

Figure 12.2 **Frost Action Working to Weather Rocks on the Face of the Niagara Escarpment at Hamilton, Ontario**

Thermal Expansion

Rock with crystals of many different colours and sizes expands and contracts at different rates when heated and

Figure 12.3 Fractured Granitic Rocks in a Desert Area of California Thermal expansion is responsible for breaking up the rocks in this stretch of desert in southern California.

cooled. This process is particularly effective in tropical and mid-latitude deserts, where temperatures may range from 40°C during the day to near 0°C at night, causing the crystals to expand and contract day after day. After decades or centuries, exposed rocks develop weaknesses and disintegrate, forming a jagged carpet of rocks over the desert surface. Physical geographers refer to this weathering process as insolation weathering or **thermal expansion**.

Pressure Release

Erosion over millions of years may strip off overlying rocks to expose intrusive igneous rocks such as granite, formed under high pressure below the earth's surface. The release in pressure allows the rock to expand, causing cracks or fractures that run roughly parallel to the rock surface. The layers separate and fall away, releasing even more pressure. New cracks form in the recently exposed rocks and the process continues. The result is the formation of mountains with rounded slopes and tops, as shown in Figure 12.4. The terms "**exfoliation**" and "**sheeting**"are also applied to this type of rock disintegration.

Figure 12.4 Pressure Release of Granitic Rocks Pressure release, or exfoliation, is responsible for peeling these flat boulders from underlying solid granite in Yosemite National Park, California.

Animals and Plants

Plants and animals are constantly at work within the upper layers of rock. Under a humid climate, small fractures can be widened and the rock split apart by the gradual pressure created by growing roots. Animals dig burrows and expose more materials to weathering processes.

Crystal Growth

As water evaporates in a desert climate, it leaves behind small salt crystals in rock pores and fissures. As these salt crystals grow, they expand and exert high pressures on the surrounding materials. Soon rock fragments break off and fall away. This process is common in desert areas where springs wet the rocks at the base of sandstone cliffs. Caves, like those housing Mesa Verde, were formed in otherwise solid rock cliffs through this process.

Figure 12.5 **Plant Root Widens Rock Fracture** A tree root penetrating a small crack in the limestone bedrock of the Niagara Escarpment will soon split apart the rock.

Figure 12.6 **Caves Created in Sandstone Rock, Mesa Verde National Park, New Mexico**

Figure 12.7 **Solution Ridges in Rock** Hollows created by the dissolving of soluble minerals pit the surface of this limestone rock in the Stone Forest near Kunming, China.

Chemical Weathering

Chemical weathering occurs where rocks suffer decay or alteration because of a change in the chemical composition of the minerals that make them up.

Solution

As rainwater falls through the atmosphere, it dissolves small amounts of carbon dioxide (CO_2) gas, turning the droplets into very weak solutions of carbonic acid. The solutions react with soluble minerals, dissolving them and carrying them away in solution. The most soluble elements in rocks are calcium, magnesium, sodium, and potassium. Limestone, composed largely of calcium carbonate ($CaCO_3$), is one rock which is subject to solution weathering.

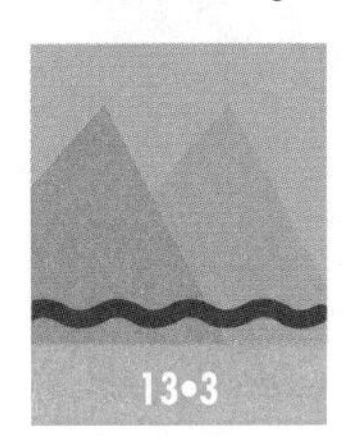

Oxidation

The process of combining oxygen with a mineral is called **oxidation**. Iron, a common element in rocks, often combines with oxygen to form a rust-coloured mineral called iron oxide. It shows up as a reddish brown stain. As iron oxide forms, it expands between other minerals, causing the rock to crumble and eventually fall apart.

Hydrolysis

Hydrolysis is a chemical reaction in which ions of water replace ions of a mineral. The reaction produces a new

Figure 12.8 **Hydrolysis of Granitic Rocks in the Sierra Nevada Mountains, Yosemite National Park, California**

clay mineral and soluble mineral compounds. These compounds are dissolved and carried away, leaving behind a weathered clay, known as kaolinite. This chemical reaction occurs in warm, moist climates in rocks containing feldspar or other silicate minerals.

Identifying Types of Weathering

It is possible to deduce the dominant weathering processes by careful observation of the rock fragments found in an area. Angular, broken rock fragments, with little change in colour or internal strength or consolidation, are typical products of physical weathering. On the other hand, rock materials produced by chemical weathering are often rounded, have altered surface colours compared to unweathered rock interiors, are less consolidated, and are likely to crumble when put under pressure.

QUESTIONS

5. a) In a group, make a list of the ways in which weathering affects your life, directly or indirectly.
   b) Organize your items into a three-column chart with the headings Positive Effect, Negative Effect, and Neutral Effect.
   c) Categorize each of these effects as either physical or chemical weathering.
   d) Identify the type of physical or chemical weathering each effect falls into, using the descriptions on pages 243 to 246.
   e) What conclusions can you draw from this analysis of weathering?
6. a) Devise and carry out an experiment to demonstrate one of the weathering types described on pages 243 to 246. Record the materials and the steps you use to carry out this experiment, along with the conclusions that you feel can be drawn from it.
   b) Indicate which of the examples of weathering you feel your experiment demonstrated. Give reasons for your choice.
   c) Evaluate the importance of this type of weathering in your local area, and its effects.
7. a) Find three photographs that show the effects of weathering.
   b) Name the types of weathering at work to create the effects.

It's a Fact...

An increase in soil temperature of 10°C doubles the rate of chemical reactions and thus the speed of chemical weathering. This is why soils in tropical areas develop faster and are deeper than soils in cooler, temperate latitudes, such as southern Canada or the northern United States.

12·3 Weathering: Variations Over Time and Space

The rates, depths, and degree of weathering vary from place to place and over time. The two most important causes of such variations are temperatures and precipitation: temperatures because they control the intensity of weathering activity and precipitation because it provides water, an effective agent in breaking down rocks and minerals.

Rate and Depth of Weathering

Weathering is greatest where high temperatures make abundant energy available and high levels of precipitation provide a

constant source of water. Such ideal conditions are found in tropical rainforest environments where weathering is rapid and deep and rocks are almost completely broken down chemically.

As temperatures and/or precipitation decrease, the rate and depth of weathering decrease. This is shown on Figure 12.9 by the shallow depth of weathering in both cold and hot desert climates. Even under the moist climates of mid-latitude forests, the depth of weathering is far less than in the moist, tropical climates, largely due to lower temperatures.

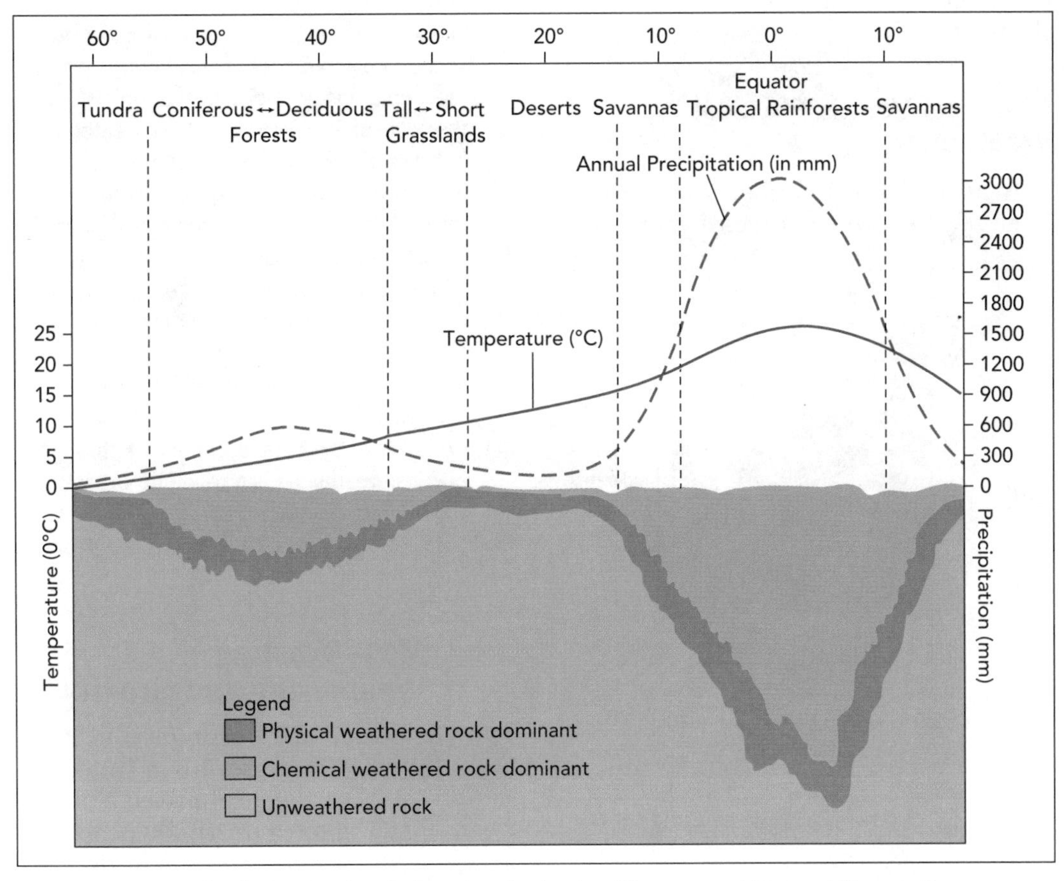

Figure 12.9 **Depth of Weathering Related to Latitude, Climate, and Natural Vegetation**

A second influence on the rate of weathering is the number of joints and bedding planes in rocks. Almost all rocks at the earth's surface have fractures known as **joints**. Joints are created by stresses and strains caused by tectonic processes such as folding, faulting, and tilting, and by the release of pressure when overlying rock layers are eroded. **Bedding planes** are parallel surfaces separating different layers of sedimentary rocks. The greater the number of joints and bedding planes, the higher the rate of weathering. This is because of the increased surface area available for weathering processes to operate and for water to penetrate below the rock surface.

Chemical Versus Physical Weathering

It is difficult to find places where only chemical or physical weathering is at work on exposed rock surfaces; nevertheless, there are environments where one or the other dominates. As Figure 12.10 shows, chemical weathering favours warm, moist climates; where temperatures are lower and water less available, chemical weathering works at a slower rate. Physical weathering, particularly frost shattering, is most effective in polar and high mountain regions where temperatures are low and precipitation moderate. Most environments have some combination of chemical and physical weathering.

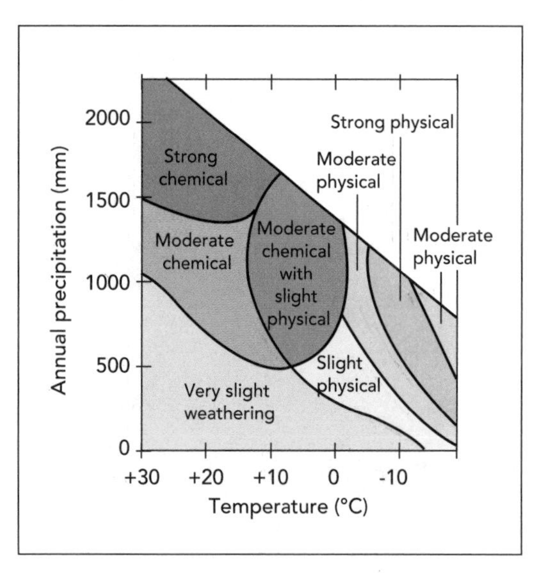

Figure 12.10 The Type of Weathering Related to Annual Temperatures and Precipitation

QUESTIONS

8. Explain how you, as an individual, contribute to the rate and extent of weathering.
9. Draw up a poster or display that points out the costs of an increasing rate of weathering because of human pollution of the world's atmosphere. Some themes you might research in your poster or display are: sources of pollution, economic impacts, human health, and destruction of the natural environment.
10. a) Describe the conditions under which weathering will be greatest.
    b) Under what conditions will weathering be weakest?
11. Explain how the breakdown of rock through physical weathering can speed up the rate of chemical weathering.

12•4 Mass Wasting: Weathering and Gravity

A landslide in Peru in 1970, triggered by a massive earthquake, buried Yungay, a town of 20 000 people, in a matter of seconds. The city of Armero, Colombia, and its 25 000 inhabitants were destroyed by a mudslide, set off by an eruption of Nevado del Ruiz volcano in 1985. A snow avalanche in the Swiss Alps almost killed Prince Charles, the heir to the British throne, in 1988. These are spectacular examples of **mass wasting**, a process defined as the downhill movement of weathered materials resulting from the pull of gravity. Besides the spectacular events described above, mass wasting includes less noticeable processes, like the slow, downhill creep of soil on a steep hillside, or the slumping of a cliff of weathered materials. The common forms of mass wasting are creep, flows, slides, and falls.

Creep

Especially on slopes in humid areas, there is a generally slow and unspectacular movement of loose rock and soil downslope. This movement is difficult to observe directly, but is shown by the downslope tilt of fences and stone walls and by tree trunks with a concave upward bend, as illustrated in Figure 12.11. **Soil creep** is helped along by expansion and contraction of the soil through freezing and thawing and wetting and drying.

Flows

Flows are common in humid climates and move like a thick liquid. The term "**solifluction**" is used to describe a slow, downhill flow of water-saturated rock and soil materials. Solifluction is common in

colder climates where the soil is ice-saturated and the ground beneath is permanently frozen. Flows occur in the narrow zone near the surface that melts during the summer months.

Where large volumes of rock fragments accumulate on mountain slopes and begin to move downhill, **rock glaciers** are formed. As Figure 12.12 shows, they flow much like ice glaciers, except they are composed largely of rock. They occur in

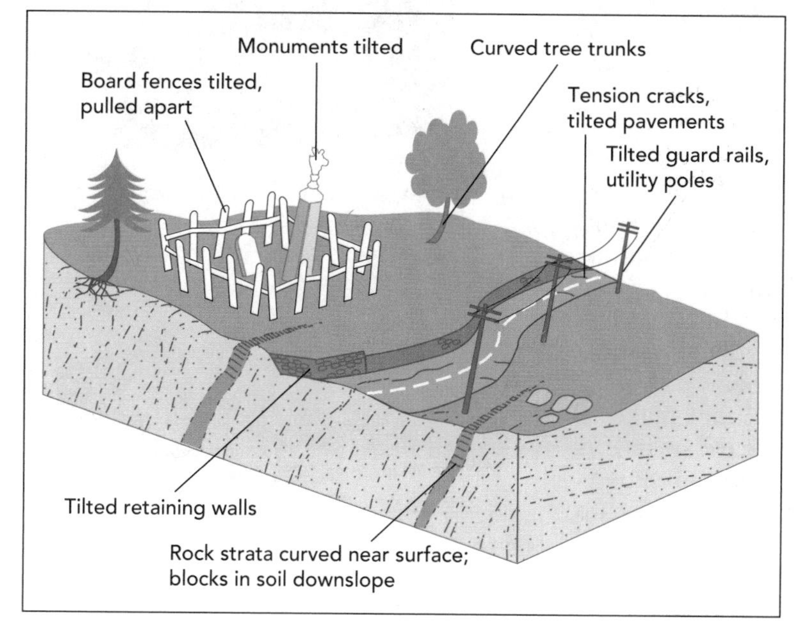

Figure 12.11
Soil Creep

Figure 12.12 A Typical Rock Glacier, Garhwal, Himalayas, India

the high mountains of humid regions where large volumes of rock are produced by frost shattering activity. Rapid flows of snow, ice, earth and/or rock masses are called **avalanches**.

Slides

The simplest form of slide is the **rock slide**. Rock slides occur in mountainous areas when large sections of rock are weathered along bedding planes. The rocks break loose and rapidly slip down to the base of the steep slope.

A block of soil and rock that makes a rotational slip along a concave surface is known as a **slump**. This form of slide is common where thick clay soils make up a cliff face. Figure 12.13 shows the appearance of a typical slump.

Debris slides are caused when loose rock and soil in steeply sloping terrain are shaken loose by an earthquake, or are undercut by river or glacial erosion. Such slides move rapidly downhill.

Falls

In high mountains, where frost shattering is very effective, individual rocks fall regularly. As they tumble down the steep

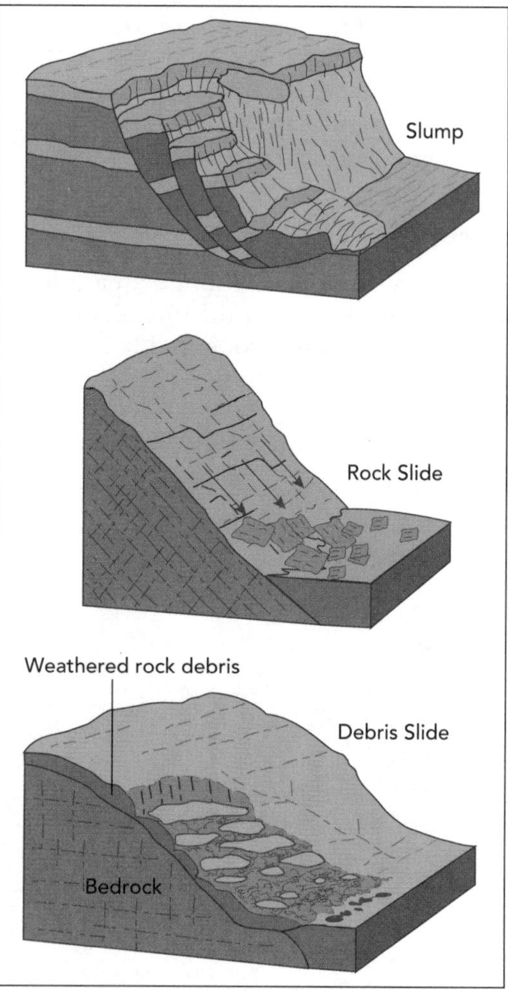

Figure 12.13 A Typical Slump, Rock Slide, and Debris Slide

slopes, they accumulate in fan-shaped piles of rubble known as **talus** or **scree**. An example of talus is shown in Figure 12.14.

Figure 12.14 A Talus Slope in the Rocky Mountains

Factors Affecting Mass Wasting

Mass wasting of weathered materials is related to two factors: friction and gravity. Any mass of rock or soil has friction — an internal resistance to movement. Water and ice have a lubricating effect that can decrease frictional resistance. When water and ice fill pores between grains of soil, the cohesion between the grains is reduced, allowing them to slide past one another. Once the resistance to movement is less than the pull of gravity, the mass of weathered material will begin to move downslope. Figure 12.15 shows the relationship between the intensity of mass wasting processes and climate.

The effectiveness of gravity is determined by the steepness of slope and the thickness of loose materials. The steeper the land surface, the greater the force available to move objects downslope. The greater the thickness of the moving mass of rock or debris, the greater the speed at which it is likely to move, everything else being equal.

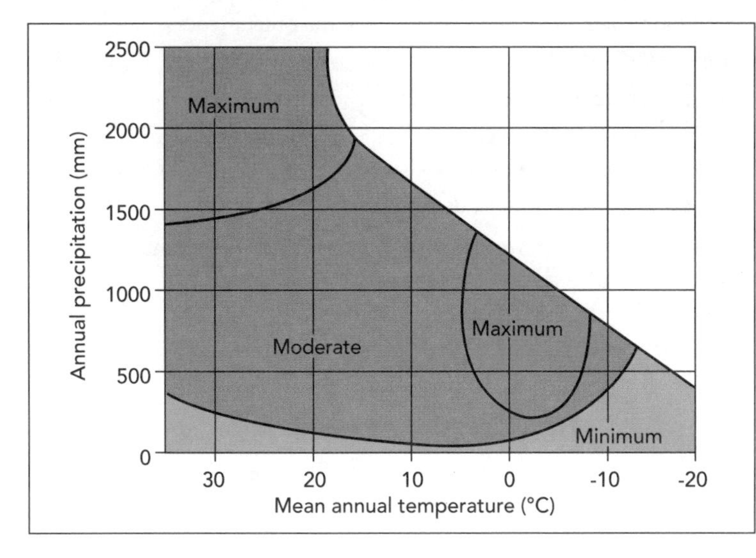

Figure 12.15 Climate and the Intensity of Mass Wasting Activities

QUESTIONS

12. Mass wasting activities lie somewhere between weathering and erosion processes rather than being a part of either of these two. Explain why.
13. a) Devise an experiment to illustrate the effects of water on the ability of soils to resist the pull of gravity.
    b) Carry out this experiment.
    c) Draw conclusions based on your observations.
14. a) Identify locations within your community or elsewhere where three to four examples of mass wasting might be observed. You might use topographic maps of your area to help identify possible locations.
    b) Point out the conditions used in identifying such locations.
    c) Where possible, photograph or sketch examples of mass wasting features in your community and include them in a display.
15. a) Identify landform and climatic conditions under which each type of mass wasting would be most active. Design and complete a chart to show this information effectively.
    b) Write a paragraph to explain the relationships among mass wasting processes and landform and climatic conditions.

Review

- Weathering is the breaking down of rocks through physical or chemical means into unconsolidated material. Erosion involves the movement of weathered rock materials to new sites.
- Weathering and erosion work to reduce the earth's surface to a base level.
- Mechanical weathering leads to fragmentation of the original rocks. Chemical weathering produces a chemical change in the original rocks, altering their chemical composition.
- The depth of weathering is influenced by such factors as temperature and the amount of precipitation.
- Weathering processes affect humans by breaking down or altering all types of materials and by modifying landscapes.
- Mass wasting processes, powered by gravity, move weathered materials downhill and are most effective in areas with steep slopes.

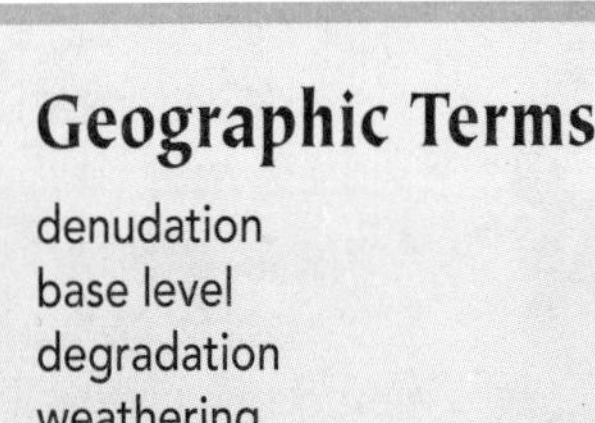

Geographic Terms

denudation
base level
degradation
weathering
erosion
transportation
aggradation
deposition
physical weathering
frost shattering
ice wedging
thermal expansion
exfoliation
sheeting
chemical weathering
oxidation
hydrolysis
joint
bedding plane
mass wasting
soil creep
solifluction
rock glacier
avalanche
rock slide
slump
debris slide
talus
scree

Explorations

1. a) Trace an outline of Figure 11.7: Natural Vegetation and Climate: How They Are Related (page 220). Mark in the various natural vegetation zones and the two scales of the graph.
   b) Using a different coloured pencil, mark in the dominant weathering types, as shown in Figure 12.10.
   c) Produce a chart to show the major type(s) of weathering found in each major natural vegetation type of the world as shown in Figure 11.5: World Pattern of Vegetation Biomes (page 219). For each type of vegetation, decide on a colour or symbol to show that type on a map.
   d) On a world outline map, draw in the patterns of dominant weathering, using your chart as a guide. Based on your world map, what type of weathering is dominant in your home region?
   e) Create a graphic that is more eye-catching than Figure 12.10 to show how weathering is related to temperature and precipitation changes.
2. Investigate ways in which your municipality combats the natural forces of weathering and mass wasting. Some topics you might wish to research are:
   - protection of historic buildings and sites
   - traffic regulations
   - construction standards
   - landscaping
3. a) Design a demonstration to show one of the following relationships:
      - the effect of angle of slope on mass wasting
      - the effect of saturation by water on mass wasting
      - the effect of fragmentation on the speed of weathering of rocks
   b) Produce a poster to show how the demonstration would work.

CHAPTER

13

DISTINCTIVE LANDSCAPES: HUMID AND ARID ENVIRONMENTS

CHAPTER 1: The Nature of Physical Geography
CHAPTER 2: Earth: Its Place in the Universe
CHAPTER 3: The Earth in Motion
CHAPTER 4: The Earth's Interior
CHAPTER 5: The Earth's Crust
CHAPTER 6: The Lithosphere in Motion: Plate Tectonics
CHAPTER 7: Solar Radiation
CHAPTER 8: Climate
CHAPTER 9: Weather
CHAPTER 10: The Hydrosphere and the Hydrologic Cycle
CHAPTER 11: Natural Vegetation and Soil Systems
CHAPTER 12: Denudation: Weathering and Mass Wasting
CHAPTER 13: Distinctive Landscapes: Humid and Arid Environments
CHAPTER 14: Distinctive Landscapes: Glacial, Periglacial, and Coastal Environments
CHAPTER 15: Natural Hazards: Disrupting Human Systems
CHAPTER 16: The Disruption of Natural Systems
CHAPTER 17: Fragile Environments

OBJECTIVES:

By the end of this chapter, you will be able to:

- use a classification system to organize the landform features created in various environments;
- recognize the impact that water has in shaping the surface of the earth in humid and arid environments;
- explain the processes that lead to erosion and deposition of materials by water and wind action;
- understand the forces that shape typical features found in humid and arid landscapes;
- apply a useful classification system to observe a cycle in river development;
- recognize the role of water in creating karst landscapes;
- develop an understanding and respect for the power of natural processes.

Introduction

Many of the features of the earth's surface seem to have little order or pattern to them. Each seems to be unique. At a local scale, this is probably the case. However, it is possible to see repetitions and patterns. In studying the buildings in a community, if the focus is on how they are different, then each building will seem unique. On the other hand, if the focus is on how the buildings are similar or related, many ways of classifying them emerge. Their uses, ages, architectural styles, shapes, building materials, heights, and so on can become the basis of a classification system.

The same is true of landscapes in Physical Geography. Order can be imposed on the chaos of individual landscapes by emphasizing common or shared, rather than unique, characteristics. It is then possible to question why such landscapes developed and what processes created them. Many landscape classification systems have been developed to do this. One of these is based on the dominant environment under which the landscape developed. This chapter will investigate the landform features created under humid and arid environments.

13·1 A World of Different Landscapes

Morphology is the study of the structure and form of phenomena. When applied to earth's landforms, it becomes **geomorphology**, meaning the scientific study of the form and structure of the landforms that make up the earth's surface.

In general, landforms fall into the morphological systems category and result from processes operating within cascading systems. All cascading systems require energy, which can come from above (solar energy) and from below (geothermal energy). Energy from above helps to shape the land through climate and, indirectly, through natural vegetation, weathering, and soils. Underlying rock types and structures and the uplift of parts of the earth's crust through tectonic activities show that energy from below is also involved in landform creation.

Landform classification schemes can be based on any number of characteristics, for example, the important forces that created the landforms, elevations, or the ages of materials. This chapter uses a classification scheme based on the major climates of the world (Figure 13.1), since climate has a major influence on the creation of landforms. A systems point of view will be used to show how climate, natural vegetation, soils, weathering, rock types, and structures interact to give rise to different landscapes.

The Major Landform Regions of the World

The major landscapes discussed in Chapters 13 and 14 are:

1. Humid Landscapes
   - Running Water and the Work of Rivers
   - Karst Landscapes
2. Arid Landscapes
3. Cold Landscapes
   - Glacial Landscapes
   - Periglacial Landscapes
4. Coastal Landscapes

In humid and arid landscapes, the amount of precipitation is the important variable. Temperatures are more important in glacial and periglacial landscapes. Coastal landscapes are unique since many variables operate on coastlines, including climate, rock types, and the size of the water

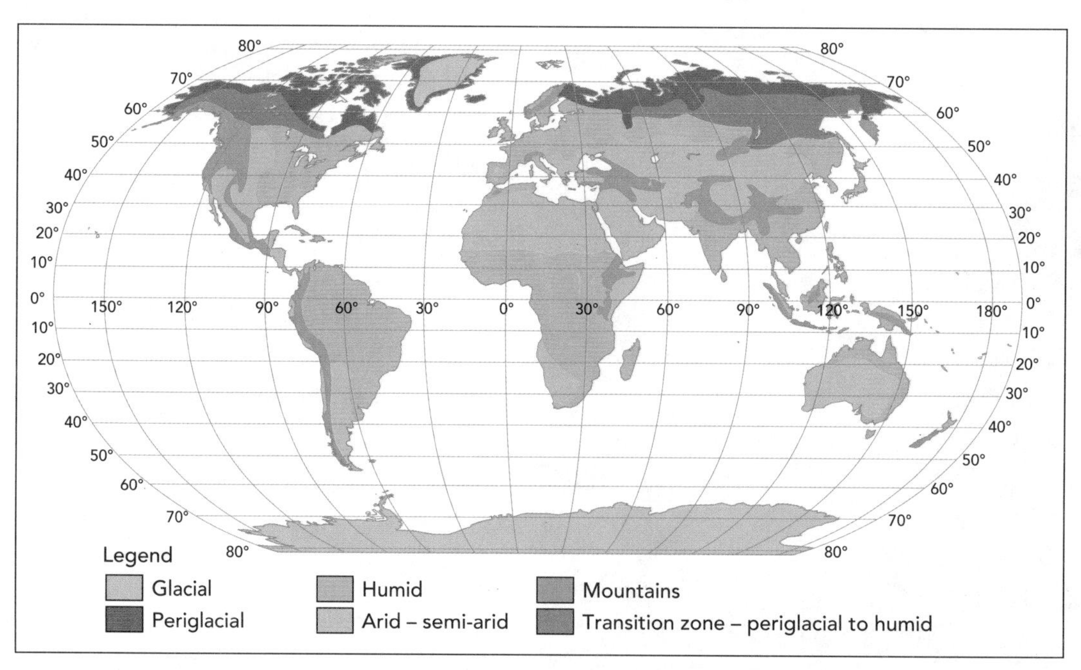

Figure 13.1 World: Landform Regions This map shows a system used to classify landforms based on the major climates of the world.

body. The special factor is the action of waves which only work on the edges of large bodies of water, particularly oceans, seas, and lakes. No other erosive agent has such a limited range of activity as wave action.

Humid and arid landscapes are the subjects of this chapter and glacial, periglacial, and coastal landscapes are considered in Chapter 14.

QUESTIONS

1. a) Make a list of special or interesting landforms that you have seen in pictures, in travels, or on film or television.
   b) Organize the items in your list into the categories that will be used to study landforms in these chapters.
   c) Keep this list handy as you work through this chapter and see how many of these features you can explain after completing your studies of the world's major landform regions.
2. a) Devise an alternative way of classifying the earth's landforms, other than by major climatic influences.
   b) Explain your classification method and describe a typical landscape for each category.
   c) Draw a diagram to show at least two cascading systems that work to influence the landforms of a region.
3. While most geomorphological processes occur slowly over a long period of time, it is still possible to observe small changes in the landscape around us.
   a) Work in groups to identify and describe three instances of landform change that people can witness.
   b) Describe how human actions can "speed up" the natural processes that wear away at the earth's surface.
4. a) Define the term "morphological system", as outlined in Chapter 1, page 8.
   b) Point out why landforms fall within this type of system.

13·2 Humid Landscapes: The Work of Rivers

The effect of water working relentlessly on the earth's surface is a major factor in landform development. Landscapes of humid climates make up a majority of the world's land surface. While all areas have abundant precipitation, there is a great variation in temperatures within this category, ranging from northern continental climates with long, cold winters to the tropical climates with year-round hot temperatures.

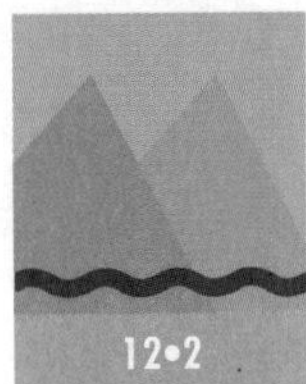

Chemical weathering is dominant due to the availability of moisture and warm summers. Most areas are underlain by weathered rock materials that are easily eroded by **run-off** where exposed. Biologic action transforms the weathered rock materials into soils containing various amounts of humus, plant roots, organic matter, and animal life. The humus layer occupies the first few centimetres of the soil surface, giving way to newly weathered rock below.

An important characteristic of humid landscapes is a surplus of water. As water droplets reach the ground surface, their fall is broken by the vegetation cover. The moisture accumulates on the leaves and falls to the ground, where it is soaked up by the soil. When the soil reaches its saturation point, the moisture begins to flow over the ground surface. Its progress is slowed by grasses, shrubs, and trees, minimizing, although not entirely preventing, soil erosion. The slow movement of run-off means the flow of streams and rivers is regulated to some extent.

Drainage Basins: Organizing the Run-off

A **drainage basin** is the area of land from which a stream gets its water supply. In mountainous areas, the "divides" between drainage basins are along the peaks and are easy to define. In lowland areas, the dividing lines between basins may be very indistinct and hard to locate accurately.

Within the basin, the run-off is collected in small, shallow pathways. These small rills in the soil are enlarged over time into rivulets and, eventually, into stream channels. These channels are interconnected and organized over the whole drainage basin, as shown in Figure 13.2; smaller rills and rivulets come together to form increasingly larger streams. The size of these channels depends on the amount of water that is collected by the series of channels upstream. The more water, the greater the ability to erode material.

The channels eroded by these streams carry the water from the drainage basin to a larger river, or to a lake, sea, or ocean.

Within a drainage basin, the shape of the land and the resistance of the bedrock to erosion influence the pattern of stream channels. In level areas with uniform bedrock, the dendritic pattern (Figure 13.3a) is most common. It has the same pattern as the veins of a leaf. Where the land surface is irregular due to differences in bedrock or mountain-building activities, stream channels develop into trellis (Figure 13.3b), rectangular (Figure 13.3c), and annular (Figure 13.3d) patterns. Where domes or volcanoes occur, the pattern often takes on a radial form (Figure 13.3e).

Frequency and Velocity of Stream Flow

Streams can be classified into three types: **perennial streams** that contain water for most of, or all, the year; **intermittent streams** that flow only part of the year and are common where there are distinct wet and dry seasons; and **ephemeral streams** that flow only rarely and are found in desert regions.

The amount of water in a stream is one factor that influences the velocity of

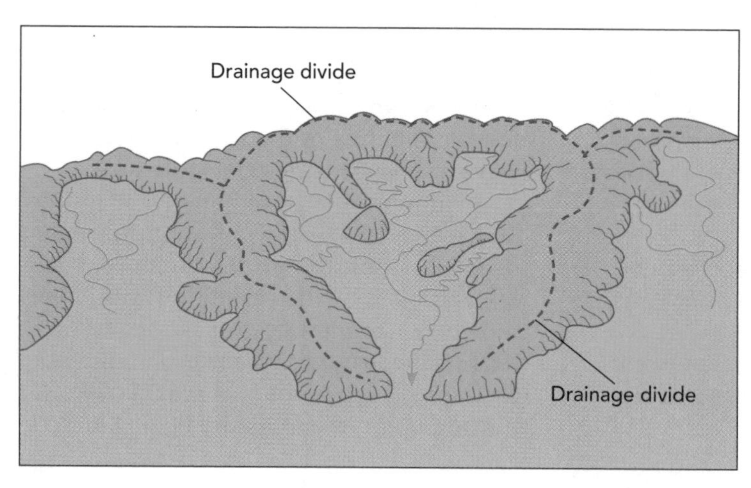

Figure 13.2
A Typical Drainage Basin

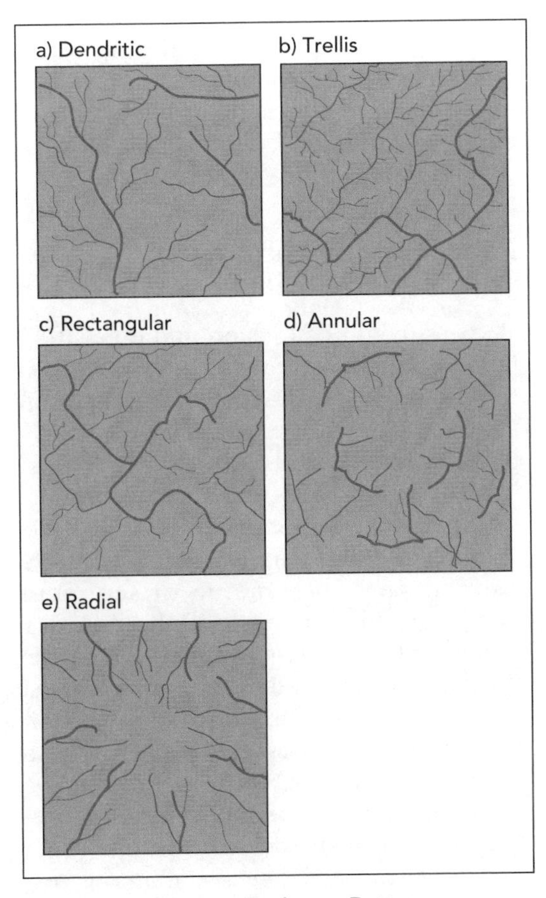

Figure 13.3 Stream Drainage Patterns

water. Other factors include the gradient or slope of the stream channel and the "friction" created by obstacles in the stream. The velocity of water will vary from one section of a stream to another. Within the channels, the water flows most rapidly in the top, centre part of the stream. This is because it is farthest from the friction and turbulence created by the bed and sides of the channel.

Stream Erosion

Streams in a drainage basin carry water to an ocean, sea, or lake. A by-product of this water movement is the erosion of soil and rock materials in the drainage basin. Water erodes material from the soil surface before it reaches a stream channel, and from the bed and banks of the river.

Water erodes through four main processes, which usually work together:

- **Abrasion** occurs when the water is "armed" with rock fragments or particles that rub, scour, or scrape rock or soil surfaces as they are carried by the stream flow. They break off or loosen rock materials that are then added to the stream's armament.
- **Attrition** occurs when rock particles within the stream collide and break one another down into smaller particles. The more rock materials that a stream contains, the more important this process becomes.
- **Corrosion** is the dissolving of soluble minerals into the stream water itself. Weak acids from dissolved carbon dioxide and other gases, from the decay of plants and other organic materials, and from acidic chemicals added to streams as the result of human action make this process more effective. Corrosion is particularly important in humid areas underlain by limestone rocks which react easily with acids.
- **Hydraulic pressure** refers to the high pressure that rapidly flowing water exerts on rock surfaces, especially in areas where rapids or waterfalls are found. This pressure weakens and eventually breaks down rocks.

Stream Transportation

The ability of a stream to carry rock and soil particles is measured in two ways: by its competence and its capacity. **Competence** is determined by the largest size of particles that can be carried, while stream

capacity is the maximum amount of particles of a given size that a stream can carry. Both of these are related to the amount of water in the stream (its discharge), water velocity (related to the slope of the stream channel), and the size of the particles available for eroding by the stream.

Streams carry their sediment loads in the ways shown in Figure 13.4. The largest materials that can be moved are rolled along the stream bed in a process called **traction**. **Saltation** occurs when sediments are bounced along the bed of the stream by the flow of the stream. **Suspension** occurs when the sediment load is carried within the stream water by turbulent flow. Dissolved minerals in the water are carried in **solution**. Soluble rocks such as salt and limestone are commonly transported in this way.

Stream Deposition

Sediments picked up by streams eventually will be deposited. The term **alluvium** is used to refer to stream-deposited sediments, such as sands, clays, silts, and gravels. Alluvial deposits are usually layered or stratified as they are laid down by streams. As the velocity of a stream decreases, its competence also decreases. As this happens, any particles above the competence level are deposited. Larger rock particles are deposited first, then smaller particles. Alluvium is found in places where the water velocity is reduced, such as where streams enter larger, slower moving bodies of water and form deltas. The individual particles of sediment in alluvial deposits have usually been rounded by the action of attrition and abrasion.

Reaching Equilibrium: The Graded Stream

The end result of erosion and deposition is the gradual lowering of the stream bed to its **base level**. Base level, the elevation at which a stream enters the ocean, sea, lake, or another stream, is the lowest level to which a stream can erode its channel. A stream constantly adjusts its channel and gradient to reach an equilibrium so that the amount of sediment being carried matches its ability to carry material. Once it reaches such a balance, it is known as a **graded stream**. For most streams in this condition, the longitudinal profile, a cross-sectional view of a stream from its source to its mouth, is a smooth, concave curve. The graded profile is usually steepest in the upstream section because the stream contains less water and is less efficient in eroding and transporting materials. Near its mouth, a graded stream is at its base level. At this time, the stream has reached a state of balance, or equilibrium, among its discharge, velocity, channel shape, sediment load, and gradient. A change in any of these factors will cause adjustments in the other factors in order to restore the balance of the stream system. For exam-

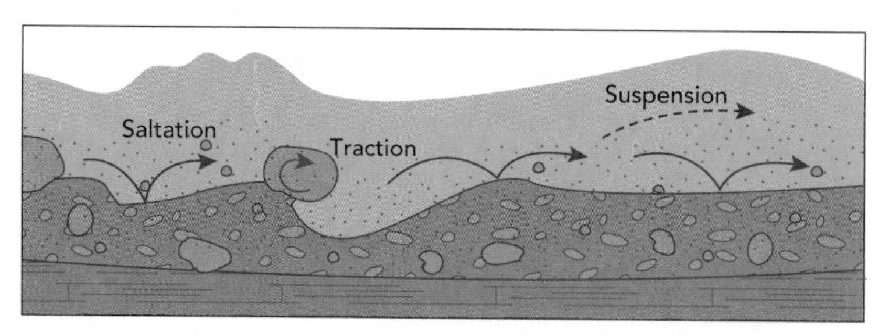

Figure 13.4
Methods of Transportation of Sediments by Streams

ple, if there is an increase in the volume of water flowing in a stream over an extended period of time, there may be a change in either the gradient, channel size and shape, and/or the velocity. These changes will allow the stream to achieve a new balance or equilibrium.

Streams with lakes or waterfalls along their courses have several local base levels, particularly in their upper reaches where they are cutting directly into rocks of different hardnesses.

Land and Water Features of Streams

The channels occupied by streams vary greatly from one part of a stream to another. There are, however, some features that are common to many streams.

Meanders

Meanders are broad, curving bends in streams. They are one of the most distinctive features of streams and usually occur where streams flow across level plains or lowlands. Meanders develop in streams of all sizes; the common requirement for development, besides a gentle slope, appears to be a bed of granular sedimentary materials, such as coarse sand or gravel. The size of meanders is related to the amount of water in the streams and the widths of the stream channels.

The process by which meanders form is shown in Figure 13.5. Because of centrifugal force, the depth and velocity of water are greatest on the outside of a river bend, and it is here that erosion occurs. On the inside of the bend, the water depth and velocity are at a minimum, so some of the sediment load is deposited to form a **point bar**. The gradual sideways growth of the river bend creates a meander curve and widens the level area on either side of the river by a process called **lateral erosion**. At the same time that this is occurring,

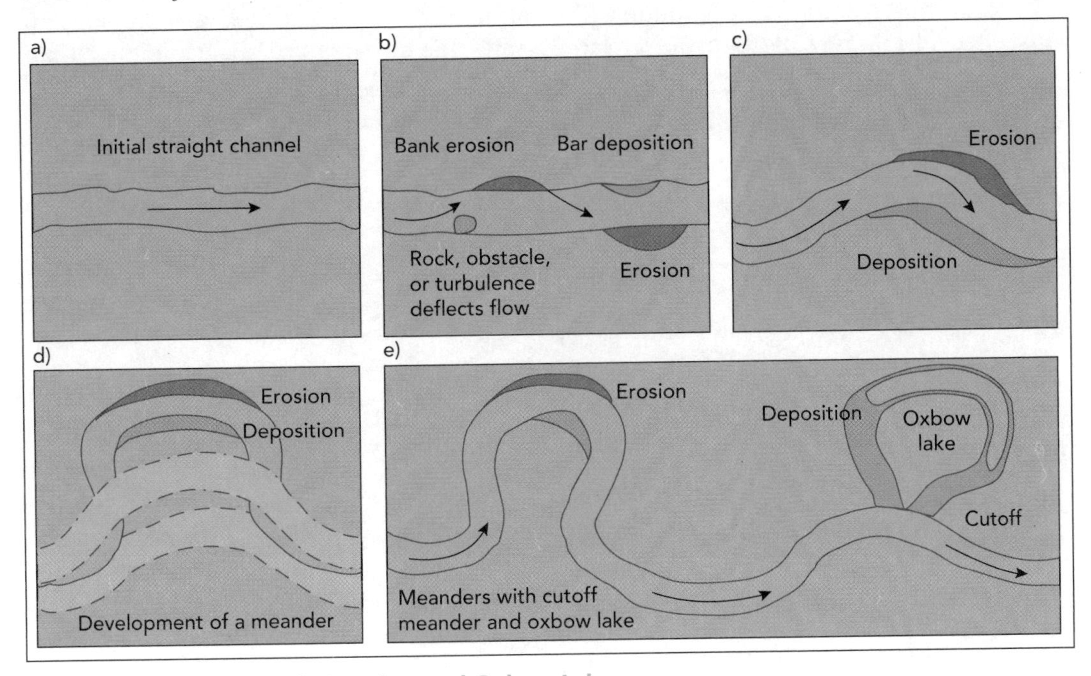

Figure 13.5 Formation of Meanders and Oxbow Lakes

erosion on the lower, outside part of the meander curve results in the gradual migration of the meander downstream towards the mouth.

As the meander continues to develop, it begins to narrow the neck of land at the base of the meander curve, as shown in Figure 13.5. Eventually, the neck of land becomes so narrow that the stream cuts through it, usually during a flood. After the cutoff has taken place, silt collects across the mouth of the abandoned stream channel, leaving the meander isolated from the main stream. Because of its shape, this body of water is known as an **oxbow lake**.

Stream Flood Plains

When streams flow over level plains or are close to base level, they often develop **flood plains**, which are level, gently sloping plains along the banks of streams that are subject to flooding. Meanders play a key role in the creation of flood plains, as illustrated in Figure 13.6. Because of lateral erosion, the meanders erode the stream banks creating a level plain, just above the elevation of the stream itself. Since the size of meander curves is related to the volume of water in the stream, the larger the stream flow, the larger the meander curves, and the wider the flood plain.

Typical flood plains develop a number of natural features, as illustrated in Figure 13.7. **Levees** are deposits of sediments along the main channels of rivers. They are formed during flooding, when the water overflows the main channel onto the flood plain. Here, since it is shallower, the velocity of the water is reduced, and the sediment load is deposited. The greatest deposition is next to the stream channels; deposition decreases as the water moves farther and farther over the flood plains. As a result, the flood plains slope very gradually downward from the stream channel to the edges of the flood plains.

In extreme cases, the levees are built up until the streams are slightly higher than

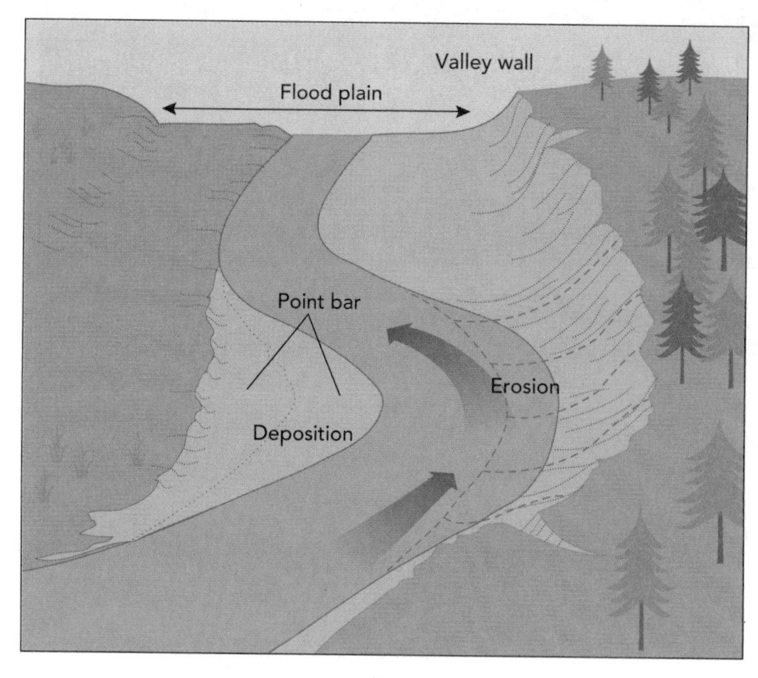

Figure 13.6
The Role of Meanders in Flood Plain Formation

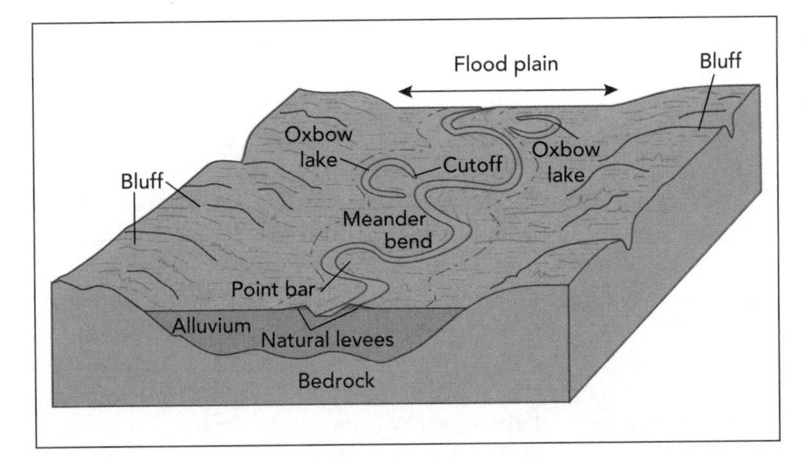

Figure 13.7
Features of a Typical River Flood Plain

large parts of their flood plains. Backswamps form at the lowest part of the flood plains, usually along the outer edges.

Braided Streams

Streams that develop a series of small, shallow, interlaced channels, separated by sand bars, are known as **braided streams**. Braiding is a product of the amount of debris carried by the river. When the stream's sediment load exceeds its capacity and competence, the sediments are deposited and the water is forced to flow through shallow channels and take a winding route. This gives the appearance of a braid.

Braided streams are common in arid and semi-arid regions of low relief. Under such climates, run-off is intermittent, only occurring during heavy rainstorms. During such storms, the sediment load is very large since there is little vegetation to reduce soil erosion. As the streams flow away from the storm areas, evaporation and seepage quickly reduce their flow. This reduces their capacity and competence to carry sediments, which are dropped in the channels, creating sand bars and shallow, interlaced channels typical of braided streams.

Streams flowing from the snouts of melting glaciers onto level plains or valleys also develop braided channels. Here, seasonal differences in flow are a result of variations in temperatures, rather than in precipitation.

It's a Fact...

When natural levees allow a river to rise higher than its flood plain, great potential for damage can occur. This is true of the lower reaches of the Mississippi River in the United States and the Huanghe (Yellow River) in China.

Deltas

Flood plains often merge into deltas where streams enter lakes, seas, or oceans. **Deltas** are level areas of alluvial deposits formed at the mouths of streams. They are created by streams that are carrying large loads of sediment, and that are flowing into relatively shallow seas that are protected from strong waves, tides, and water currents. Deltas take many shapes and forms depending on local conditions of stream deposition and water currents. The channels that carry the river water across the deltas are called **distributaries**.

Alluvial Fans

Alluvial fans are deposits of sediment formed where fast-flowing streams carrying large amounts of sediment flow abruptly onto level plains, valleys, or basins, as illustrated in Figure 13.13 (page 270). Fans develop where narrow mountain stream channels spread out over newly forming fans, causing a drop in stream velocity and depth and resulting in the deposition of sediments. Evaporation and seepage into the coarse sediments of the fan further reduce stream flow and increase deposition. Deposition in the stream bed causes the stream channel to constantly shift from one location to another across the face of the fan. Where several fans grow together, they are called **bahadas**. Alluvial fans are most commonly found in areas of arid to semi-arid climates, but occur in humid mountain environments as well.

Classification of Rivers

As discussed on pages 255-56, classification systems are useful for bringing order to natural systems so that we can understand them more easily. A common classification system for rivers and their immediate valleys was developed in the early twentieth century by a geographer, William Davis. He suggested that, like living things, landforms evolve through a series of stages from early forms (youth), to intermediate conditions (maturity), to a final stage (old age). However, unlike living things, rivers can be rejuvenated by geologic processes. Davis called this scheme the "Geographical Cycle". It gives a dynamic view of the earth's surface and the landforms that develop and change through time as a result of erosion by running water. Although many aspects of this cycle have been questioned by researchers since Davis first devised it, it is still a useful system for studying rivers.

Youth

In the initial stages of the Geographical Cycle, erosion is the dominant process. **Vertical erosion** allows the rivers to cut downwards into the land to form V-shaped valleys like those in Figure 13.8a. **Headward erosion** occurs where streams are cutting backwards into higher and higher ground, lengthening their channels. The upland areas between the narrow river valleys — the interfluves — are wide and undissected, with areas of poor drainage, such as swamps and small lakes. Stream profiles are steep, with rapids and falls where the rivers cross beds of resistant rocks. The streams are turbulent and their paths irregular, as large rocks, boulders, and other obstructions occupy the stream bed.

Maturity

By the mature stage, the drainage system has developed over much of the river basin. Rivers are eroding vertically and deepening their valleys, but meanders are beginning to form through lateral erosion, as shown in Figure 13.8b. This process widens the floors of the valleys and creates narrow flood plains, giving the valleys more of a narrow U-shape. Few of the rapids and waterfalls remain and the river profiles are smoother. Most of the interfluves are now dissected by stream channels and those that remain are very narrow. The landscape has good to excellent drainage.

Old Age

At this stage of development, according to Davis, the most important processes are lateral erosion and deposition of sediments carried from upstream. The flood plains are wide and marked with large meander curves, oxbow lakes, backswamps, and well-developed natural levees. Deposition of sediments carried from upstream dominates over erosive activity. The streams now have graded profiles and their channels are wide and smooth, with an absence of rapids and waterfalls. Well-developed deltas are found at the mouth of the streams where they enter large bodies of water.

Rejuvenated River

Davis suggested that uplift of the land could change the old-age landscape, causing the rivers to begin eroding vertically once again. Or, there could be a drop in the base level of the stream, such as a drop in sea level or lake level. This would cause streams to begin downcutting once again. In effect, the region would return to the youthful stage and work its way, over thousands or millions of years, through the various stages once again.

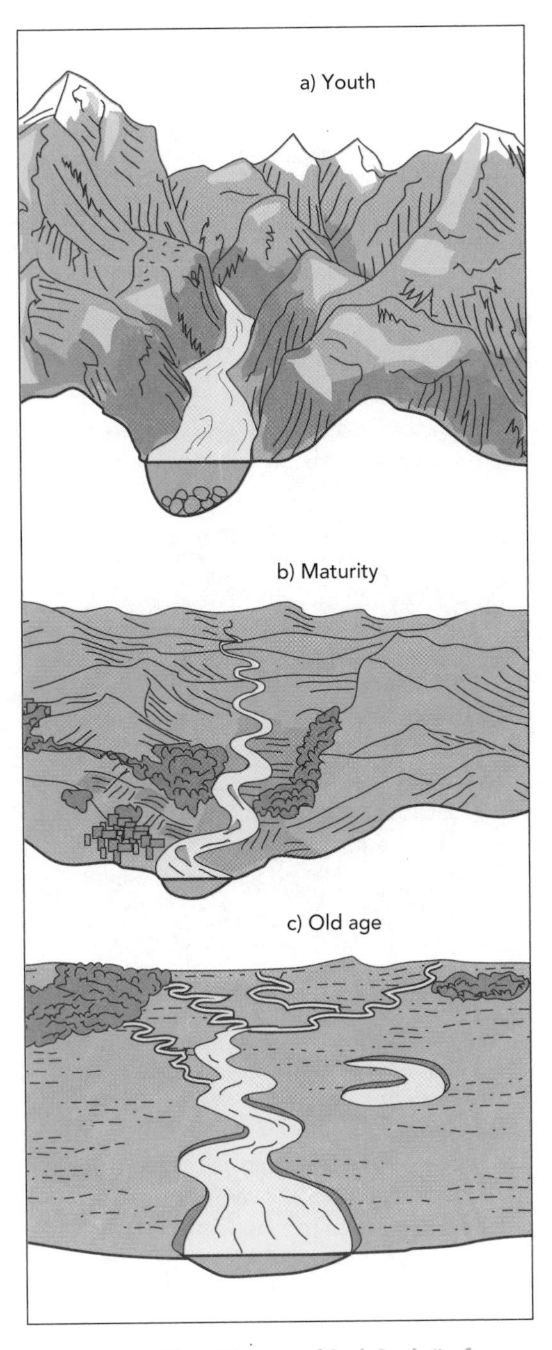

Figure 13.8 The "Geographical Cycle" of Erosion: Youth, Maturity, and Old Age

It's a Fact...

On average, the world's rivers deposit 20 billion tonnes of sediment in the oceans every year. This rate is equivalent to lowering all the earth's land masses by 3 cm every thousand years.

QUESTIONS

5. Create a diagram to show the factors that influence the amount of water that flows from a drainage basin.
6. In an atlas, find an important river in your region or country.
   a) Name the river and the political areas through which it flows.
   b) On a piece of clear plastic, trace the shape and extent of the drainage basin and the main stream channels.
   c) Calculate the length of the main river and its major tributaries.
   d) List the three most important factors that influence the size and volume of water in the river.
   e) Describe ways in which humans have influenced the river and the area through which it flows.
7. Draw a labelled diagram to show the place rivers and streams occupy in the hydrologic cycle, discussed on pages 196-198.
8. a) In your own words, describe each of the four processes by which streams erode material.
   b) Describe ways in which you use the same processes in your life.
9. Make up a chart to compare the four ways in which sediment load is carried by streams. You may choose to consider these characteristics in your chart:
   - method of movement
   - particle sizes
   - required velocity
   - amount of material moved

   Add one or two other characteristics of your choice to the chart.
10. a) Explain why deposits of alluvial materials are usually layered, or stratified.
    b) In what order would clay, gravel, and sand be deposited by a stream? Explain.
11. a) Explain why many municipalities prohibit the construction of homes and other buildings on stream flood plains.
    b) What are some appropriate land uses for flood plains?
12. Devise and carry out an experiment to illustrate how braided streams or meanders develop.
13. Draw up a chart to compare rivers in each of the stages of the Geographical Cycle. Some characteristics you might include in your chart are: size of streams, direction of erosion, dominant processes, slope of the rivers, typical landforms, and uses of the rivers.

13·3 Karst Landscapes

Humid areas with limestone bedrock develop special landform features known as **karst landscapes**, named after a limestone region in Slovenia. The conditions necessary for the development of karst landscapes are beds of thick, hard, fractured limestone at the earth's surface, a humid climate, and fairly high relief (mountains, foothills, or plateau areas).

The most characteristic surface features of karst landscapes are the many pits and hollows formed when water dissolves soluble limestone rocks. Streams often enter fissures in the rocks and disappear underground. In many karst areas, most of the drainage is underground, with few surface rivers or lakes despite the humid climates. Caves, with their spectacular stalactites and stalagmites, are part of the underground features created in karst landscapes.

Karst Landforms

Water, moving over limestone rock surfaces, creates **solution furrows** or **rills**, a few millimetres to several metres in length. The rills drain into hollows on rock surfaces. **Rainpits** form where drops of rainwater dissolve hollows into level rock surfaces. As these pits grow larger, they often hold water in small pools. Deep cracks, or **grikes**, from 15 to 60 cm in width, are created by solution along fractures or weaknesses in limestone rocks.

Where limestone is dissolved over a long period, large, enclosed, cone- or bowl-shaped hollows, known as **dolines**, form where streams drop into underground cavern systems. Others are formed by the collapse of caves below ground level. Dolines often occur in groups, giving a pitted appearance that is a characteristic feature of karst landscapes. The photo in Figure 13.9 shows a typical doline-studded karst landscape. As the dolines grow together, they create large, flat-bottomed and steep-sided depressions known as **uvalas**. Where uvalas grow into large basins and valleys, they are known as **poljes**.

Figure 13.9 **Pitted Land Surface of a Karst Landscape, Slovenia**

Many karst drainage features cannot be observed easily since they are underground. Rivers commonly sink through cracks in the limestone and flow completely, or partially, below ground, leaving behind **dry valleys**. Only during periods of heavy rainfall will water flow in these "dry" valleys. Where they occur above ground, rivers usually occupy deep, steep-sided gorges, since the water cuts down quickly into the limestone through solution and abrasion. Underground rivers and water continue to dissolve limestone as they travel through the rocks. As time passes, the fissures are enlarged to form underground **cavern systems** that can extend for kilometres.

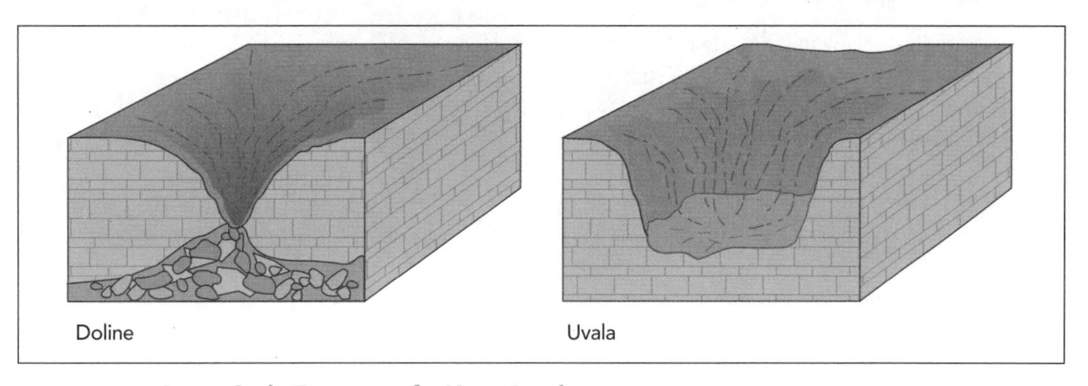

Figure 13.10 **Large-Scale Features of a Karst Landscape**

It is only in caverns that depositional features are to be found with any regularity. Deposits of calcium carbonate ($CaCO_3$) — also called calcite — occur where water seeps out of the rock into an open cavern. The two most common depositional features are stalactites and stalagmities. **Stalactites** are calcite tubes or cones that grow downward from cavern ceilings. **Stalagmites** grow upward from the cavern floor and are usually much wider than stalactites. Many famous caverns have spectacular displays of stalactites, stalagmites, and other unusual depositional features, made up of calcium carbonate. One example is illustrated in Figure 13.11.

Figure 13.11 **Underground Cavern, Carlsbad Caverns National Park, New Mexico**

It's a Fact . . .

The world's largest known natural underground room is the Sarawak Chamber, found in the Good Luck Cave in Mulu National Park in Borneo (Kalimantan), Indonesia. It is 700 m long, 400 m wide, and never less than 70 m in height and could hold 17 football fields!

QUESTIONS

14. Describe how you would determine if the region in which you live, or through which you are travelling, is a karst landscape.
15. a) Explain why humid climates are necessary for the development of karst landscapes.
    b) Explain what impact air pollution and acid rain would have on karst landscapes.
16. Explain why karst landscapes are:
    a) potentially dangerous or difficult areas in which to build urban areas;
    b) often excellent areas to develop for tourism.

13•4 Arid Landscapes

Arid landscapes develop where the rate of evaporation is greater than the annual precipitation. Precipitation, when it does come, is often a result of convectional rainfall and is brief and intense. This means that only narrow bands of the landscape receive rainfall from a given storm, and specific locations receive rainfall infrequently. In some cases, rain may not occur for a year or more. Due to a sparse vegetation cover and a hard, rocky surface, very little rain soaks into the ground. Most of it quickly runs off, creating flash floods that are responsible for much of the erosion of desert landscapes. Large amounts of material can be eroded and moved under flash flood conditions.

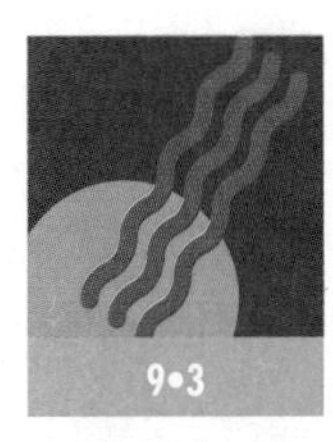

Because of limited precipitation, physical weathering is relatively more important in arid environments than in more humid climates. The main physical processes are heating-cooling, freezing-thawing, and exfoliation. The products of these weathering processes are angular rock materials strewn across the desert landscape. Recent research in desert geomorphology indicates that chemical weathering is also important in breaking down desert rocks.

Paradoxically, water is the chief agent of erosion in deserts, despite being available in very limited quantities. Careful study of desert landscapes reveals the presence of many water-formed features, including wadis, braided river channels, and alluvial fans. These are products of the heavy rains that occur during intense thunderstorms. The hard, bare ground is easily eroded by run-off, since there are few plants to break the fall of the raindrops, slow down the water's movement over the ground, and hold onto the soil.

Landforms Created by Water

Wadis are common water-formed features in arid landscapes. They are water-eroded valleys that seldom contain flowing water, except after infrequent storms. Wadis have steep sides and wide, level bottoms covered with thick beds of alluvial sands and gravels. Wadis are most commonly found where streams flow from mountains onto lowland areas. Most of the erosion of the sides and beds of wadis occurs during infrequent flash floods.

Sheet flooding is another cause of desert erosion. During very intense thunderstorms, precipitation is so heavy that water covers the whole ground surface and flows as a continuous sheet. Sheet floods occur on level surfaces, such as alluvial fans or lowlands where few stream channels have formed. They can carry

Figure 13.12
A Typical Desert Wadi, Utah

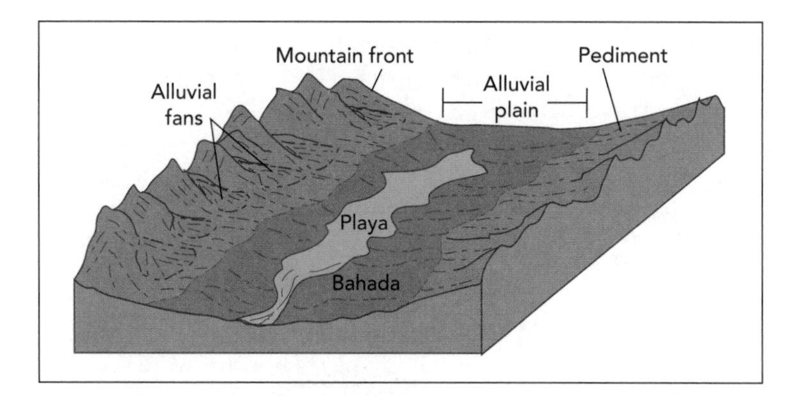

Figure 13.13
A Typical Landscape of Mountainous Desert Landforms

vast amounts of sediment into the centres of lowlands and inland basins in very short periods of time.

Where mountains and hills surround lowland basins, common water-formed features include pediments, alluvial fans, bahadas, and playas. A typical landscape made up of these features is shown in Figure 13.13.

Pediments are bedrock plains, with a thin or non-existent soil cover, that start at the base of steep mountains and slope gently down into the centres of desert basins.

Alluvial fans are made up of water-transported sands and gravels and occur where streams flow from steep, mountain valleys onto level, lowland plains. Where many alluvial fans grow together along the base of a mountain range, they form a sloping alluvial plain known as a bahada.

Playas are dry lake beds in the centres of inland basins. They are underlain by fine-grained alluvium, such as clays and silts. After storms, playas fill with water which deposits fine clays and silts. As the waters evaporate, large amounts of soluble salts are left behind as salt pans.

In level, sedimentary rock deserts, the most common features are escarpments, mesas, buttes, and badlands, as shown in Figure 13.14.

Escarpments are formed by the differential erosion of level sedimentary rocks. The top of the escarpment is a hard, resistant **cap rock**; the softer rock layers beneath have been eroded to form the escarpment face. The large, relatively flat area of cap rock above the escarpment forms a **plateau**.

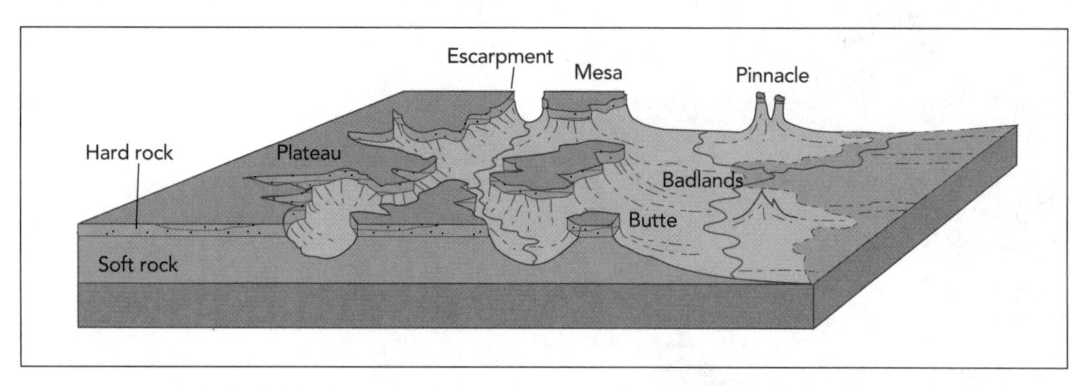

Figure 13.14 **A Typical Arid Landscape in Sedimentary Rock Areas**

Mesas and **buttes** form along the margins or edges of plateaus. For one reason or another, parts of the plateau are eroded more quickly than others. The more slowly eroding sections are soon separated from the main plateau by a gap or valley, leaving them isolated as mesas and buttes. The major difference between the two is that mesas are much larger in size than buttes.

Badlands form in arid areas where deep beds of soft rocks are exposed to erosion. Infrequent, heavy rains gouge out stream channels easily since vegetation is sparse. The result is often a spectacular array of gullies, carved hillsides, and tall spires of resistant rocks.

Landforms Formed by Wind

Wind erosion plays an important role in creating arid landscapes. Even though water is the major influence, wind has more impact here than in any other landscape type. In the great sand deserts, which cover approximately 10 percent of all arid environments, wind action is actually the dominant erosive and depositional agent. But wind is much less dominant in the remaining 90 percent of arid landscapes. The action of wind in eroding, transporting, and depositing sediments is often called **eolian** or **aeolian erosion**.

Wind erodes rock surfaces by using sand as an abrasive agent. The fine sand particles that are carried by the wind wear away rocks up to about half a metre above the ground surface. The effectiveness of wind blasting is influenced by the wind speed, the hardness and amount of sand carried by the wind, and by the hardness of the rocks being attacked. Softer rock layers are soon hollowed out, while harder rocks are polished to a smooth surface. Undercutting can result in large boulders being perched on narrow, wind-eroded bases to form **pedestal rocks**, as shown in Figure 13.15.

Figure 13.15 **Pedestal Rock, Arches National Park, Utah**

The fine sand used to blast rocks comes from the surface of arid regions. It is picked up by winds, creating other interesting landforms. **Desert pavements** are similar to the cobblestone streets once common in many older towns and villages of Europe and North America. These pavements are formed when wind action removes the finer clay, silt, and sand particles, leaving behind the larger stones and boulders. Figure 13.16 shows the steps in the formation of such pavements, which often form a continuous

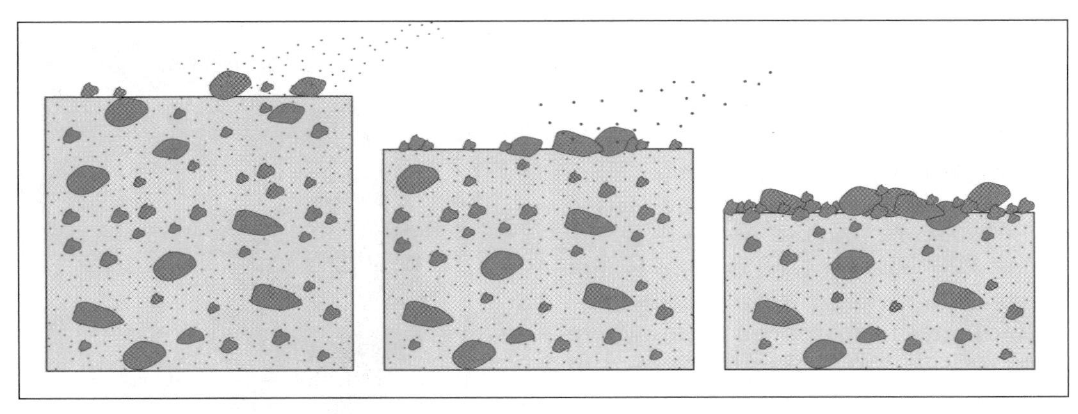

Figure 13.16 The Formation of Desert Pavements

ground cover. The individual rocks are sandblasted and heated and cooled by the sun until they develop a black, shiny surface known as **desert varnish**. Desert pavements are known as **regs** in Algeria, **serir** in Libya, and **gibbers** in Australia.

Deflation hollows are important landforms created by wind action. Such hollows form when fine-grained particles from easily weathered rocks are carried away by wind action. As the hollows deepen, they collect water during infrequent rains. The water helps speed up rock weathering in the hollow, creating more fine particles to be carried away by the wind. Because of this, the hollow deepens at a faster rate than surrounding higher and drier areas.

Wind also forms depositional features. Wind can be seen as a river of air, and sand is the sediment that it carries. As with rivers, when the wind velocity is reduced, so is its ability to carry sediment. Depositional features of wind vary tremendously in size.

The greatest depositional features are the major **ergs** or **sand seas** (Figure 13.17) covering thousands of square kilometres of Saudi Arabia, Libya, and Algeria. Despite opinions to the contrary, sand seas cover only about 25 to 30 percent of the desert areas of the world. The sand grains are largely quartz, a hard, abrasive, chemically resistant mineral formed by the weathering of sandstone rocks.

Dunes are the most common and familiar landforms created by wind action. In general, dunes develop an asymmetrical shape. The windward sides of dunes have gentle, upward slopes to steeper, leeward slip slopes. In the lee of dunes, wind speeds are reduced, causing the deposition of sand on the slip slope. The result is moving sand dunes that can advance across the desert landscape at speeds from 50 m per year for smaller dunes to 1-2 m per year for large dunes.

Figure 13.17 A Sand Sea, or Erg, in the Western Sahara

QUESTIONS

17. a) Describe what your image of a desert was before reading this section of the book.
    b) How has this study altered your views on what arid landscapes are really like?
    c) How would you develop an arid landscape to make it an attractive tourist destination?
18. Arid regions are estimated to cover approximately 20 percent of the earth's surface. Using an atlas, rank the world's continents based on the percentage of their areas that have arid landscapes.
19. Draw sketches to illustrate the key ways in which arid landforms differ from those in humid areas.
20. Water is important in forming arid landscapes. Why is this the case, despite the arid climate of such landscapes?
21. Sketch a series of diagrams to show the processes involved in the formation of any two of the following features:
    a) an escarpment
    b) a butte
    c) a wadi
    d) an alluvial fan
22. a) Devise an experiment that you could carry out to show the difference in rates of erosion between a slope covered with vegetation (humid area) and one with very sparse vegetation cover (arid area).
    b) If possible, carry out this experiment and summarize your results in an illustrated report.
23. Describe some of the factors you would investigate before buying or building a home in a desert landscape in such areas of Arizona or Nevada.
24. Explain why the sandblasting action of wind erosion is confined mostly to the half metre or so above ground level.
25. Suppose you are a farmer in Egypt whose farmlands are being threatened by approaching sand dunes. What actions could you take to stabilize the dunes?

Review

- The world's landforms can be organized under the categories of humid, arid, cold, and coastal landscapes.
- The work of running water dominates the processes of erosion where precipitation exceeds evaporation.
- The drainage basin is the basic unit of the landscape of humid climates.
- Rivers erode, transport, and deposit materials to create distinct landforms and to reduce the landscape to base level.
- The Geographical Cycle is one system of classifying rivers.
- Karst landscapes develop in limestone regions under humid climates.
- Karst landscapes are characterized by a unique set of surface and underground features.
- Arid landscapes develop in places where evaporation exceeds precipitation.
- Precipitation in arid landscapes is often sudden, intense, and very effective in eroding the soil.
- Most desert landforms are created by water erosion.
- Wind is an effective erosive agent in arid landscapes.

Geographic Terms

geomorphology
run-off
drainage basin
perennial stream
intermittent stream
ephemeral stream
abrasion
attrition
corrosion
hydraulic pressure
competence
capacity
traction
saltation
suspension
solution
alluvium
base level
graded stream
meander
point bar
lateral erosion
oxbow lake
flood plain
levee
braided stream
delta
distributary
alluvial fan
bahada
vertical erosion
headward erosion
karst landscape
solution furrow or rill
rainpit
grike
doline
uvala
polje
dry valley
cavern system
stalactite
stalagmite
wadi
sheet flooding
pediment
playa
escarpment
caprock
plateau
mesa
butte
badlands
eolian (aeolian) erosion
pedestal rock
desert pavement
desert varnish
reg
serir
gibber
deflation hollow
erg
sand sea

Explorations

1. a) Collect a series of photographs or maps of rivers and attempt to classify them into the stages in the Geographical Cycle.
   b) Find photographs or maps showing the major streams or rivers in your region and attempt to classify them in the same manner.
2. Create a series of landscape diagrams or sketches, or a 3-dimensional model, to illustrate the various surface features associated with humid, karst, and arid landscapes.
3. a) Draw a set of contour maps which would show a typical humid, karst, and arid landscape.
   b) Label the important landform features that appear on each of these maps.
4. Point out how human activities are increasingly affecting the operation of natural processes in humid, karst, and arid landscapes.

CHAPTER

14

DISTINCTIVE LANDSCAPES: GLACIAL, PERIGLACIAL, AND COASTAL ENVIRONMENTS

OBJECTIVES:

By the end of this chapter, you will be able to:

- recognize that understandings about glaciers, glaciation, and permafrost are relatively recent;
- understand the processes by which ice and waves move, erode, transport, and deposit materials;
- describe the unique landforms and processes that result from the action of glaciers, frost, and wave action;
- appreciate the impacts glaciation, frost, and wave action have on the landscapes where they operate;
- develop an understanding and respect for the power of natural processes.

CHAPTER 1: The Nature of Physical Geography
CHAPTER 2: Earth: Its Place in the Universe
CHAPTER 3: The Earth in Motion
CHAPTER 4: The Earth's Interior
CHAPTER 5: The Earth's Crust
CHAPTER 6: The Lithosphere in Motion: Plate Tectonics
CHAPTER 7: Solar Radiation
CHAPTER 8: Climate
CHAPTER 9: Weather
CHAPTER 10: The Hydrosphere and the Hydrologic Cycle
CHAPTER 11: Natural Vegetation and Soil Systems
CHAPTER 12: Denudation: Weathering and Mass Wasting
CHAPTER 13: Distinctive Landscapes: Humid and Arid Environments
CHAPTER 14: Distinctive Landscapes: Glacial, Periglacial, and Coastal Environments
CHAPTER 15: Natural Hazards: Disrupting Human Systems
CHAPTER 16: The Disruption of Natural Systems
CHAPTER 17: Fragile Environments

Introduction

The cold landscapes of the world have only slowly released their secrets to explorers and scientists. Because they have resisted the spread of human settlement far more than any other environment, they contain many of the only remaining undisturbed natural regions on earth. Even so, the lure of untapped resources and the development of various earth resource satellites have speeded up scientific investigations of these frozen landscapes.

Coastal landscapes represent a frontier region between land and sea, and are one of the most densely inhabited zones on earth. The power of waves beating upon a rocky headland and the thrill of surfing on a magnificent beach are images that characterize coastlines for many people. They are also zones of hazard and risk when the storm surge running before a hurricane strikes a settled coastline.

This chapter will focus on the various processes that work to mould the surface features of the earth in cold climates and along the contact zones between land and water.

14·1 The Theory of Glaciation

The idea that parts of Europe and North America had been covered by huge ice sheets was first suggested in 1837 by Louis Agassiz, an earth scientist. His "glacial theory" was greeted with ridicule by scientists of the day. It was simply unthinkable that parts of Europe and North America had been buried beneath ice sheets up to three kilometres thick.

As evidence accumulated from both sides of the Atlantic Ocean, scientists saw that the only satisfactory explanation for the facts was the spread of ice sheets over the northern parts of the continents during what came to be known as the Pleistocene Ice Ages. When Agassiz died in 1873, his glacial theory was widely accepted.

Continuing research has revealed the Pleistocene ice sheets advanced and retreated several times. Based on evidence from drill cores of deep sea sediments and ice layers in Antarctica and Greenland, it is believed that up to ten major glacial advances (glacials) and retreats (interglacials) have occurred over the past two million years.

Earth scientists now believe the earth has not emerged from the ice ages; that it is only in an interglacial period. The fact that ice sheets still survive on Greenland, Antarctica, and in high mountain areas is evidence that the ice ages may not be over!

The areas affected by the Pleistocene glacial advances in the northern hemisphere are shown in Figure 14.1. Unglaciated areas also felt the impact of the glacial periods. The ice ages caused a global lowering of sea level due to the huge volumes of water that were locked up in the ice sheets. These changes altered the pattern of ocean currents as well as the salinity and temperatures of ocean waters. There were also changes in global climatic patterns since the increased albedo of the snow and ice cover reflected large portions of incoming solar radiation directly back into space, cooling the polar regions. It is believed that tropical climates were much the same as today, except they did not extend as far from the equator. In the middle latitudes, however, the changes were considerable. The various vegetation zones, and the climates to which they were closely related, were squeezed in towards the equator.

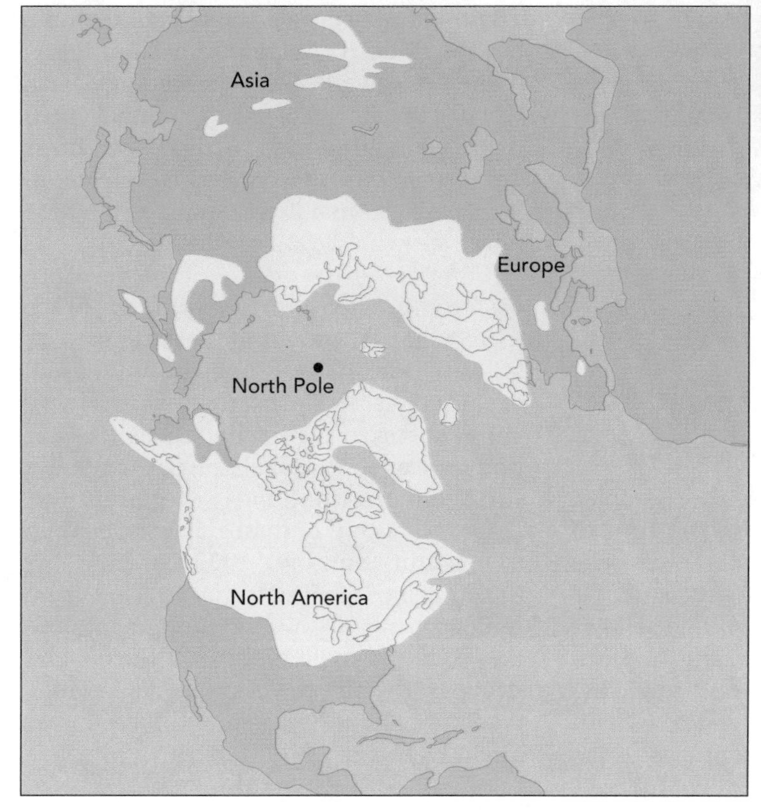

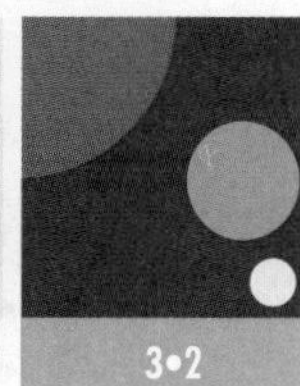

Figure 14.1
Northern Hemisphere: Maximum Extent of Continental Pleistocene Glaciers and Ice Sheets

It's a Fact...

The term "Pleistocene" comes from the Greek word meaning "most recent". The Pleistocene is a subdivision of the Quaternary Period on the Geological Time Scale that began approximately 2 million years ago as the ice ages began and lasted until 10 000 years ago, when the glaciers had largely disappeared.

QUESTIONS

1. a) Describe the image you have of cold landscapes.
   b) From what sources did you get this image? How might your information sources have affected your image?
2. Many earth scientists believe the world will eventually enter a new glacial period. Why might they believe this could occur?
3. a) Suggest reasons why scientists found it so difficult to accept Agassiz's glacial theory.
   b) What are some other theories that were widely criticized when they were first proposed but eventually became widely accepted?
4. Brainstorm a list of reasons that might account for glacial advances. Rank the reasons from most to least plausible.

14·2 The Formation of Glaciers

The formation of massive sheets of ice requires special conditions. The first of these is obvious — cold temperatures. The cooling of the climate during the summer months is a critical ingredient that allows winter snows to survive through the summer and add to the snows of the following winter. A summer blanket of snow, with its high albedo, further contributes to the cooling of the climate since more solar radiation is reflected back into space, without adding much heat energy to the atmosphere. This, in turn, means less melting of the snow and more cooling of the atmosphere.

But not all of the northern areas of the continents were covered with ice sheets. These places were certainly cold enough, but they lacked a second important ingredient — snowfall. Glaciers can only form where snowfalls exceed the annual melting over long periods of time. Ice sheets are formed when the pressure of accumulating layers of snow compresses the base of the snowfield to form glacial ice.

Compacted ice crystals that have not yet been pressed into a solid mass of ice form what is known as **firn**. There are still air pockets between the individual crystals. As the pressure of overlying snow and firn layers increases, the air pockets are squeezed out, forming a solid, impermeable mass of **glacial ice**. Under very cold climates, it may take up to a century and over 100 m of overlying firn and snow to create glacial ice. Where temperatures are warm enough in summer to allow some melting of the ice and snow in the firn layers, glacial ice may take only one to two years to form under a relatively thin 10-15 m of firn.

Ice: A Moving Solid

The formation of glacial ice beneath a snowfield is just the first stage in the story of glaciation. The pressure that turns snow to ice is also responsible for starting the solid ice flowing downslope. The flattening of ice crystals allows them to slide over one another in a process known as **plastic deformation**. Also, **shearing** within the ice sheet or glacier takes place. Miniature faults create surfaces along which different blocks of glacier ice can slip over, past, and beside each other. A third contributor to ice movement is **basal slippage**. The heat of friction allows the ice to slip over the surface of the bedrock on a microscopic film of water that lubricates the contact surface. The result of these processes is a solid mass of moving ice.

As a glacier moves downslope, friction at the base and sides of the glacier slows the movement in these parts, while the top centre flows most quickly.

As ice moves outwards from a snowfield, it will advance in all directions unless channelled by landforms to flow in certain ways. This outward flow will continue until the ice front reaches warmer areas where melting begins to sap the strength of the advancing ice front. When the amount of melting balances the rate of outward movement of the ice, the front or **snout** of the glacier will become stationary. It is important to note that, although the ice front is stationary, the ice is still advancing from the snowfield towards the snout. It is just that the amount of melting matches the rate of outward movement so the ice front appears to stand still. A glacier that is stationary is in equilibrium with its surrounding environment and provides an example of a cascading system (Figure 14.2).

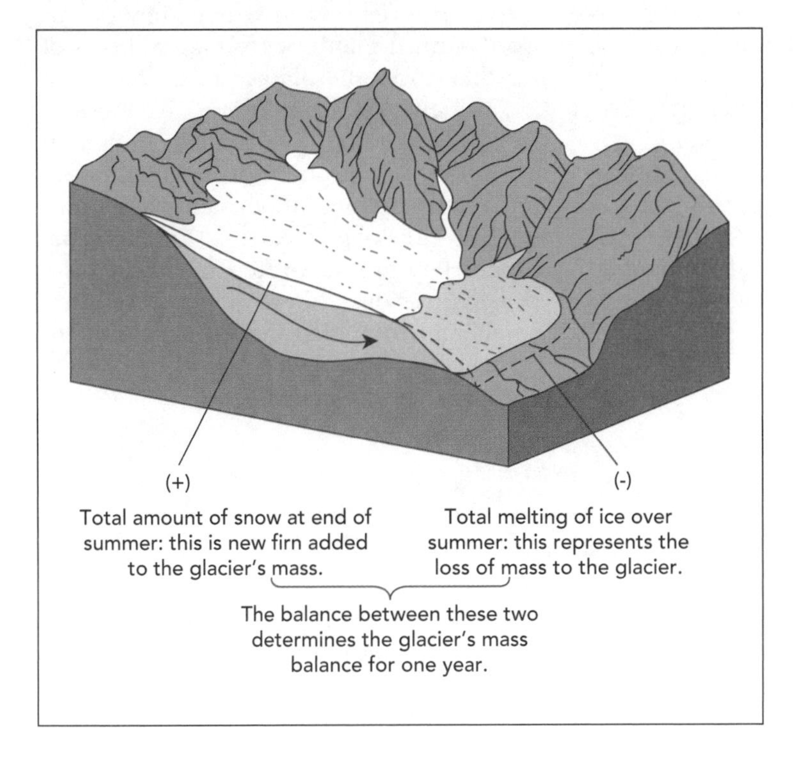

Figure 14.2
The Glacier as a Cascading System

How Ice Erodes

Ice erodes rock in a variety of ways. Abrasion occurs at the base of the glacier where it comes into contact with the bedrock over which it is moving. As the ice moves over the ground surface, rocks, boulders, and sand particles trapped in the ice literally scour away at the rocks on the ground like sandpaper. Weaker bedrock is gouged out to create future lake basins. Large, hard-rock boulders gouge grooves in rocks, sometimes over hundreds of metres in length. Where smaller particles grind directly against the bedrock, it may be polished and covered with myriad scratches, known as **striations**. Striations and grooves are very useful because they reveal the direction of ice movement.

A second method of erosion occurs as ice moves slowly over bedrock. Summer meltwater seeps into the bedrock below, freezing around it. As the ice moves forward, the glacier plucks the rocks from the surface, leaving behind holes that are smoothed and deepened by later ice movement. This is known as **glacial plucking**.

A third method of glacial erosion comes from the sheer weight of the ice. Small obstructions, such as hills, rock outcrops, or ridges, are literally ploughed out of the way by the weight and power of the ice.

QUESTIONS

5. a) Describe the changes that occur as snow becomes glacial ice.
   b) Devise a demonstration that would model the process by which glaciers move.
6. a) Figure 14.2 shows a glacier in equilibrium with its surrounding environment. Draw a second diagram to show what would happen to the glacier if:
      i) the climate warmed up;
      ii) more annual snowfall occurred.
   b) Explain the changes in the glacier that result from these two changes.
7. What effect do you suppose the temperature of ice in a glacier has on the speed of glacial movement? Explain your answer.

14·3 Continental Glaciation

Physical geographers divide glaciers into two major groups based on the topography of the regions in which they occur. **Continental glaciers** are major sheets of ice that cover the plains and lowlands of a continent as a series of massive lobes of ice that spread out like pancake dough on a griddle. **Alpine** or **mountain glaciers** form in upland and mountainous regions, within the larger valleys and basins. Since they are separated by many high peaks and ridges, these glaciers are often smaller in total size than their continental counterparts. This section deals only with continental glaciers.

It's a Fact...

Today, ice covers about 10 percent of the world's land surface. It is estimated that glaciers covered 29 percent of the land surface during the peak of the last ice age.

Erosional Features

Erosional features of continental glaciers vary tremendously in scale. The largest, most visible erosional features are the millions of **rock basin lakes** found in Canada and the northern United States, and in Finland and the surrounding countries of Europe. These lakes occupy basins, gouged out of the bedrock by advancing ice sheets. The structure of the bedrock influenced the size and shape of the basins. For example, the pattern of lakes in Figure 14.3a reveals fault and fracture lines in the rocks of the Canadian Shield, whereas the lakes and rivers in Figure 14.3b follow diagonal beds of softer rocks. Landscapes consisting of rounded rock hills and rock basin lakes are called **rock knob topography**.

Near their origins, glaciers are thick and powerful, capable of eroding huge amounts of material. They strip off most of the soils and fragmented rocks to form **bedrock plains**, where bedrock is completely exposed. Such bedrock plains cover large portions of the Canadian Shield and the Baltic Shield of Europe. These areas are surrounded by arc-like regions of glacial deposition along the southern boundaries of the shield areas.

Depositional Features

As glaciers move farther and farther from their source, they carry increasingly heavier loads of rock debris picked up from the places across which they have already moved. In addition, they become steadily thinner due to melting. As this

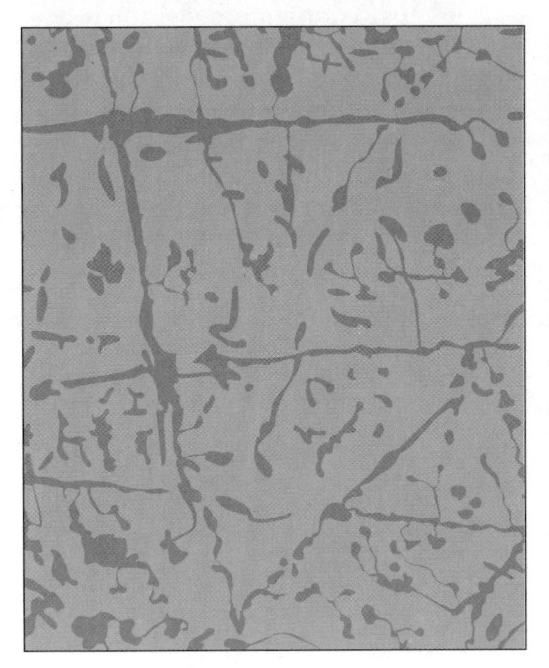

a) Faulting

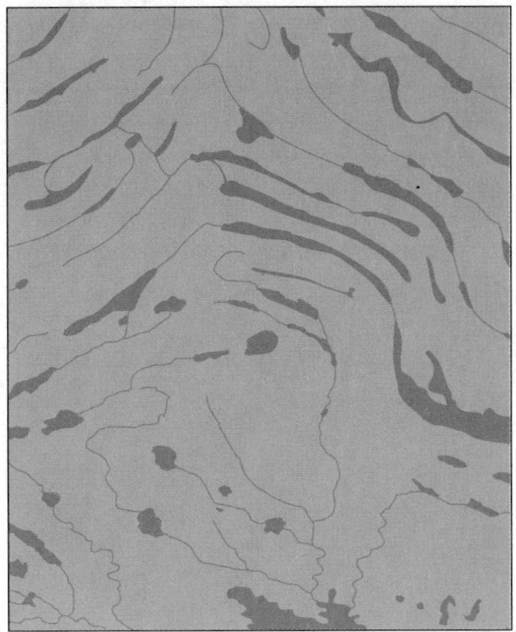

b) Folded hard and soft rock layers

Figure 14.3 Lake and River Patterns on the Canadian Shield: An Example of Glacial Erosion Revealing Rock Structures

happens, the capacity of the ice to carry sediment decreases. Eventually, the ice sheets begin to deposit many of the materials. Also at work at this time are huge volumes of glacial meltwater that transport sediments. Figure 14.4 shows the major features formed by ice and water action near the edges of typical continental ice sheets.

Ice-deposited materials are made up of a jumble of clay, sand, silt, gravel, and boulders, known as **glacial till**. The materials are usually unsorted and unlayered. The glacier deposits them in the same way that a bulldozer pushes materials in front of it into a pile. The larger rocks and boulders in ice-deposited materials are often angular or irregularly shaped.

Materials deposited by glacial meltwater usually consist of sorted and layered sediments. The velocity of the water determines what size materials will be dropped in its bed. Fast-moving streams deposit only larger boulders and stones; as the velocity of the water slows, gravels, sands, then silts are dropped to the stream bottoms. Clays, the finest-sized sediments, are only deposited in the calm waters of lakes and ponds. In this way, the sediments are sorted. A second characteristic of water-deposited materials is that they are layered. The volume and velocity of the streams change from season to season, resulting in the deposition of different-sized sediments. In addition, water-deposited materials are often rounded, rather than angular, in shape. These characteristics help identify glacial landform features and their methods of origin.

Ice-Deposited Materials

Glacial erratics are rocks that have been carried from their areas of origin to distant locations and deposited by glaciers. Granitic erratics eroded from the Canadian Shield are easy to identify when they are deposited in regions of sedimentary rocks, such as southern Ontario or the Prairies.

The most common depositional features of continental glaciers are **moraines** (Figure 14.5a). Moraines are formed when debris is pushed up in front of, or along the bottom of, an advancing glacier. They are composed of unsorted and unlayered glacial till, varying in texture from gravels to sands to clays. Moraines typically form a series of rolling hills that extend in an arc-

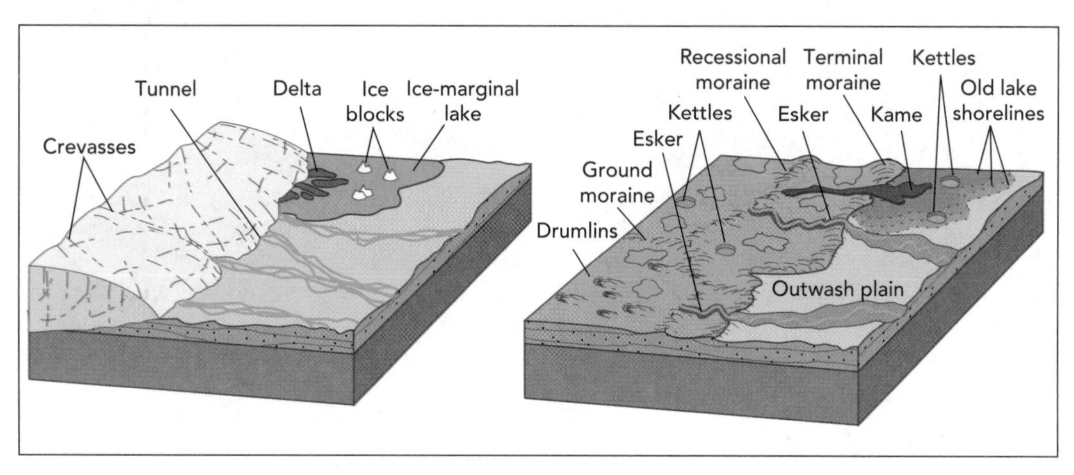

Figure 14.4 Depositional Features of Continental Ice Sheets

like shape across the landscape. In some cases, their surfaces are marked by small depressions, which may be filled with water. These depressions are formed when ice blocks, trapped beneath the till materials, melt and leave small lakes, known as **kettles**, as shown in Figure 14.5b.

Moraines formed at the furthest extent of glaciers are known as **terminal moraines**. Glacial till deposited along the sides of lobes of ice form **lateral moraines**, while those formed where two lobes of ice met are **interlobate moraines**. Since the climate occasionally warmed during glacial periods, the continental ice sheets sometimes retreated a few kilometres, only to stall when the climate cooled. Moraines deposited during halts in the retreat of the glacier are known as **recessional moraines**. These are the most common moraines in Ontario, the Prairie Provinces, and the adjacent states of the United States. A final type of moraine was formed from till dropped beneath ice sheets as their capacities to carry materials were diminished. Such deposits of till are called **ground moraines**.

Drumlins are streamlined, tear-shaped hills, with a wide and rounded front end (or stoss) and a longer, tapering tail (Figure 14.5c). They are formed when ice re-advances over an existing moraine. Overloaded with till, the glaciers sculpt what they cannot carry into a series of smooth, tear-shaped hills composed of unsorted till. They often occur in huge numbers in what are known as drumlin fields. The area surrounding Peterborough, Ontario, is one of the largest drumlin fields in the world. Since the tails point in the direction of ice movement, drumlins are used to determine the direction of flow of ice sheets in areas of glacial deposition.

Figure 14.5a Rolling hills of a typical recessional moraine, central Alberta

Figure 14.5b Kettles dot this moraine landscape near Saskatoon, Saskatchewan.

Figure 14.5c This side view of a drumlin in southern Ontario shows its steep side (stoss) and gently sloping tail. The glacier that created this drumlin moved from right to left across this area.

Water-Deposited Materials

Most depositional activity occurs near the snouts of glaciers where melting is most active. Glacial meltwaters produce a wide variety of depositional features, the most extensive of which are the various types of plains formed in glacial lakes. Wherever glacial streams empty into glacial lakes, extensive **sand plains** are formed. The Norfolk Sand Plain in Ontario is one example of this type of feature. Where the waters are calm over long periods, **clay plains** are created, as fine clay particles settle out of the calm waters.

When meltwaters flow out from glacier fronts in sheets, covering large areas, they deposit sands and silts to form **outwash plains**. These plains usually slope gently downward from the original position of the glacier. If ice blocks are buried under such plains, kettle lakes are formed; the name **pitted outwash plain** is used when such lakes are common.

At times, meltwaters concentrate into large stream channels where they carry heavy loads of silts, sands, clays, and other sediments. As these rivers flow over the landscape, they erode wide, shallow valleys known as **spillways** (Figure 14.5d).

Streams flowing on, inside, or beneath the ice also contain many sediments. Occasionally, the streams carry meltwater from several kilometres inside the glaciers to the snouts, through ice tunnels. The sediments form beds over which the streams flow. When the glaciers retreat, long, snake-like ridges of sorted sands, gravels, and boulders are left behind. Such features are known as **eskers** (Figure 14.5e).

Wherever meltwater streams empty into ponds or lakes, deltas form. As the ice melts, these deltas, composed of sorted sands and gravels, are left behind as small, conical hills, known as **kames**.

Figure 14.5d The wide, level floor of this spillway, south of Maple Creek, Saskatchewan, is typical of such glacial features.

Figure 14.5e An esker winds its way across a kettle lake near Whitehorse, Yukon. The lighter coloured areas outline the places where the esker dives below lake level.

Many of the water-deposited glacial features are important sources of sands and gravels needed for building roads, sidewalks, buildings, and other structures. As a result, many of these natural features of glacial landscapes are fast disappearing in the vicinity of major urban areas. This is especially true of eskers, kames, and spillway deposits.

QUESTIONS

8. a) Using a two-column chart with the headings Ice-Deposited Features and Water-Deposited Features, indicate the glacial depositional features that fit into each of these categories.
   b) Indicate how you could tell if a glacial feature was deposited by ice or water. Add this information to the chart you completed in part a) above.
9. Suggest how scientists could work out the direction of flow and paths of flow of the glaciers that formed over North America during the last ice age.
10. With the help of simple sketches, describe the processes that lead to the formation of two of the following glacial features:
    a) kettle lakes
    b) interlobate moraines
    c) drumlins
    d) eskers

14•4 Alpine Glaciation

Alpine glaciers form in upland or mountainous regions. They originate in **ice-fields**, or **snowfields**, in the highest areas of mountains where falling snows can collect to great enough depths to start ice formation. Snow avalanches are common ways in which snow is pulled downward by gravity and is concentrated in the snowfields to form glacial ice.

In most mountain ranges, the largest alpine glaciers, or evidence of their former existence, are found on the windward sides of mountains where snow, the basic ingredient needed in ice formation, is more abundant. Many glaciers survive in mountain areas of mid- and low latitudes where snowfalls are heavy. Wherever the snowfall is light, even in the coldest mountains of the high latitudes, few, if any, glaciers are found.

Alpine glaciers generally occupy a series of connected valleys and flow down into surrounding valleys and mountain foothills where they combine to form continuous ice sheets. Most present-day glaciers outside of Greenland and Antarctica are alpine or mountain glaciers, surviving due to their high elevation and/or high latitude locations.

Erosional Features

In alpine areas, erosion is somewhat different than for continental glaciers. In high sloping areas, the ice is channelled into rivers of ice rather than into massive and continuous continental ice sheets. In addition, alpine glacial action roughens and steepens the terrain, widening valleys and sharpening peaks instead of smoothing off features. The tremendous erosive power of the ice moves vast amounts of rock and debris from the mountains onto surrounding plains or ocean basins.

As glacial ice moves outwards from its collection area, it erodes the sides, bottoms, and back of its source, creating and enlarging a depression in the rock. The bowl-shape of such depressions, known as **cirques**, is only revealed when a glacier has completely melted. The bottoms of cirques are often occupied by small lakes, known as **tarns**.

Where many cirques erode into the side of a single mountain, knife-edge ridges — **arêtes** — are formed between the cirques. A rectangular, sharp-pointed **horn peak** is created when several cirques erode back into a mountain, as shown in Figure 14.6. The Matterhorn peak in the Alps is a famous example of this glacial feature.

As the ice moves down from a cirque, it deepens, widens, and straightens out the valley, giving it a typical U-shaped cross-section. Figure 14.7 shows an example of a typical, glacially eroded **U-shaped valley**. Such valleys have much wider and gentler curves than river valleys in non-glaciated areas. Wherever U-shaped valleys have been cut down below sea level, and are flooded by the sea, **fiords** are formed. Fiord coasts occur in Norway, British Columbia, Labrador, New Zealand, southern Chile, and Ellesmere and Baffin Islands.

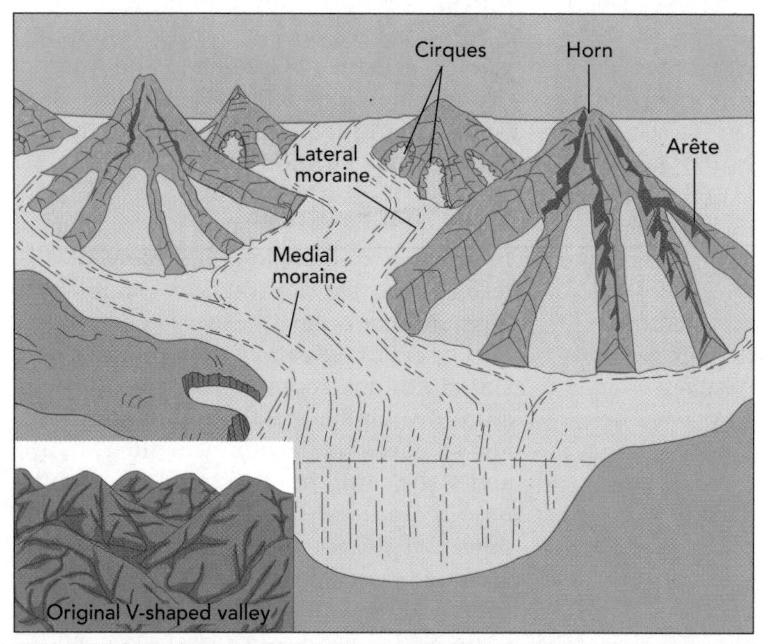

a) During alpine glaciation

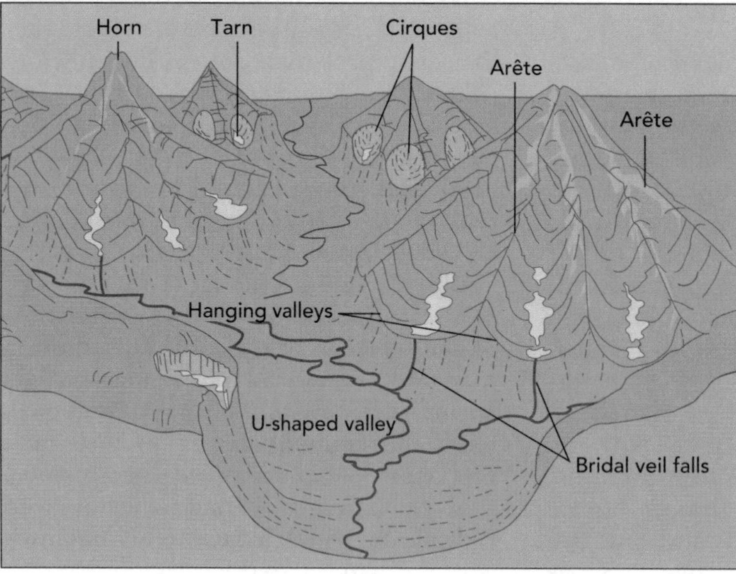

b) After alpine glaciation

Figure 14.6 **Features of a Glaciated Mountain Landscape**

Figure 14.7 Typical U-Shaped Valley of Glacial Origin This U-shaped valley was cut by glacial action into the Long Range Mountains, Gros Morne National Park, Newfoundland.

Figure 14.8 Medial and Lateral Moraines Dark stripes of debris mark the lateral (side) and medial (centre) moraines on the Kaskawulsh Glacier, Kluane National Park, Yukon.

Where glaciers of different sizes come together, the larger glaciers usually have deeper valley floors. When the ice recedes, the valleys of the smaller glaciers are often left perched high above those of the larger glacier to form **hanging valleys**. **Bridal veil falls** are created where streams flow over such hanging valleys.

Depositional Features

Glaciers are constantly eroding the sides and bottoms of the valleys over which they flow. In addition, where ice does not completely cover the rocky slopes of mountains, heavy frost action and snow avalanches bring tonnes of rock debris down onto the glaciers in the valleys below. The result is a massive amount of debris carried on the bottom, sides, and top of the rivers of ice. Where the amount of debris is greater than the ability of the glaciers to carry it, or where the glaciers are gradually retreating, many sediments are laid down. Most of these deposits are similar to those created by continental ice sheets, while others are unique to alpine areas.

As with continental glaciers, ice-deposited materials are unsorted. Those deposited along the sides of alpine glaciers are known as lateral moraines. Where two rivers of ice join into a single ice sheet, glacial till trapped between the newly joined glaciers is deposited as **medial moraines**. Figure 14.8 shows the lateral and medial moraines of the Kaskawulsh Glacier of Kluane National Park in the Yukon. Like continental glaciers, terminal moraines form from till pushed up at the furthest extent of glaciers. Recessional moraines are located where the ice retreated, then stalled, then retreated again in a series of steps.

Depositional features are also formed by the meltwaters of alpine glaciers. These features include eskers, created by rivers that flowed inside or below the glaciers. Where meltwaters empty out of glaciers into **ice-dammed lakes** or ponds, kames of sorted sands and gravels are formed. The rivers are usually heavily loaded with sediments, to such an extent that they have the colour of chocolate milk. Meltwater streams often occupy the whole width of valleys during the spring and summer melting season and become **braided** in their flow pattern. Over time, such braided rivers form narrow outwash plains across the whole of the valley bottoms. The rivers also carry tonnes of debris that form deltas and alluvial plains far from glacial areas.

Figure 14.9 **A Typical Periglacial Landscape near Tuktoyaktuk, Mackenzie River Delta**

QUESTIONS

11. "Glaciated mountains are much more spectacular and attractive to tourist development then unglaciated mountains." Agree or disagree with this statement, giving reasons for your opinion.
12. Compare alpine and continental glaciation using a chart. The characteristics you should include in your chart are: Erosion Processes, Erosion Features, Deposition Processes, and Deposition Features.
13. Explain how each of the following conditions would affect the size and erosive power of alpine glaciers:
    a) the direction a mountain faces
    b) summer temperatures
    c) steepness of slopes

14•5 Periglacial Landscapes

Periglacial landscapes develop in glacier-free polar regions where the climates are very cold but snow does not accumulate. The prefix "peri" means around or surrounding. The actions of frost and snow are the most important forces shaping the earth's surface in these regions. Three characteristics make these lands different from other landscapes:

- the presence of permafrost; **permafrost** is ground that has been frozen for more than two years. In effect, periglacial landscapes are "lands under refrigeration".
- the development of landforms resulting from the constant freezing and thawing of the upper layer of the soil; and
- the predominance of mechanical over chemical weathering.

A typical periglacial landscape is shown in Figure 14.9. The development of this type of landscape requires very long, cold winters with below-freezing temperatures for at least ten months of the year. Combined with a shallow snow cover, the long, cold period allows deep frost penetration of the soil to create permafrost. In some places, permafrost extends hundreds of metres below ground level. The greatest depth known is 1450 m in Russia. Canada's deepest known permafrost depth is 700 m.

Depending on the latitude, permafrost can be continuous or discontinuous. Approximately 12 percent of the earth's surface is underlain by continuous permafrost, much of it in the northern hemisphere. It is believed that up to 24 percent of the earth's surface was underlain by continuous permafrost during the height of the last ice age. The present permafrost areas of the northern hemisphere are shown in Figure 14.10.

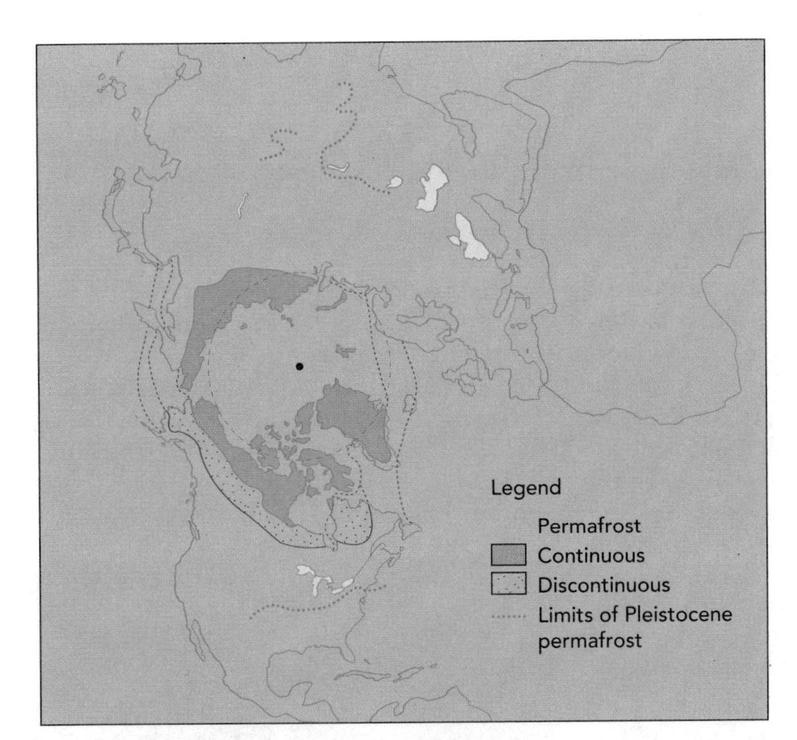

Figure 14.10
Permafrost Areas of the Northern Hemisphere

Periglacial landscapes are either devoid of vegetation or have only shallow rooted plants such as grasses, mosses, and lichens. Where soils exist, they are made up largely of undecayed organic materials and peats. Most of the landscape is very poorly drained with a high water table, waterlogged soils in the summer, and many shallow lakes, ponds, and swamps. This occurs despite the low annual precipitation because the permafrost layer is just below the ground surface and there is little evaporation in the cold, polar climate.

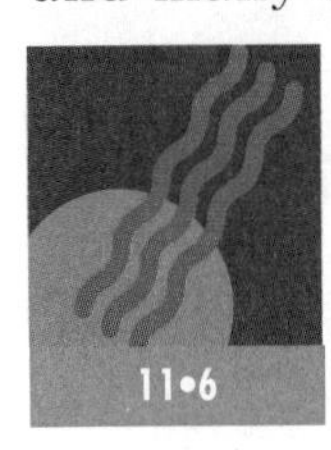

Frost action is the main process at work in periglacial landscapes. Such action involves the freezing of water-saturated ground and the formation of ice wedges which crack the ground surface. The annual thawing of the surface layer — the **active layer** — during short, cool summers is another significant process. This layer is largely responsible for the distinctive surface features of periglacial areas. **Solifluction** is a third process at work in the summer melting period. Solifluction is the slow, downslope movement of water-saturated soil and rock under the force of gravity. This process is responsible for a large part of the movement of rock and soil materials in periglacial areas.

Periglacial Landforms

A typical periglacial landscape is dotted with a myriad of small lakes, ponds, and swamps occupying shallow depressions amid low, rolling hills. The depressions form where deeper summer melting of the permafrost, due to disturbances of the soil and vegetation by natural or human actions, causes the surface to sag. Water from the summer melting of the active layer collects in the depressions.

It's a Fact...

It is estimated that continuous permafrost underlies about one quarter of Canada and of Alaska.

The most distinctive landforms created by periglacial processes are **pingos**. Pingos are "hills" with cores of ice, water, and mud that can rise up to 90 m above the plains on which they occur (Figure 14.11). They form when water seeps down to the permafrost layer, freezes, and expands. As water continues to freeze, the growing ice domes push up the topsoil to form hills. Once the icy interiors of pingos are exposed, they will melt during the summer months, leaving behind circular rims of soil with small lakes inside. Such lakes can be up to 100 m in diameter.

Patterned ground is another unique feature of periglacial areas. As Figure 14.12 illustrates, patterned ground forms over permafrost where intense winter freezing causes the ground to crack into polygon shapes. During the summers, water seeps into the cracks. In the following winters, ice wedges form and widen the cracks to as much as three to four metres. As the cracks are widened, the soil is pushed out and raised around the edges of the polygons to form patterned ground landscapes.

A third distinctive feature of periglacial landscapes is **stone circles** which surround dome-shaped areas of fine soil. The first step in their formation is the doming of a soil area by frost heaving.

Figure 14.11 **Pingo Hill: A Unique Periglacial Landform Along the Harding River, Northwest Territories**

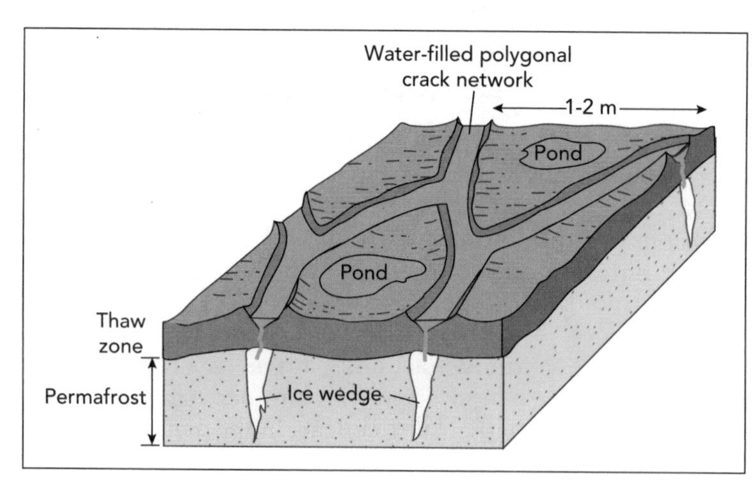

Figure 14.12 **The Formation of Patterned Ground**

This also brings stones and boulders to the ground surface. When the active layer melts, the soil becomes very moist, allowing the stones to slip downhill to form a ring around the base of the dome. Often, plants will grow in the protected areas between the stones to make the ring even more prominent. The area within the circle is usually bare of vegetation due to the frequency of soil heaving.

Rivers and streams of periglacial landscapes are also distinct. Due to heavy melting in the summer season and the limited natural vegetation cover, the streams carry very heavy loads of silt. The stream beds fill with such sediments, causing the streams to develop a braided pattern. Deltas often develop where the rivers enter lakes, seas, or oceans.

In steeply sloping areas of high mountains where permafrost exists, a different set of landforms is created. Here, the action of frost on exposed rock surfaces and the downslope movement of water-saturated soil and rock materials are most important. Where the movement of weathered materials is relatively slow, it is known as solifluction; where it moves rapidly, it is an avalanche. Some of the features resulting from such flows include rock glaciers, talus and screes, and solifluction lobes.

QUESTIONS

14. a) Why can periglacial landscapes be described as "lands under refrigeration"?
    b) Physical geography textbooks written a generation ago seldom contained information on periglacial landscapes. Why might this have been the case?
    c) Why are periglacial studies becoming increasingly important in Canada?
15. Explain why periglacial landscapes are dotted with many swamps, lakes, and ponds during the summer season, despite having "cold desert" climates.
16. Choose a typical feature of the periglacial landscape and describe the processes by which it was formed.
17. Describe the types of landscapes where braided streams develop and point out, using sketches and diagrams, the common causes of such stream channels.
18. Write a short note describing the problems of transportation and the construction of buildings in periglacial landscapes.

14·6 Coastal Landscapes

Coastal landscapes are quite distinct and are not associated with special climatic or bedrock conditions like the other distinctive landscapes. They span all of the world's climates and occur in all types of bedrock. The common theme is the contact zone between the continents and the oceans where wave and tidal actions create distinctive land and seascapes.

The coastal landscape is different in another way from the other landscapes because it extends only a short distance inland and out into the ocean, sea, or lake. It is also very limited in its range above and below the level of the waves that meet the land's edge.

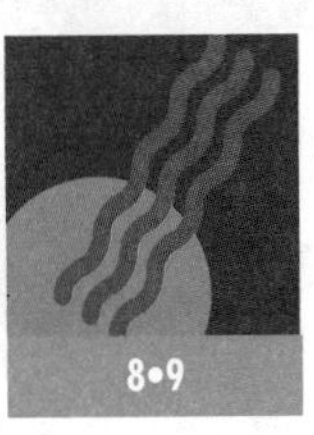

The Coastal Energy System

The sun is the source of most of the energy that drives coastal processes. This energy takes a rather long path to reach shorelines, beginning as solar energy entering the atmosphere. This creates differences in heating and thus pressure from one location to another. The resulting winds transfer some of their energy to the surface of the water to form waves and currents. As this energy moves through the water, as shown in Figure 14.13, it causes water molecules to move in circular orbits, with the molecules remaining essentially in the same place in the water. It is only the energy that moves through the water from one point to another. The energy waves move outwards in all directions from windy areas or storms like ripples on a pond. On a calm day, both the amount of energy and the amount of work carried out by wave action are small. However, when driven by storms or hurricanes, the force exerted by waves is enormous.

Waves on a Shoreline

As waves approach a shoreline, changes begin to occur when the water depth is approximately half of the wave length, as

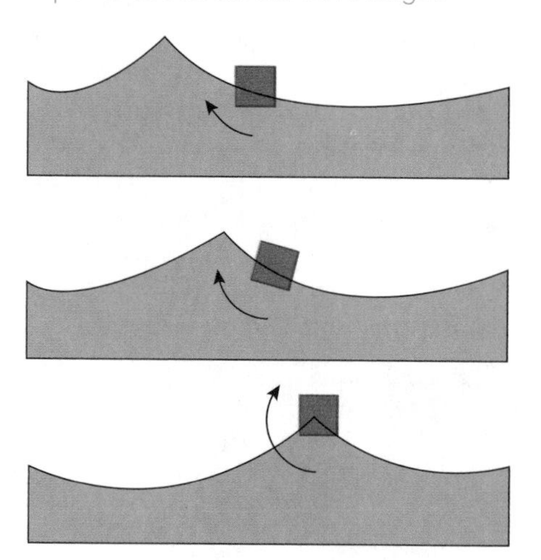

a) The orbital motion of water in a wave decreases with depth. It dies out at a depth equal to about half the wave length.

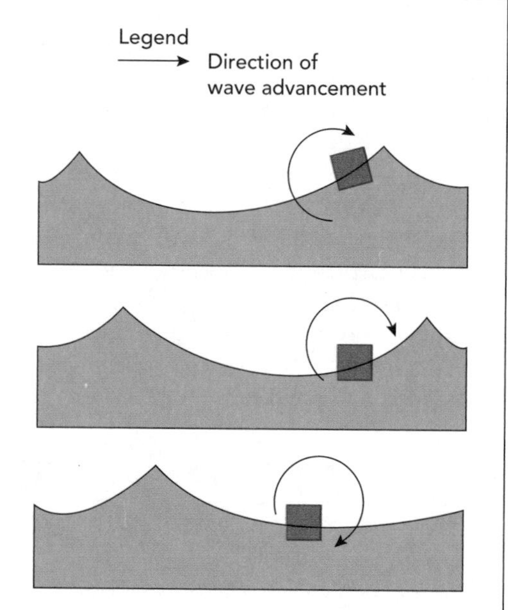

b) The motion of a water particle as a wave advances is indicated by the movement of a floating object. As the wave advances (from left to right), the object is lifted up to the crest and then returns to the trough. The wave form advances, but the water particles move in an orbit, returning to their original position.

Figure 14.13 Energy and Wave Movement Through Ocean Waters

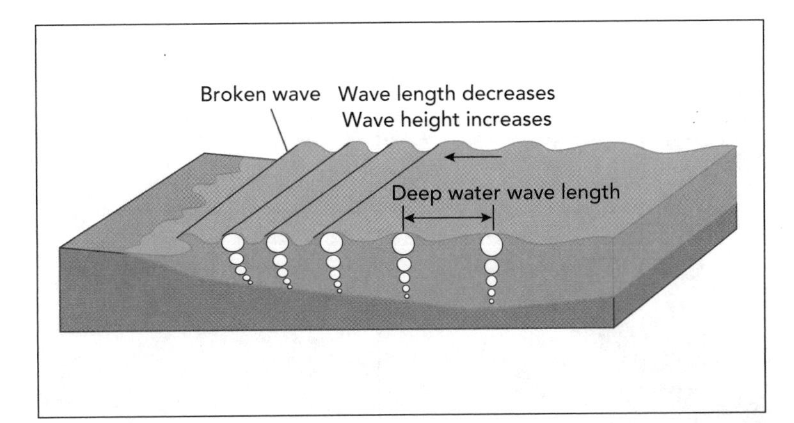

Figure 14.14
The Effect of Shallow Water as Waves Approach a Shoreline

shown in Figure 14.14. The bottom of the wave is slowed by the frictional drag of the shallowing bottom, while the top of the wave continues on at its former pace. Once the wave "leans over" far enough, it breaks, spending its energy against the shoreline. The term "**breaker**" is applied to such waves.

Viewed from above, waves bend as they hit a shallowing bottom. On an irregular coast, the waves bend so that they become almost a reflection of the shape of the coastline, bending around the headlands and looping into the bays. Wave crests approaching a straight coastline are slowed by frictional drag on the bottom and begin to bend as the outer sections of the wave, still in deeper water, continue at their normal speed.

In both cases, the waves move onto the shore at an angle. On a straight coastline, this angular arrival path means the water moves down the coastline, creating a **longshore drift** within the shallow breaking zone next to the beach. On an irregular coastline, longshore drifts are set up along the headlands leading into the bays. In both cases, sand, pebbles, or other debris will be carried by the longshore drift to be deposited on a beach further along a straight shoreline or at the head of the bay on an irregular coastline.

QUESTIONS

19. Suggest why coastlines are fascinating to people of all ages.
20. Point out how coastal landscapes differ from the other types of landscapes discussed in Chapters 13 and 14.
21. Draw a diagram to illustrate the steps involved in transferring energy from the sun into waves on an ocean or lake surface.
22. a) Using diagrams if necessary, explain why waves approaching a straight coastline are bent or refracted.
    b) In what other situations is energy bent or refracted in much the same way?

14·7 Erosion and Deposition by Waves

The major processes of erosion by wave action are abrasion, corrosion, and hydraulic pressure. **Abrasion** occurs as waves hurl bits of rock and sand against cliffs or **headlands**. In addition to weakening and wearing away the rocks, the abrading material is also broken down, making it more transportable by longshore drift. **Corrosion** occurs when soluble minerals are dissolved by water. Such action is common on limestone headlands, especially when the water is

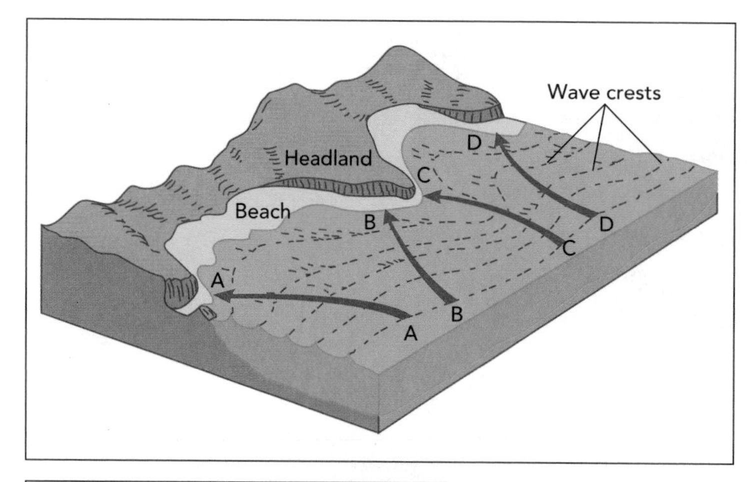

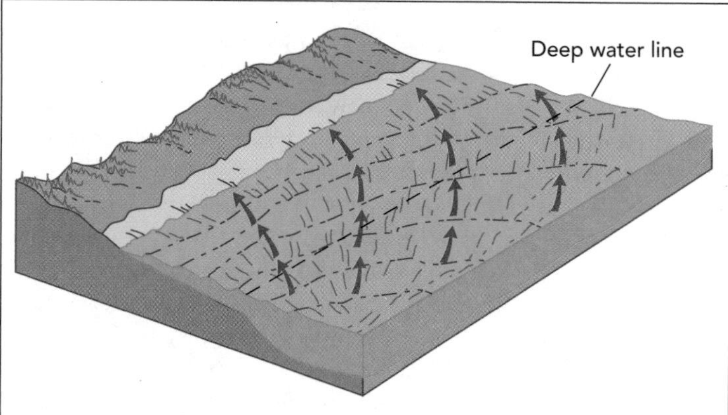

Figure 14.15
Wave Refraction on an Irregular and Straight Coastline

a) Irregular Coastline
Each segment of the unrefracted wave, AB, BC, and CD, has the same amount of energy. As the wave approaches shore, segment BC encounters the sea floor sooner than AB or CD and moves more slowly. This difference in the velocities of the three segments causes the wave to bend, so that the energy contained in segment BC is concentrated on the headland, while the energy contained in AB and CD is dispersed along the beach.

b) Straight Coastline
Along a straight coast, waves refract evenly; thus, wave energy is evenly distributed.

slightly acidic. **Hydraulic pressure** is not a major erosive process under normal weather conditions, but does become important during severe storms. Here, water hurled at great speeds against rock surfaces puts pressure on fractures or other weak points in the rocks, breaking off fragments. These broken rocks, in turn, increase the abrasive effect of the waves. All three processes operate at the same time.

Figure 14.16 shows the sequence of events that turn a rocky headland into a sea **cave**, an **arch**, then a **sea stack**. The process is carried out by the three processes of erosion attacking weaknesses in the rocks of the cliff face. In some cases, a sea stack may be joined to the shore by a sand bar. This bar may be built up in the calmer waters protected by the sea stack itself. The name "**tombolo**" is given to such features.

Wave action operates only within a narrow range above the water surface. However, through the process of undercutting, waves can wear away cliff faces far above wave heights. Undercutting causes landslides and slumping of cliff materials, which are then broken up and removed by wave action.

On the other hand, wave action can only erode a few metres below the lowest water level. Since it is limited in its ability to cut down further than this, the

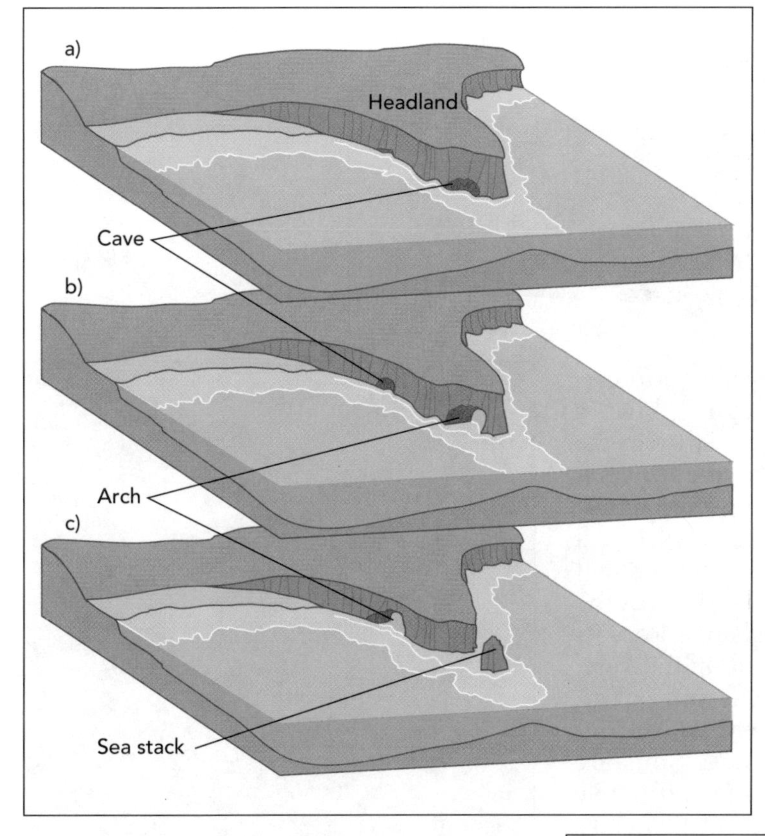

Figure 14.16
Sequence of Erosion on Rocky Headland

a) Wave energy is concentrated on a headland as a result of wave refraction. Zones of weakness, such as joints, faults, and nonresistant beds, erode faster, so sea caves develop in those areas.

b) Sea caves enlarge to form a sea arch.

c) Eventually, the arch collapses, leaving a sea stack. A new arch can develop from the remaining headland.

result of headland erosion is the formation of **wave-cut platforms** or terraces, as shown in Figure 14.17. Such wave-cut platforms are often raised above sea level (by a rise in the land or a drop in sea levels) where they can be clearly identified as former shorelines.

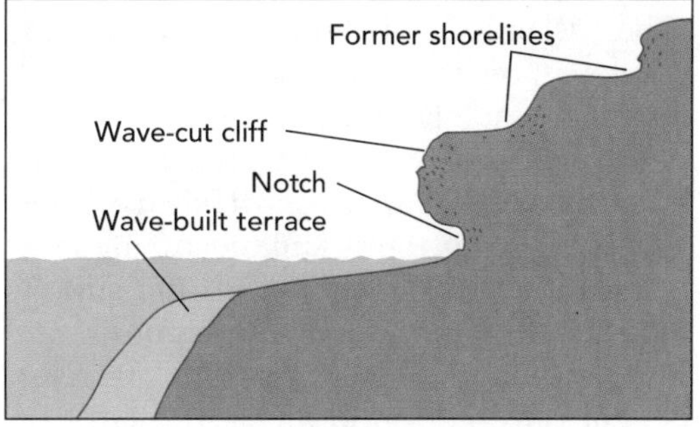

Figure 14.17 **A Wave-Cut Platform or Terrace** Wave action cuts at the base of the cliff, undermining the cliff until it collapses. Wave action soon removes the debris, and undercutting continues. Continued erosion causes the cliff to recede farther, leaving a gently sloping wave-cut platform. Some of the sediment eroded from the shore can be deposited in deeper water, forming a complementary wave-built terrace.

Deposition by Waves

Beaches are the most common depositional feature of coastal landforms. On irregular coastlines, the materials eroded from headlands are carried by short longshore drifts on each side of the headlands into bays and dropped to form **bayhead beaches**.

On straight coastlines, sand picked up from beaches farther upshore, or deposited by rivers, is carried along long-

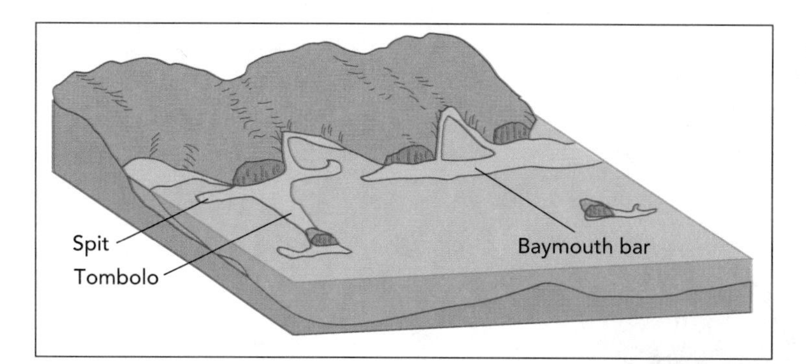

Figure 14.18
Depositional Features on an Irregular Coastline

shore drift. In fact, the longshore drift carries a constant "river of sand" down a coastline. If there is an irregularity in the coastline, such as a bay, a **baymouth sand bar** may be formed, cutting the bay off from the ocean or lake (Figure 14.18). If the longshore drift reaches a headland that juts out, the longshore drift may be deflected into deep water where it loses its source of energy and dies out. In this case, the sand builds up into a long **spit** or bar. It may also sink into submarine canyons along mountainous coastlines. Spits are attached to land at only one end, while baymouth bars are attached at two ends. Where islands lie offshore, a sand bar may develop in the lee of the island and connect it with the mainland to form a tombolo.

Many coastlines bordering lowlands have **barrier islands** or **offshore bars** separated from the mainland by lagoons (Figure 14.19). Depending on the supply of sediments from rivers, the **lagoons** can be open water or very marshy. Barrier beaches are common along the Gulf of Mexico in the United States and along the Gulf of St. Lawrence in New Brunswick and Prince Edward Island. Such shorelines are thought to originate in several different ways. One way is that the offshore bars are spits that have extended long distances along a shoreline from a headland or promontory over thousands of years. A second method of formation

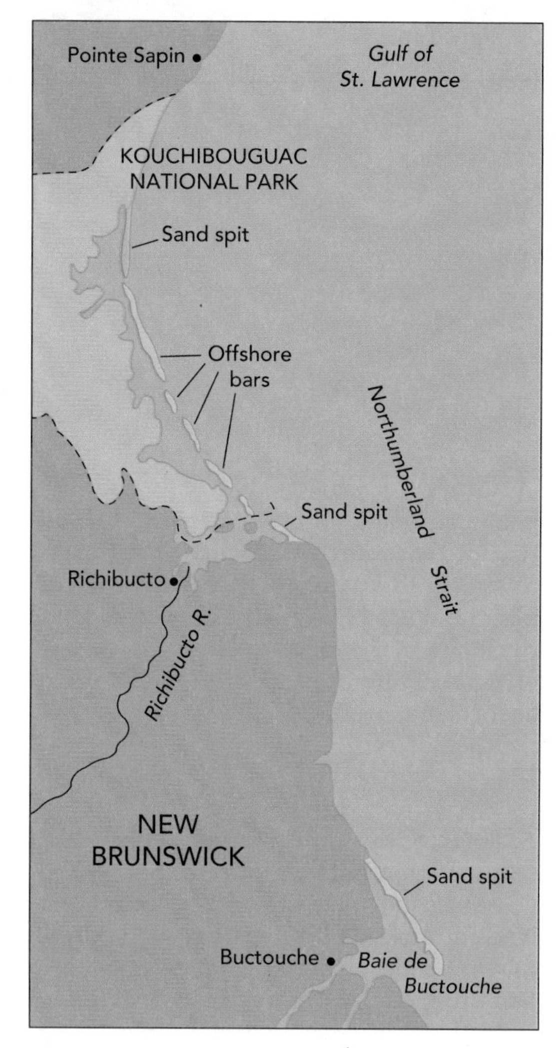

Figure 14.19 **Spits and Offshore Bars (Barrier Beaches), New Brunswick**

involves a rise in sea level following the melting of ice age glaciers. As the sea rises, it floods a sand ridge formed on a former beach by wind action, creating an offshore bar or barrier island, separated from the mainland by a lagoon.

QUESTIONS

23. Write an explanation of the sequence of events associated with the erosion of the headland shown in Figure 14.16.
24. a) Describe an experiment in a wave tank that would illustrate how sand spits and bayhead bars form.
    b) Either carry out the experiment or describe the sequence of events that would occur in forming such a feature.
25. a) Describe the formation of an offshore bar or barrier island.
    b) Use an atlas to locate examples of barrier islands along the coastline of the United States or Atlantic Canada.
    c) Why are many of these barrier islands the sites of tourist facilities?
26. Will sediments on a beach be well sorted or poorly sorted? Explain your answer.

14·8 Tides

Wave action, the main process in coastline formation, is powered by solar energy. **Tides**, on the other hand, are a product of the gravitational attraction of the moon and the centrifugal force of the earth-moon system.

Tides, the regular rise and fall of sea level that occur on most of the world's seacoasts, are caused by the combination of forces shown in Figure 14.20. The moon's gravitational attraction causes two bulges of water on the world's oceans. The bulge on the side nearest the moon results from gravitational attraction. The bulge on the farthest side results from the centrifugal force of the earth's rotation. As the earth rotates, the bulges are aligned with the moon, so the tides rise and fall twice every 24 hours.

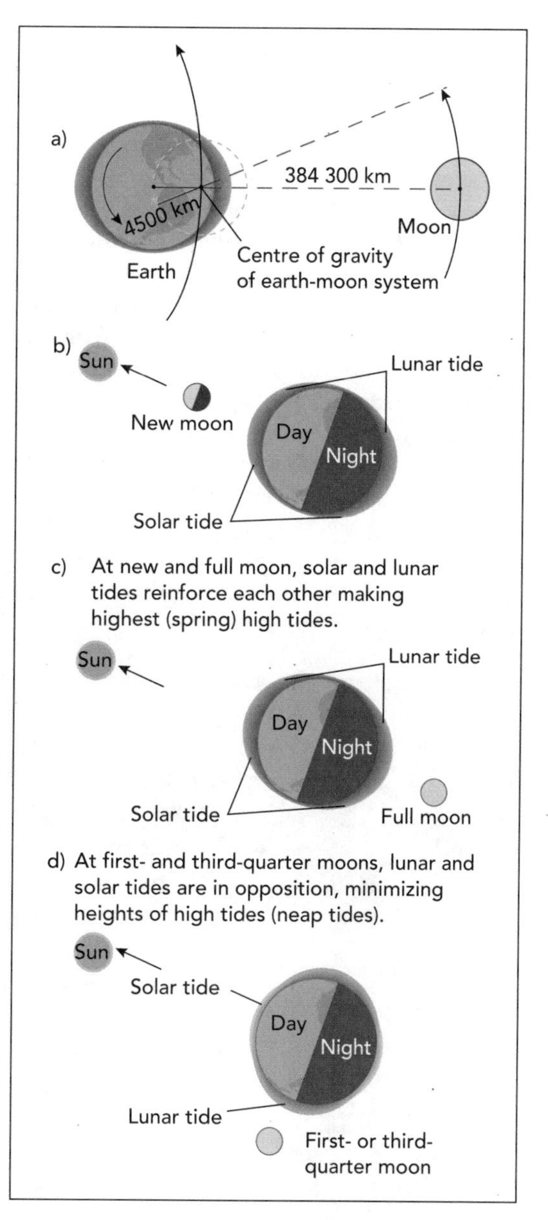

Figure 14.20 The Creation of Tides

When the sun, moon, and earth line up, the sun adds its gravitational attraction to that of the moon, creating even higher tides, known as **spring tides**. Such tides come every two weeks when there is a full or new moon. When the sun and moon are at right angles to one another, the lowest tides, or **neap tides**, occur.

The effects of tides are the rise and fall of the water level and the generation of tidal currents. Where tidal fluctuations are great, tidal currents may surge in and out of inlets, bays, and river estuaries, moving large quantities of sediment along the coast or over the shallow ocean floor. Tidal ranges up to 21 m have been recorded; the volumes of water movement are so great that they influence the shape of the coastline.

QUESTIONS

27. a) In what ways do tides affect people along the shore of an ocean?
    b) Describe a situation where tides can pose serious hazards for visitors to an ocean shoreline.
28. a) Explain why there are two high tides and two low tides each day.
    b) Why do the tides not occur at the same time each day?
29. A friend of yours is planning to purchase a home along an ocean shoreline. What advice would you give to your friend?

Review

- Glaciers have advanced across the higher latitudes several times over the past two million years. The earth appears to be in an interglacial period.
- Glaciers form where snow accumulation is greater than snow melting over an extended period of time.
- Glaciers are responsible for the creation of unique erosional and depositional features.
- Continental glaciation occurs over continental plains, while alpine glaciation takes place in mountainous areas.
- Periglacial landscapes develop under cold, dry climates, where glaciers do not form due to a lack of snowfall.
- Periglacial landscapes are underlain by permafrost, a condition that produces interesting and unique surface features.
- The energy that creates waves is derived from the sun.
- Wave action is limited to a very narrow range above and below sea level.
- Longshore drifts, created by the angular arrival paths of waves, are movements of water, sand, and gravel along shores.
- Erosion by waves carves features such as sea stacks and arches.
- Deposition of eroded materials in calmer areas creates features such as tombolos, spits, and offshore bars.
- Tides are created by the gravitational pull of the moon and sun on the ocean's waters.

Related Careers

- oceanographer
- engineer
- geologist
- archaeologist
- soil scientist
- civil engineer
- tour guide
- outfitter
- botanist
- surveyor
- conservationist
- statistician
- park ranger
- travel agent
- hydrographer
- geomorphologist
- planner
- biologist

Geographic Terms

firn
glacial ice
plastic deformation
shearing
basal slippage
snout
striation
glacial plucking
continental glacier
alpine or mountain glacier
rock basin lake
rock knob topography
bedrock plain
glacial till
glacial erratic
moraine
kettle
terminal moraine
lateral moraine
interlobate moraine
recessional moraine
ground moraine
drumlin
sand plain
clay plain
outwash plain
pitted outwash plain
spillway
esker
kame
icefield or snowfield
cirque
tarn
arête
horn peak
U-shaped valley
fiord
hanging valley
bridal veil fall
medial moraine
ice-dammed lake
braided
periglacial landscape
permafrost
frost action
active layer
solifluction
pingo
patterned ground
stone circle
breaker
longshore drift
abrasion
headland
corrosion
hydraulic pressure
cave
arch
sea stack
tombolo
wave-cut platform
bayhead beach
baymouth sand bar
spit
barrier island or offshore bar
lagoon
tide
spring tide
neap tide

Explorations

1. Draw separate hydrological cycles (see Chapter 10) to show reasons for sea level changes as the continental ice sheets (a) expanded and (b) retreated or disappeared. Include arrows of differing thicknesses to show the flow of water through the cycle, especially the flow from the continents to the oceans.

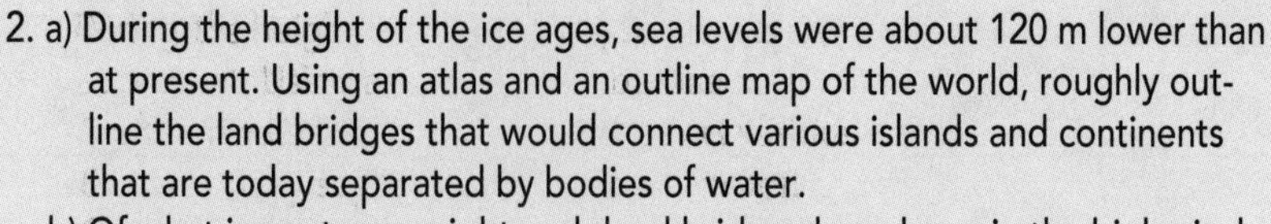

2. a) During the height of the ice ages, sea levels were about 120 m lower than at present. Using an atlas and an outline map of the world, roughly outline the land bridges that would connect various islands and continents that are today separated by bodies of water.
   b) Of what importance might such land bridges have been in the biological history of the world? How might this have changed the present world?
3. What impact has glaciation had on your local area? Research an answer to this question and record information about types and locations of features, the time since the last glaciation, and the benefits to present inhabitants.
4. Conduct research to find the major coastal processes operating to produce one of the following landscapes.
   a) Toronto Islands
   b) the white cliffs of Dover in Britain
   c) the Frisian Islands of the Netherlands
   d) the offshore bars and spits of the Prince Edward Island or New Brunswick coasts

   Describe the distinctive features of the place you research.
5. Oil and natural gas and other mineral exploration activity is a fact of life in periglacial landscapes in Canada, Alaska, and Russia. Prepare an environmental impact statement on the effects that such exploration might have on a periglacial landscape, including suggestions on how to adapt to, and minimize the effects on, the natural environment.
6. Research the special techniques necessary to construct buildings, roads, and other structures in a periglacial landscape.
7. Describe the impact of human activities on the processes operating in coastal areas. Some modifications you should consider are:
   a) the construction of dams and the diversion of rivers
   b) poor agricultural practices
   c) construction of breakwater barriers and harbour facilities

   Produce a three-page report to summarize your conclusions. Include sketches and diagrams where appropriate in your report.

Integrative Studies

UNIT

6

Human beings are an integral part of the ecosphere. We influence and are influenced by the earth's physical systems. The planet's natural systems are being altered in many planned and unplanned ways. Many of these "altered systems" are now threatening life on our planet. We need to understand the earth's physical systems and our impact on them if Planet Earth is to continue as we know it. The final three chapters of the text examine the impact of nature on people, important physical systems that we are altering, and some of the earth's fragile biomes.

As you work through this unit, consider these questions:

- In what ways can humans protect themselves against natural hazards?
- What similarities and differences exist among the various environmental problems?
- What role can the individual play in helping to solve environmental problems?
- What steps need to be taken to ensure that we leave a healthy planet for future generations?
- In what ways are most environmental problems also political and economic in nature?

CHAPTER

15

NATURAL HAZARDS: DISRUPTING HUMAN SYSTEMS

CHAPTER 1: The Nature of Physical Geography

CHAPTER 2: Earth: Its Place in the Universe

CHAPTER 3: The Earth in Motion

CHAPTER 4: The Earth's Interior

CHAPTER 5: The Earth's Crust

CHAPTER 6: The Lithosphere in Motion: Plate Tectonics

CHAPTER 7: Solar Radiation

CHAPTER 8: Climate

CHAPTER 9: Weather

CHAPTER 10: The Hydrosphere and the Hydrologic Cycle

CHAPTER 11: Natural Vegetation and Soil Systems

CHAPTER 12: Denudation: Weathering and Mass Wasting

CHAPTER 13: Distinctive Landscapes: Humid and Arid Environments

CHAPTER 14: Distinctive Landscapes: Glacial, Periglacial, and Coastal Environments

CHAPTER 15: Natural Hazards: Disrupting Human Systems

CHAPTER 16: The Disruption of Natural Systems

CHAPTER 17: Fragile Environments

OBJECTIVES:

At the end of this chapter, you will be able to:

- define "natural hazard";
- classify the various types of natural hazards that affect human beings;
- show the causes and impacts of natural hazards on human systems and societies;
- understand reasons for the increasing toll being exacted by natural hazards on humans;
- learn how predictable such natural hazards are and how their destructive effects can be reduced by long-range planning;
- explain the great loss of life in developing nations resulting from natural hazards;
- appreciate efforts to reduce the loss of life due to natural events in all nations, and especially in the developing nations of the world.

Introduction

A tropical storm lashes the islands of the western Pacific; an earthquake destroys villages in the Andes Mountains of Peru; in Kansas, a tornado cuts a path of destruction through a quiet farming town. These are examples of natural events that occur frequently on this planet bringing great suffering and loss of life and property.

15·1 Natural Disasters: The Environment as Hazard

The idea that the natural environment can be hazardous is an important area of study in Physical Geography. Geographers involved in such studies explore how natural hazards affect people and property, how inhabitants of places view the hazards that occur there, how people adjust to such hazards, and how they attempt to prevent or reduce the impact of the hazards on their lives and property. All hazards involve some degree of human choice. For example, if people know a coastline is subject to hurricanes, they are making a choice if they continue to live there. Reasons for making choices that appear to be risky vary from person to person.

A **natural hazard** is an **extreme natural event** that can potentially affect people in negative ways. Common natural events that are hazards to human beings are snow avalanches, coastal erosion, droughts, earthquakes, floods, fog, frost, hail, landslides, lightning, snow, tornadoes, tropical cyclones (hurricanes), volcanic eruptions, and severe winds. When one of these events actually affects people, it is known as a natural **disaster**.

Almost daily, somewhere in the world, disasters bring death and destruction.

These events are particularly devastating in areas that have not tried to predict, prepare for, and/or prevent them. Failure to take precautions might result from shortages of money and resources, lack of knowledge, the difficulty of predicting a time of occurrence, or a lack of preparation for a predictable event. In cases where precautions are taken, disasters may still occur because the magnitude of the natural event was greater than expected, or the event occurred in an unexpected place.

Figure 15.1 Mount St. Helens Eruption, May 18, 1980

The Scale of Disasters

In an effort to analyse hazards, researchers have devised ways of measuring their impacts. Probably the most commonly used measure of the size or scale of a disaster is the loss of life and property it caused. The media usually quote these numbers in headlines or news bulletins when reporting disasters.

A second measure of the scale of disasters is the amount of energy they release. For example, the Richter scale of earthquake intensity uses a numeric scale to indicate the amount of energy released by an earthquake.

A third measure is the length of time disasters last. The natural event may endure for only a few seconds, minutes, months, or even years. A meteor impact would fit into the first time scale. On the other hand, a major drought might last years, as was the case in Ethiopia in the 1980s. Even where the event lasts only a few seconds, its impact may last for months, years, or decades.

Another measure considers how often disasters occur. In 1980, Mount St. Helens erupted for the first time in 123 years. On the other hand, the western Pacific Ocean from Japan to northern Australia averages 40 tropical cyclones a year, many causing death and destruction.

Each of these measures of the magnitude of disasters has strengths and weaknesses, but they do allow comparisons to be made.

Research points to an increasingly high loss of life in the poorer nations of the world, while in the wealthier nations, the trend is to lower number of fatalities but increasingly high property damage totals.

Analysing Natural Disasters

Studies of natural disasters usually focus on specific features or events, which often include:

- location,
- relationships to other similar disasters,
- length of the natural event causing the disaster,

- general and specific causes that brought the disaster,
- immediate effects on people, property, and the landscape,
- disruptions caused by the disaster to natural or human systems,
- long-term effects of system disruptions,
- predictability of such disasters, and
- possibilities for reducing losses of life and property.

This list of characteristics can be used to analyse case studies of natural disasters.

Overall Rank	Event	Grading of Characteristics and Impacts[a]							
		Degree of Severity	Length of Event	Total Areal Extent	Total Loss of Life	Total Economic Loss	Social Effect	Long-term Impact	Suddenness
1	Drought	1	1	1	1	1	1	1	4
2	Tropical Cyclone	1	2	2	2	2	2	1	5
3	Regional Flood	2	2	2	1	1	1	2	4
4	Earthquake	1	5	1	2	1	1	2	1
5	Volcano	1	4	4	2	2	2	1	3
6	Mid-latitude Storm	1	3	2	2	2	2	2	5
7	Tsunami ("tidal wave")	2	4	1	2	2	2	3	2
8	Bushfire	3	3	3	3	3	3	3	2
9	Sea-level Rise	5	1	1	5	3	5	1	5
10	Iceberg	4	1	1	4	4	5	5	2
11	Dust Storm	3	3	2	5	4	5	4	1
12	Landslide	4	2	2	4	4	4	5	2
13	Beach Erosion	5	2	2	5	4	4	4	2
14	Debris Avalanche	2	5	5	3	4	3	5	1
15	Creep & Solifluction	5	1	2	5	4	5	4	2
16	Tornado	2	5	3	4	4	4	5	2
17	Snowstorm	4	3	3	5	4	4	5	2
18	Ice at Shore	5	4	1	5	4	5	4	1
19	Flash Flood	3	5	4	4	4	4	5	1
20	Thunderstorm	4	5	2	4	4	5	5	2

[a]Hazard characteristics and impacts are graded on a scale of 1 (largest or greatest) to 5 (smallest or least).

Figure 15.2 **The Top Twenty Natural Hazards**

It's a Fact...

The United Nations has estimated that 3 million people have died as a result of natural hazards over the period from 1970 to 1990 and that 1 billion people have been permanently injured or left homeless by catastrophes.

QUESTIONS

1. a) Which of the natural hazards listed in Figure 15.2 do you think are likely to occur in the region where you live?
   b) Add any hazards that are likely in your area that are not listed in Figure 15.2.
   c) Compare your list of local natural hazards with others in the class and discuss reasons for the similarities and differences among the lists produced.
2. a) Formulate your own definitions of the terms "extreme natural event", "natural hazard", and "natural disaster". In a group, share your definitions with others and attempt to reach a consensus on the meaning of each term.
   b) Give examples of common natural hazards.
   c) Search through a daily newspaper to find examples of natural disasters. Summarize the ways in which people were affected by such disasters.
3. Decide which two of the measures of natural disasters given in Figure 15.2 are most useful, in your opinion. Explain your choices.
4. Prepare an argument to show that our society is more, rather than less, susceptible to disruption by natural events as our level of technology increases.

15•2 Atmospheric Hazards: Droughts and Floods

Atmospheric hazards occupy a number of positions at the top of the hazards' list. These hazards range from droughts and floods through various types of wind storms, including tornadoes, dust storms, and tropical cyclones. Other atmospheric hazards include storm surges along exposed coastlines, flash floods, snowstorms, blizzards, freezing rain, and bush and forest fires. In the oceans and seas, hazards related to the atmosphere include sea ice (icebergs), high waves, sea-level changes, and shoreline erosion.

Droughts and floods are discussed together in this section of the chapter because they often occur in the same regions. This may seem unusual at first, but the combination of drought and flood is common in many semi-arid regions. China is an excellent example. The country has experienced 1029 floods and 1056 droughts since records of such events were first begun over 2100 years ago. Many of these events were centred on the North China Plain, where millions have died as a result of the extremes of precipitation associated with these two hazards.

Droughts as Hazards

Droughts are extended periods of low rainfall during which the growth of food and other crops is drastically reduced or even completely halted. They are the most damaging natural hazard because they are often severe, widespread, and long lasting. It is difficult to get an accurate number of deaths that result from droughts since they last for such a long time that human systems, such as governments and record keeping, often break down under the stress.

Droughts are common in semi-arid to sub-humid climates where precipitation is irregular or unpredictable. A decrease in precipitation is not the only reason for the tremendous loss of life during such droughts. In most cases, human actions make the consequences of these droughts much greater than they might normally have been. This process, called **desertification**, is discussed in greater detail in Chapter 17.

Floods as Hazards

Floods are one of the most destructive of natural hazards. Between 1947 and 1987, more than 180 000 people died from flooding, mainly in the developing countries of the world. Floods result from a number of natural causes, including high precipitation over short periods, rapid snowmelt, coastal flooding during severe storms, and **tsunamis**. Tsunamis are ocean waves generated by large displacements of water caused by undersea earthquakes or volcanic eruptions.

Of the different types of floods, those caused by higher than normal rainfall are most common. But, a number of other factors, including:

- the steepness of slopes,
- the nature of the soils,
- the natural vegetation cover, and
- the degree of human modification of the area,

also help determine the severity and frequency of floods.

Under heavy rainfall conditions, the amount of surface run-off collected by the river system may exceed the capacity of the stream channel and the river will overflow its banks, inundating the flood plain. Where the flood plains are wide, such floods can affect thousands of square kilometres of land. Once the flooding subsides, the water slowly returns to the river channel and drains away.

Case Study: China's Sorrow

Over the past several thousand years, the Huanghe River has caused the deaths of millions of people who live on the densely settled, level flood plain at its mouth. It has truly earned the name "China's Sorrow".

"Huanghe" comes from the Chinese word for "yellow earth". The river was given this name because of the tremendous load of yellow-coloured sediment that it carries as it moves onto the North China Plain. This sediment builds up in the river bed as the river slows in its passage across the level plain. In order to keep the river from flooding the farms and villages along its banks, the natural levees have been raised and are maintained by human labour. Now, the river flows above its own flood plain. When flooding occurs, a large area is inundated by the river and millions of people risk drowning. The people also face famine resulting from the loss of crops in a region that can just barely feed its people.

Since 602 B.C., the river has broken its banks 1573 times. Twenty-six of these breaks have caused huge loss of life. One of the most spectacular occurred in 1852 when the Huanghe broke through its banks, changed course, and shifted its mouth from the Yellow Sea to the

Gulf of Chihli, 600 km to the north! The many channels occupied by the river across the North China Plain since 600 B.C. are shown in Figure 15.4.

The two most devastating floods of the Huanghe occurred in 1887 and 1931. In 1887, heavy rains fell over North China, causing the river to burst through its dikes and flood an area of 130 000 km^2. The floodwaters destroyed over 2500 villages and vast areas of farmland were flooded. In some places, it took up to two years for the floodwaters to disappear, making huge areas unfit for cultivation.

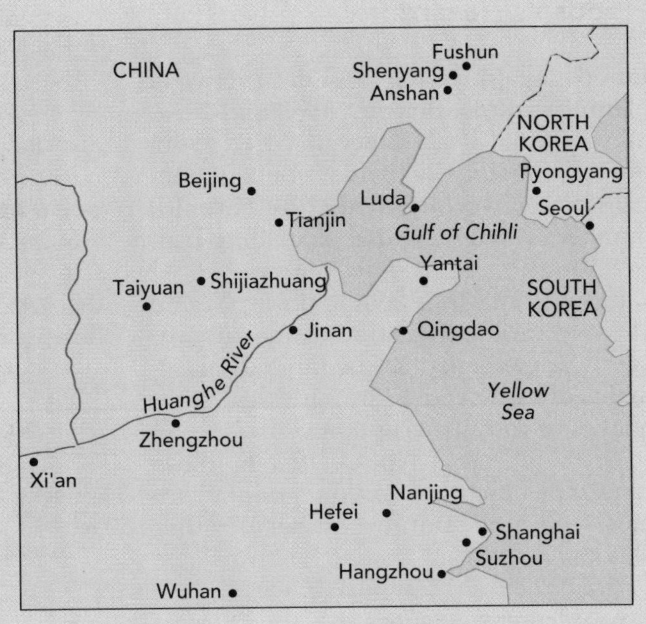

Figure 15.3 **Location Map of Huanghe River**

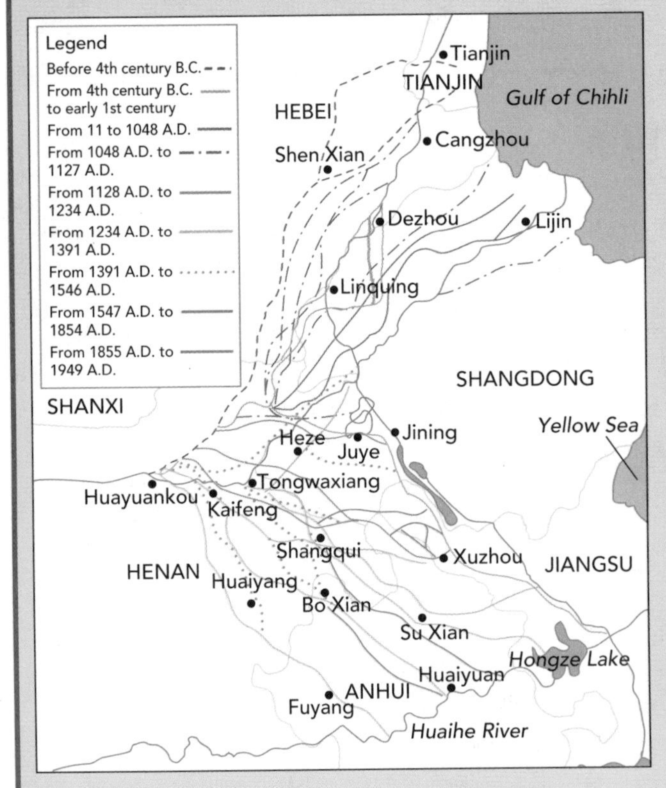

Between two and seven million people died in the flood and associated famine.

Similar disasters occurred in 1931 and 1939. The death toll reached an estimated 4 million people in 1931 and approximately 10 million people were made homeless and nearly 200 000 died in the famine that followed the 1939 floods.

Figure 15.4
The Changing Channels of the Huanghe River on the North China Plain

It's a Fact...

"Tsunami" is a Japanese term to describe large waves that inundate coastlines as a result of the displacement of the sea floor by underwater earthquakes. Tsunami is the correct term for what is incorrectly called a "tidal wave". Tsunamis occur most frequently in the Pacific Ocean since it is ringed by earthquake and volcanic belts. The waves can travel at speeds of 800 km/h and reach heights of 75 m.

QUESTIONS

5. a) List the various types of atmospheric natural hazards.
   b) Identify potential natural hazards in the region where you live, based on your own perceptions.
   c) From your list, rate the hazards you perceive as having the highest risk in terms of:
      i) frequency
      ii) danger to life
      iii) danger to property
      iv) magnitude of total effect
   d) Compare your perceptions with others in your group or class.
6. For what reasons are droughts ranked as the highest natural hazard in Figure 15.2?
7. Suggest reasons why it is so difficult to estimate the numbers of people who have died as a result of droughts.
8. a) List the factors that influence the likelihood and extent of flooding in a river system.
   b) Explain briefly how each of these factors might influence the flood potential of a river.

15·3 Atmospheric Hazards: Tropical Cyclones

Tropical cyclones are intense storms that originate over warm tropical seas. Since such storms get most of their energy from the evaporation of water from warm oceans, they are confined to tropical oceans and the coasts, as Figure 15.5 shows. Once over land, these storms quickly lose their strength and die. Tropical cyclones are called by different names throughout the world. They are known as hurricanes in North America, typhoons in east and south Asia, and willy willys in Australia.

Tropical cyclones are ranked as the second most important natural hazard. They were responsible for approximately 25 percent more deaths than earthquakes from 1960 to 1987. Such storms cause high death tolls because they occur frequently and often strike densely populated regions such as east, southeast, and south Asia. Rarely a year passes without reports of extensive death and destruction caused by tropical cyclones in either the Philippines, Taiwan, Japan, or coastal China.

Tropical cyclones cause death and destruction in three ways:

- through **storm surges** that are sudden rises in water levels caused by high winds and low atmospheric pressure. Storm surges can bring extensive flooding to low-lying coasts, destruction of coastal facilities through powerful wave action, and death by drowning.
- through extremely high winds, which can topple tall structures, rip roofs off

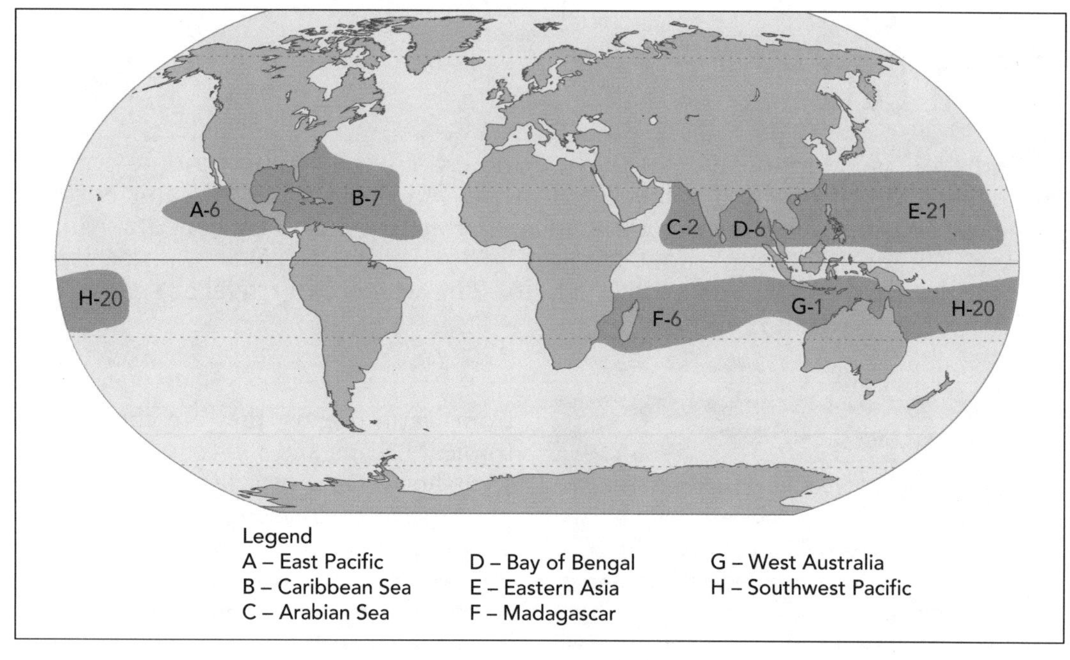

Figure 15.5 **World: Areas of Origin and Average Annual Number of Tropical Cyclones**

buildings, tear up trees by the roots, and fling debris through the air.

- through heavy rainfall in very short periods of time, which results in flooding and the contamination of water supplies, with a spread of diseases. Heavy rains can also trigger landslides and mudflows that may disrupt transportation and communications facilities, preventing relief efforts from reaching disaster zones.

With the potential increase in the earth's average temperature as a result of the greenhouse effect, it is expected that tropical cyclones will increase in strength and frequency as a result of the greater energy available for their formation.

It's a Fact...

There have been four tropical cyclones in recorded history that killed over 300 000 people, all of them in east and south Asia.

QUESTIONS

9. Suggest reasons why tropical cyclones are the second ranked of all the world's natural hazards.
10. Using Figures 9.11 (page 181) and 15.5, draw a world tropical cyclone hazard map. Divide the world into four zones: Little or No Risk, Low to Moderate Risk, Moderate to High Risk, and Extreme Risk. On the back of your map, summarize the criteria you used to establish the zones.

15•4 Atmospheric Hazards: Tornadoes

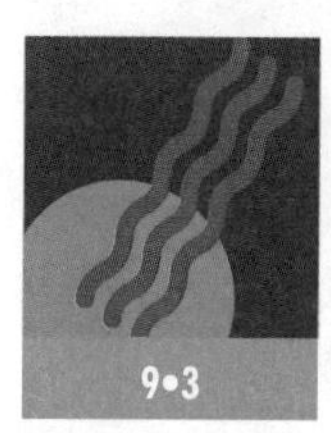

Tornadoes are among the most feared natural hazards in those parts of the world where they occur. Most tornadoes occur in the United States, but they also occur in the Great Lakes-St. Lawrence region and Prairie provinces of Canada, in Russia, the Ukraine, China, India, Bangladesh, Argentina, and Australia, where warm and cold air masses collide on a regular basis.

On the average, between 600 and 700 tornadoes occur each year, with an annual average death toll of about 160 people, worldwide. Despite the seemingly large number that form each year, tornadoes are relatively rare events that affect very limited areas. They also exist for much shorter periods than tropical and mid-latitude cyclones. Tornadoes threaten about 45 million people in the Midwest, Great Plains, and Gulf States of the United States. The average annual tornado death toll in the U.S. of 125 exceeds that of hurricanes (75 deaths). The unpredictability and higher frequency of tornadoes are two reasons why the annual average death rate from tornadoes, although decreasing over recent decades, is higher than that of hurricanes in the United States.

Case Study: Southern Ontario, May 31, 1985

A cold front was moving south across Georgian Bay and Lake Huron on May 31, 1985. It was pushing against a very warm, humid, tropical air mass that spread over the southern reaches of the province. Along the cold front between the two air masses, a line of thunderstorms developed, intensifying as the front moved over the southern end of Georgian Bay. All of this activity was monitored through radar and weather balloons by the Ontario Weather Centre. The meteorologists recognized that the atmospheric conditions were right for the formation of severe thunderstorms and tornadoes, and issued a Severe Weather Watch for the areas around Georgian Bay.

By late afternoon, the thunderstorms had moved into central Ontario, north of Toronto. The first tornado was sighted after 4:00 p.m. northwest of Toronto. Once news reached the Weather Centre, a tornado warning was issued for the areas north and east of Toronto.

As reports came in, it became clear that several tornadoes had developed; these became known as the "Black Friday" tornadoes. The paths of the tornadoes across central Ontario

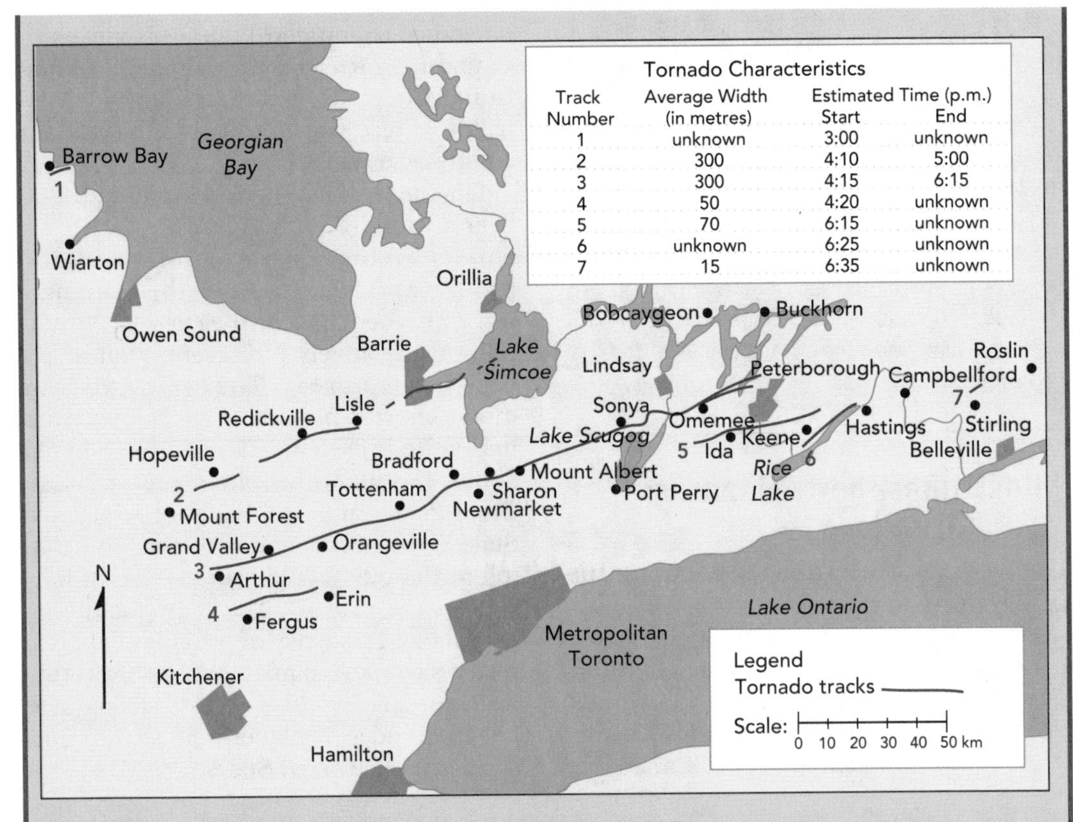

Tornado Characteristics

Track Number	Average Width (in metres)	Estimated Time (p.m.) Start	End
1	unknown	3:00	unknown
2	300	4:10	5:00
3	300	4:15	6:15
4	50	4:20	unknown
5	70	6:15	unknown
6	unknown	6:25	unknown
7	15	6:35	unknown

Figure 15.6 **Southern Ontario Tornado Map, May 31, 1985**

are shown in Figure 15.6. The longest path was 190 km long and about 300 m wide, for a tornado that lasted approximately two hours and killed four people. A second tornado lasted about 50 minutes, killed eight people, and cut a path 85 km long and 300 m wide. By 9:20 p.m., the front had moved east out of southern Ontario and the tornado warnings were cancelled.

Figure 15.7 **Damage Caused by Tornadoes, Southern Ontario, May 31, 1985**

When all reports were in, twelve people were dead and property damage amounted to over $100 million. Barrie, due north of Toronto, was the largest centre struck. A tornado cut a swath of destruction through the town, demolishing houses, stores, industrial buildings, a race track, and anything else in its path (Figure 15.7).

QUESTIONS

11. After reviewing the information on tornadoes in Chapter 9 on pages 183-86 and in this chapter, write a description of what it might be like, and how you would feel, if a tornado touched down in your community.
12. What characteristics of tornadoes make them such a feared hazard in regions where they occur?
13. Suggest reasons why tornadoes are not ranked as high in the list of natural hazards in Figure 15.2 as hurricanes, even though they are responsible for more deaths in the United States than hurricanes.
14. a) In a group, work out some possible rules or procedures to follow in your home or school to reduce the loss of life due to tornadoes.
    b) Share these ideas with other groups to make up a combined list of procedures.
    c) Point out the areas of Canada where such measures would be most needed.

15•5 Geological Hazards: Earthquakes

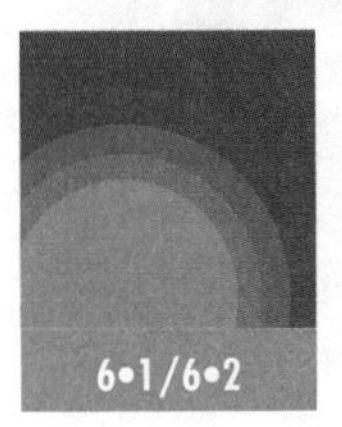

Geological hazards are largely related to movements of the earth's crust because of plate tectonics. Included in this group are earthquakes, volcanic eruptions, tsunamis, and various types of landslides and avalanches. Most of these hazards occur along the boundaries of tectonic plates.

Geologists explain **earthquakes** as the release of stresses and strains built up between massive tectonic plates. The plates are moving slabs of the earth's crust powered by convection currents, caused by the differential heating of the mantle by the decay of radioactive elements. This simple model gives earth scientists a powerful set of concepts to help predict the occurrences of most of the world's geological hazards.

The most frequently used measure of the magnitude of earthquakes is the Richter scale. Developed by seismologist Charles Richter, it gives an estimate of the amount of energy released by an earthquake, based on the amplitude of the waves recorded by seismographs and its distance from the seismograph. The method used to calculate the magnitude of an earthquake is shown in Figure 15.8.

This section will focus on the effects of earthquakes on human activities and ways of reducing the impacts of these hazards, particularly in densely populated areas.

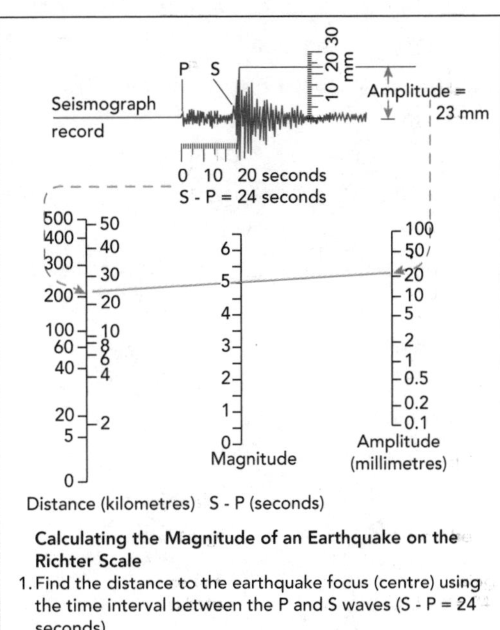

Calculating the Magnitude of an Earthquake on the Richter Scale

1. Find the distance to the earthquake focus (centre) using the time interval between the P and S waves (S - P = 24 seconds).
2. Measure the height of the maximum wave recorded by the seismograph (23 mm).
3. Draw a straight line between the points on the distance scale (left) and the amplitude scales (right) to obtain a magnitude of 5.0 on the Richter scale.

Figure 15.8 Calculating Earthquake Magnitude on the Richter Scale

Case Study: San Francisco, California, October 17, 1989

It was 5:04 p.m. on Tuesday, October 17, 1989, just 30 minutes before the start of the third game of the World Series. Sixty thousand baseball fans were packed into San Francisco's Candlestick Park to see the Oakland Athletics play the San Francisco Giants. Millions more were watching the pre-game show on television.

Suddenly, the stadium began to shake as shock waves from an earthquake rippled through a 300 km stretch of the coast of California. The quake was a result of crustal movements along the San Andreas fault, activity that was centred in Nisene Marks State Park, 90 km southeast of San Francisco and 22 km northeast of Santa Cruz. The quake measured 6.9 on the Richter scale of earthquake intensity, making it one of the most powerful earthquakes recorded in California since the devastating San Francisco earthquake of 1906.

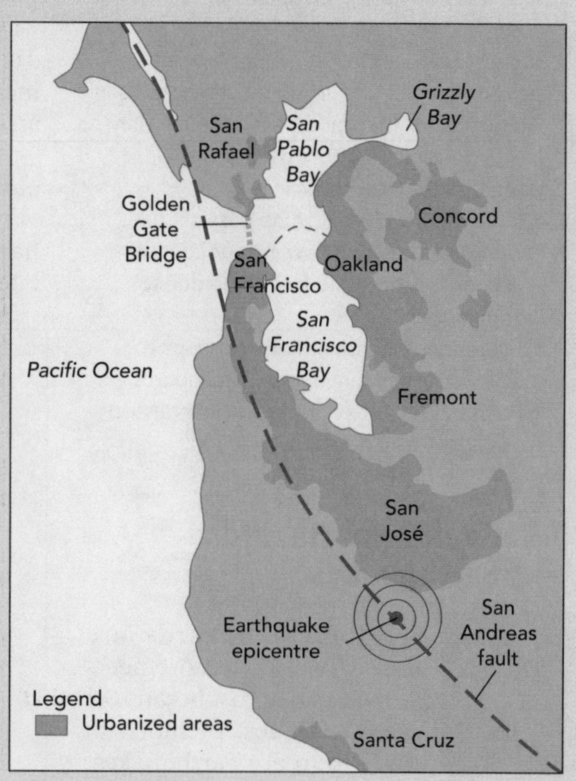

Figure 15.9 The "World Series" Earthquake Location Map

The ballgame was cancelled. The baseball stadium and its 60 000 fans emerged relatively untouched. Other parts of the San Francisco area were not so lucky. When it was over, 272 people were dead; 35 of them dying when a 2.3 km stretch of elevated freeway in Oakland collapsed onto rush hour traffic (Figure 15.10). A 15 m section of the Bay Area Bridge gave way, cutting one of the links between Oakland and San Francisco. Buildings constructed on unstable landfill in San Francisco's Marina District also collapsed. In Santa Cruz County, near the epicentre of the quake, 12 500 people were left homeless while flames spewed out of broken gas lines. Total property damage was estimated as high as $6 billion.

This was only one of approximately 15 000 earthquakes that occur each year in California. Most are so minor that they can only be detected by the most sensitive seismographs.

Such was the case with most of the 1400 aftershocks that followed the "World Series" quake. **Aftershocks** are lesser earthquakes that occur in the same area following a major earthquake. Even though they are less intense, they are dangerous because they might cause the collapse of buildings weakened, but left standing, by the major quake.

Figure 15.10 Damage Caused by the "World Series" (Loma Prieta) Earthquake

On the positive side, San Francisco's strict building codes and technological innovations in construction techniques saved high-rise office skyscrapers and apartments from structural damage. Some even swayed a metre or more during the height of the quake. The results proved that strict building controls can reduce losses of both life and property even though occurrences of earthquakes cannot be predicted with any certainty.

Contrasts with the 1988 Armenian earthquake, which also registered 6.9 on the Richter scale, are great. In Armenia, poor construction methods and an almost total disregard of the earthquake threat in designing many of the high-rise buildings resulted in their total collapse, even those only recently completed. This alone accounts for the great differences in the number of deaths and the degree of property damage. (In Armenia, an estimated 55 000 people were killed and property damage was in the $14-15 billion range.)

The San Andreas fault is one of the most closely monitored fault lines anywhere in the world. The U.S. Geological Survey spends millions of dollars and invests the time of many highly trained seismologists to collect and interpret data about earthquakes along the fault lines. Scientists are trying to discover ways to predict the timing, size, and locations of future major earthquakes. So far, they can suggest areas along the San Andreas fault where major earthquakes are likely to occur, what magnitudes they might have, and, in a very general way, when they will occur. They are a long way from being able to predict earthquakes with enough accuracy to warn people to evacuate unsafe areas and buildings; and they may never be able to do this.

It's a Fact . . .

- An estimated 1 million people died from earthquakes between 1970 and 1990.
- Many of Canada's largest earthquakes occur where the Pacific plate meets the North American plate. The boundary between the two plates is marked by the Queen Charlotte Island fault, which is located 150 km offshore in the Pacific Ocean.

QUESTIONS

15. Explain why wealthier parts of the world should be better able to reduce death rates resulting from earthquakes.
16. By examining Figure 15.2, discuss three ways in which earthquakes differ in nature from atmospheric natural hazards.

15·6 Geological Hazards: Volcanic Eruptions

After earthquakes, volcanic eruptions are the most spectacular geological hazards. Most of the world's volcanic activity occurs either along plate boundaries or above hot spots within the plates. Because of the nature of the magma, eruptions of volcanoes along separating boundaries at the mid-ocean ridges are less violent and dangerous than those along converging boundaries associated with island arcs and continental margins.

Case Study: Nevado del Ruiz, Colombia, November 13, 1985

By November of 1985, Nevado del Ruiz, a 5389 m volcano in the Andes Mountains of Colombia (Figure 15.11), had been rumbling and shaking for many months. In September, there had been a small steam explosion in the crater at the summit of the volcano. Still, the 25 000 townspeople of Armero did not listen to geologists' warnings of the danger of an eruption. They also forgot that 1000 people had died in mudflows from the same volcano 140 years before, and that Armero was actually built on top of those mudflows.

On November 13, a relatively small eruption of hot pumice and ash melted part of the large ice and snowfield that occupied the top of the volcano. The water quickly produced massive mudflows that picked up ash, soil, rocks, trees, and anything else in their path. These mudflows rushed down the canyons draining the steep sides of the mountain at speeds of up to 60 km/h. One started down the canyon of the Lagunillas River, which led to the town of Armero, 50 km from the volcano. Two hours after the start of the eruption on

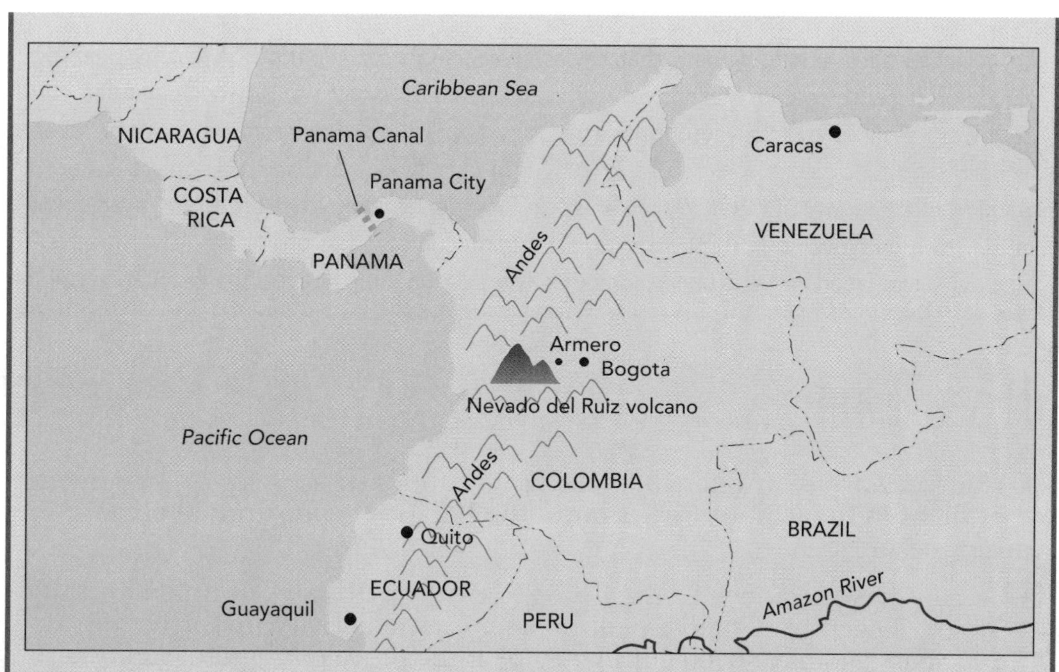

Figure 15.11 Location of Armero and Nevado del Ruiz Volcano, Colombia

Nevado del Ruiz, a 30 m high mudflow instantly swept away the town of Armero and all its inhabitants (Figure 15.12). The townsite and its surrounding farmlands were left covered with a one-metre-thick layer of sulphurous volcanic mud, studded with boulders up to ten metres in diameter.

After the disaster, survivors searched the debris for bodies of their loved ones. Relief supplies from around the world were slow to reach the victims. Much of this can be blamed on confusion and a lack of co-ordination among the various relief agencies in the area. In addition, damaged roads and communications facilities also hampered relief workers.

Looking back, it is clear that the tragedy could have been prevented. Nevado del Ruiz has a history of volcanic activity and mudflows going back at least to 1595.

Figure 15.12 Destruction of Armero caused by Nevado del Ruiz Volcano

The volcano gave ample warning that it was awakening from a period of rest by belching clouds of ash and steam for almost a year before the fatal eruption. However, alarmed citizens were assured by local authorities that they had little to fear from the volcano, even when the volcano became increasingly active in early November and geologists issued a warning of the threat of mudflows. Even as an emergency committee urged the evacuation of the town on November 13, the local radio station continued to urge calm. Given conflicting advice and faced with the uncertainty of leaving, few people heeded the evacuation call.

It's a Fact . . .

- There are approximately 650 active volcanoes in the world; 75 percent of them in the Rim of Fire around the Pacific Ocean, and 20 percent in Indonesia.
- Canada has volcanoes in the Coast Ranges of British Columbia and the Yukon. Legends of native peoples of volcanic activity in Terrace, B. C., 200 years ago have been confirmed by radioactive dating of lava flows in the area.

QUESTIONS

17. Suggest measures that could be used to reduce the death toll from volcanic eruptions.
18. What lessons learned from Nevado del Ruiz give hope that future losses due to volcanic eruptions can be reduced?

15•7 Natural Hazards: An Overview

Natural disasters continue to take their toll of human life and property. In response to the growing number of deaths and the rising loss of property, the United Nations General Assembly declared the 1990s as the International Decade for Natural Disaster Reduction. The focus on disaster reduction takes the form of coordinating research, data gathering, and information sharing among all the world's nations. Much of the effort centres on the key natural hazards such as floods, hurricanes, landslides, earthquakes, and droughts. The report that suggested the establishment of such an international effort praised the emergency relief that is provided by many of the world's nations to help victims of catastrophe. On the other hand, it expressed concern that much of this effort was directed at the results of the disaster rather than trying to reduce the vulnerability of people to natural hazards.

The reduction effort is made possible by the growing range of electronic **remote-sensing** devices mounted on satellites and placed on the earth's surface that can gather enormous amounts of information quickly, inexpensively, and accurately, despite their locations. High speed supercomputers can process massive amounts of data to make more accurate climatic and weather forecasts for the entire globe. The intention of this U.N. program is to give even the poorest nations access to information that can make the difference between life and death for many of their citizens.

A United Nations specialist was quoted as saying: "Remote-sensing offers the first and vital early warning, while flood control technologies, safer building techniques and land surveying give planners the option to resettle populations in less vulnerable areas." As this quotation indicates, air and land based instruments can pick up the warning signs of impending disasters with great accuracy. In addition, inexpensive microcomputers and communications technology can make the processing and sharing of these warnings so convenient and easy that it is no longer defensible "to leave poor and vulnerable populations defenceless in the face of predictable disasters."

One example of the effectiveness of such technologies was described by a Canadian scientist. He pointed out that satellites equipped with sensing devices readily available today could easily determine the depth of snow cover on the Himalayas. This data could then be used to predict the volume of run-off during the spring melt period, making it possible to give early warning of a flood hazard in the densely populated, flood-prone Ganges-Brahmaputra delta of India and Bangladesh.

If humanitarian arguments are not enough, a U.N. official said: "the cold, hard cost of cleaning up after floods, earthquakes, and other natural disasters should convince politicians . . . that a satellite dish antenna and a small computer in high-risk communities would be cheap insurance" against many natural hazards.

QUESTIONS

19. A Japanese proverb says: "A natural calamity will strike at about the same time the terror of the last one is forgotten."
    a) Suggest reasons why this is a proverb with some degree of truth to it.
    b) If this proverb continues to be true, suggest reasons why the worldwide toll on human life and property will continue to increase into the twenty-first century.
20. a) Point out the differences between natural hazards and their impact on the developed as opposed to the developing nations of the world, as shown by the case studies in this chapter.
    b) What reasons account for these differences and what might be done about them?
    c) Do the developed nations of the world have some obligation to help change this situation? Outline your own views on this subject and discuss them with others in a group.
21. Suggest reasons why many of the world's wealthiest nations support the United Nations International Decade for Natural Disaster Reduction.

It's a Fact . . .

Since the late 1960s, cyclones, floods, volcanic eruptions, and other natural disasters have caused an estimated $100 billion damage to property.

Review

- The environment, while it supports human existence on earth, must be viewed as a hazard at times in certain locations.
- Natural hazards become natural disasters if there is loss of life and property.
- Droughts and floods often occur in the same places, usually semi-arid parts of the world.
- Other atmospheric hazards are tropical cyclones and tornadoes, which are more typical of mid-latitude locations.
- Developing parts of the world do not have the resources to adequately warn residents of natural hazards, or to respond rapidly and effectively should disasters occur.
- Earthquakes and volcanoes are considered geological disasters. They are most commonly located along the boundaries between crustal plates.
- The United Nations is attempting to focus research efforts on natural hazards to reduce the costs of disasters in terms of loss of life and property damage.

Geographic Terms

natural hazard
extreme natural event
disaster
atmospheric hazard
drought
desertification
flood
tsunami
tropical cyclone
storm surge
tornado
geological hazard
earthquake
aftershock
remote-sensing

Explorations

1. Keep a diary of natural disasters that occur in the world for a month or more and record their key features as outlined under "Analysing Natural Disasters" on pages 304-305. As an addition, or alternative, consult the vertical files or use the yearbooks of encyclopedias in the library/resource centre to compile a list of such disasters. Once collected, analyse the results to:
   a) point out additional types of natural hazards/disasters that were not discussed in the case studies in this chapter;
   b) describe what pattern of location they had over the earth's surface that was noteworthy;
   c) decide whether they fit the pattern of higher loss of life in the developing nations of the world and higher property damage in the developed nations;

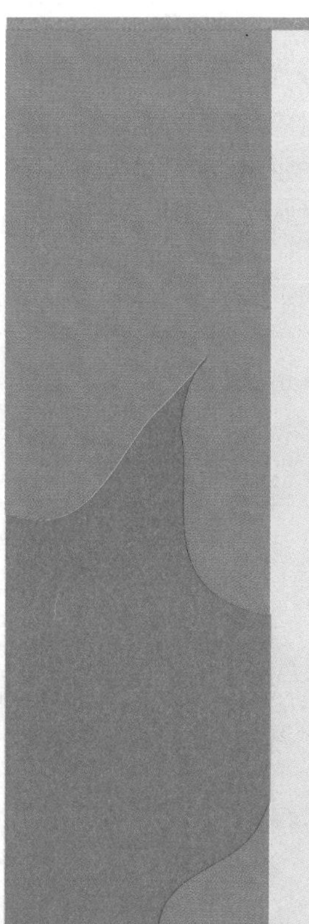

d) identify what preventative measures were, or might have been, taken to reduce the loss of life and property;

e) point out any other interesting or noteworthy findings that your study brought out.

2. a) Research a specific natural hazard and find out the types of predictive and preventative measures that might be, or have been, taken in some places to reduce the loss of life and/or property.

   b) Analyse your findings to draw conclusions about how much it is possible for prediction and prevention to reduce losses of life and/or property.

   c) Comment on the cost-effectiveness of continued investments in prediction and prevention technologies and methods.

3. Research the differences in predictability between atmospheric and geological natural hazards.

4. Prepare a report on the risks presented by natural hazards in your local community or region. In the report, indicate the hazards, their likelihood of occurrence (the risk), rank them according to their potential impact, and identify measures that might/should be taken to predict them and to reduce the toll of human life and property that they might inflict.

5. Research one of the natural hazards listed in Figure 15.2 that was not covered by the case studies in this chapter. Prepare a case study of a disaster that resulted from such a natural hazard, including in your report its key features, as listed on pages 304-305.

CHAPTER

16

THE DISRUPTION OF NATURAL SYSTEMS

OBJECTIVES:

At the end of this chapter, you will be able to:

- use a systematic approach to investigate changes in natural systems that are caused by humans;
- understand the causes and effects of selected environmental problems that have arisen as a result of human interference in natural systems;
- appreciate the seriousness of human intervention in natural systems;
- clarify your own values as they relate to the environment;
- understand the different solutions that can be employed to lessen the impact of humans on the environment.

CHAPTER 1: The Nature of Physical Geography
CHAPTER 2: Earth: Its Place in the Universe
CHAPTER 3: The Earth in Motion
CHAPTER 4: The Earth's Interior
CHAPTER 5: The Earth's Crust
CHAPTER 6: The Lithosphere in Motion: Plate Tectonics
CHAPTER 7: Solar Radiation
CHAPTER 8: Climate
CHAPTER 9: Weather
CHAPTER 10: The Hydrosphere and the Hydrologic Cycle
CHAPTER 11: Natural Vegetation and Soil Systems
CHAPTER 12: Denudation: Weathering and Mass Wasting
CHAPTER 13: Distinctive Landscapes: Humid and Arid Environments
CHAPTER 14: Distinctive Landscapes: Glacial, Periglacial, and Coastal Environments
CHAPTER 15: Natural Hazards: Disrupting Human Systems
CHAPTER 16: The Disruption of Natural Systems
CHAPTER 17: Fragile Environments

Introduction

Acid rain, water pollution, endangered and extinct species — these are some of the results of the disruption of natural systems. Some disruption is inevitable as human beings produce food, build cities, and manage the planet on which we live. Our environmental problems today are a result of many planned and unplanned alterations of the earth's natural physical systems. These alterations have brought about changes that we did not anticipate and that we do not fully understand. The changes threaten our economic, social, and physical health. To understand and begin to cope with the negative results of change, we must understand the natural physical systems and examine how these systems have been altered by human activity.

Natural systems are very complex. We are only just beginning to understand the processes involved in these systems and the interrelationships among the systems themselves. In order to help us investigate environmental problems, we will use the following methodology in this chapter:

a) Identify the problem.
b) Review the natural system being affected.
c) Discuss how the natural system is being altered.
d) Analyse the ways in which the altered system is affecting humans and Planet Earth.
e) Generate possible solutions to the problem.

The solutions to most environmental problems can be broken down into five areas — prevention, treatment, knowledge, adjustment, and philosophy. *Prevention* and *treatment* of the problem can move altered systems back to more natural ones. Ongoing research and study are necessary to increase our *knowledge*, which will make our prevention and treatment techniques more efficient. Education is an important component of this research and study, as informed citizens will support and understand the treatment and prevention policies. *Adjustment* to the altered systems produced by human activity might be necessary to some degree. Ultimately, the solution to environmental problems lies with developing an environmental *philosophy* that places the care of Planet Earth as a top priority in all economic and developmental decision making.

This chapter examines three natural systems and how they have been altered by human intervention.

16•1 Case Study: The Greenhouse Effect

Problem

There is increasing evidence that the earth is experiencing a "greenhouse effect". The greenhouse effect refers to the warming of our planet as a result of an increase in longwave radiation, and thus also the net radiation received at the earth's surface. This increase is caused by the increase in the amount of several gases in our atmosphere. These gases are known as the **greenhouse gases**.

The Natural System

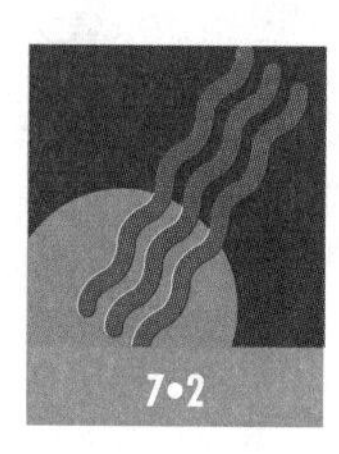

The solar radiation balance was discussed in Chapter 7 and illustrated in Figure 7.5. The sun provides energy to the earth in the form of electromagnetic waves. This energy, called solar radiation, fuels most of the natural systems operating on Planet Earth. Because the sun is so hot, it emits shortwave radiation. Some of this radiation travels through the earth's atmosphere to the earth's surface directly. Some of the energy is scattered and reflected by the earth's atmosphere and by the surface of the earth. Some of the energy is absorbed by the earth's atmosphere and surface which emit longwave radiation. When all these radiation fluxes are taken into account, we are left with net radiation, an amount of energy available for use at the earth's surface. This energy is used to heat the earth and the atmosphere, to evaporate water, and to allow photosynthesis to occur.

The Altered System

Over the past few hundred years, human activities have been altering the natural system indicated in Figure 7.5. This alteration has occurred due to the introduction into the atmosphere of the greenhouse gases. Figure 16.1 outlines the most important of the approximately thirty gases which are now designated as greenhouse gases. Figure 16.1 also indicates the sources and increasing magnitudes of these gases in our atmosphere.

The greenhouse gases prevent outgoing longwave radiation from leaving the earth-atmosphere system in the same manner in which glass in a traditional greenhouse allows shortwave radiation to penetrate into the greenhouse but prevents longwave radiation from escaping. The greenhouse gases are particularly efficient at absorbing longwave radiation in the 7 to 19 micron wavelengths. These wavelengths correspond to the peak wavelengths of the longwave radiation emanating from the earth's surface. The

Gas	Sources	Magnitudes	Relative Contribution to the Greenhouse Effect as of 1990 (estimated)
Carbon dioxide CO_2	Carbon dioxide is produced by the burning of fossil fuels such as coal, oil, and natural gas. CO_2 is thus a by-product of most industrial processes.	From a "norm" of 280 ppm in pre-industrial times, CO_2 is now at 350 ppm and, if the present trends continue, will reach 500 ppm by the year 2040.	60%
Methane CH_4	Methane is produced by the bacterial decomposition of organic matter. Cow dung is a major source, as are landfill sites.	The methane level in the earth's atmosphere has been rising at a rate of 1% per year since 1980. Methane concentrations have doubled over the last 500 years and are now at 1.4 ppm.	15%
Chloro-fluoro-carbons CFCs	These chemicals are used as refrigerants, cleaning solvents, and in the making of plastic foam. The gases are released as these products are used, burned, or decay.	Levels of chlorofluorocarbons have been rising at the rate of 5%–6% a year. CFCs are very stable. They can last for 100 years in the atmosphere. They take various forms and are also known as the freon gases.	11%
Ozone O_3	Ozone occurs naturally in the stratosphere but is present at lower altitudes due to the burning of fossil fuels.	Although O_3 is extremely variable in the lower layers of the atmosphere, the concentrations of this gas have been rising steadily.	7%
Nitrous Oxide N_2O	This gas is produced by the burning of fossil fuels and by the increased use of nitrogen-based fertilizers.	Since World War II, the levels of nitrous oxides in the atmosphere have increased rapidly. The amount of N_2O in the atmosphere will triple by the year 2025 if present trends continue.	4%

* *Note:* The chart contains those greenhouse gases whose concentrations are being altered by human intervention. Water vapour is a greenhouse gas also; however, we are not sure to what degree we are altering its atmospheric amount, nor are we sure of its contribution to the overall warming of the planet.

Figure 16.1 The Major Greenhouse Gases

gases absorb outgoing longwave radiation and radiate much of the energy back to earth as incoming longwave radiation. The net effect is to increase the amount of incoming longwave radiation and thus the amount of energy available for use at the earth's surface. This excess energy is changing the earth's climatic, hydrologic, and biological systems. Figure 16.2 indicates the altered radiation system as a result of the increase in greenhouse gases in our atmosphere. The longwave fluxes have been highlighted as these are the exchanges that are being altered.

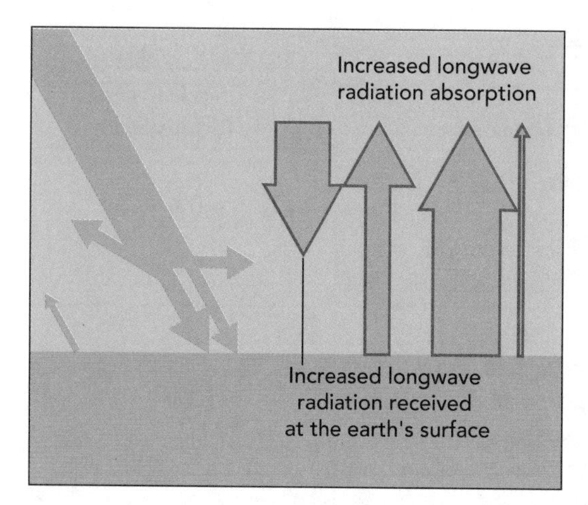

Figure 16.2 **The Greenhouse Effect**
Note: Compare this figure with Figure 7.5 on page 124.

Compounding the increase in these greenhouse gases is the increase in the rate of deforestation on Planet Earth. The forests of the world represent a **carbon dioxide "sink"**. They absorb atmospheric carbon dioxide and give off oxygen. As we deforest the planet, we eliminate a means by which to rid the atmosphere of excess carbon dioxide. In addition, especially in the tropics, much of the cut forests are burned, thus directly adding carbon dioxide into the atmosphere.

Early studies on the greenhouse effect concentrated on carbon dioxide; however, it is now believed that the combined effects of all the other greenhouse gases will be at least equal to the effects of the increasing carbon dioxide concentrations by the year 2020. Scientists are particularly concerned with the increase in the **chlorofluorocarbons**, also known as the freon gases or CFCs. As indicated in Figure 16.1, the concentration of these gases is growing at a rate of 5 to 6 percent annually. If this rate continues, CFCs could become the dominant greenhouse gas by the early twenty-first century. A molecule of a CFC gas is some 10 000 times more efficient at absorbing longwave radiation than is a molecule of carbon dioxide. In addition, CFCs remain in our atmosphere for decades, much longer than the other greenhouse gases. CFCs are also causing the breakdown of the atmospheric ozone layer.

Effects

Studies over the past decade have led to a growing consensus by most scientists on the following two statements:

1. The greenhouse effect is not a "future" event, it is happening now; and
2. The greenhouse effect cannot be stopped.

The net radiation received from the sun is used by the earth in one of four ways. In Chapter 7, we called the equation that illustrates how the earth uses solar radiation the "Energy Balance". An increase in the net radiation available at the planet's surface means that all the systems that depend upon the input of solar radiation will and are being changed. Climatic, hydrologic, and biological systems are being influenced by the increase in the net radiation available on the planet.

The fundamental change that is occurring is an increase in the heating of the earth's surface. This change will bring about many other changes. There is uncertainty concerning the degree to which the earth will heat up. Estimates are based on computer models of the earth's climate and on existing trends. The sophisticated models now developed by many countries can simulate past and present climates with a high degree of accuracy. By changing the terms in the models based on present trends, we can predict future climates. For example, by changing the level of carbon dioxide from 350 ppm (parts per million), its present level, to 500 ppm, its predicted level in the year 2040, the models will generate future climates based on this new data. The models predict increases of from 1.5°C to 5.0°C on a global scale. The uncertainty occurs because so many different variables are considered in the models. Some of the major variables are the trends in the uses of all of our fossil fuels, the ability of the oceans to absorb excess atmospheric carbon dioxide, the rate of deforestation on the planet, the sensitivity of the atmosphere to increases in the greenhouse gases, and the natural variations of the earth's climate.

The implications of a warmer planet are so great that it is difficult to predict and categorize all of the changes that might occur. Some of the most significant effects that just one country might experience are summarized in Figure 16.3.

Coastal Changes

As Figure 16.3 indicates, all countries with ocean coasts face the possibility of much higher sea levels. As temperatures rise, the rate of melting of the ice at the

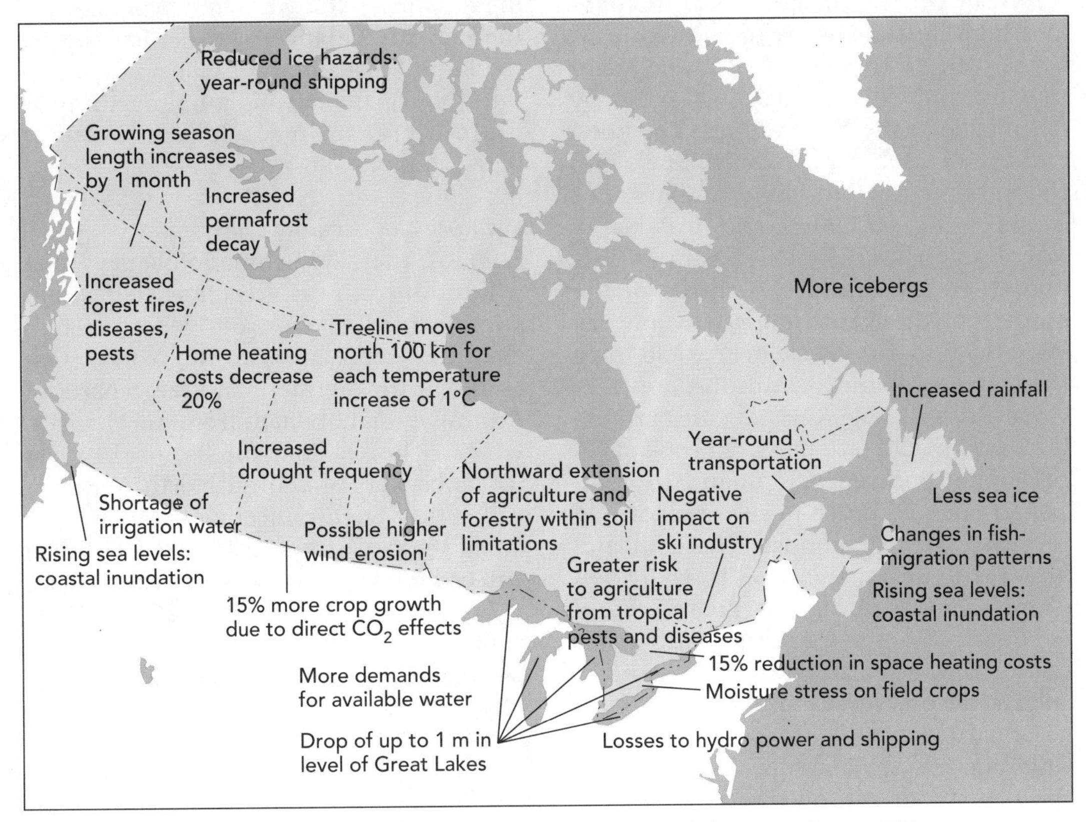

Figure 16.3 **Possible Consequences for Canada as a Result of the Greenhouse Effect**

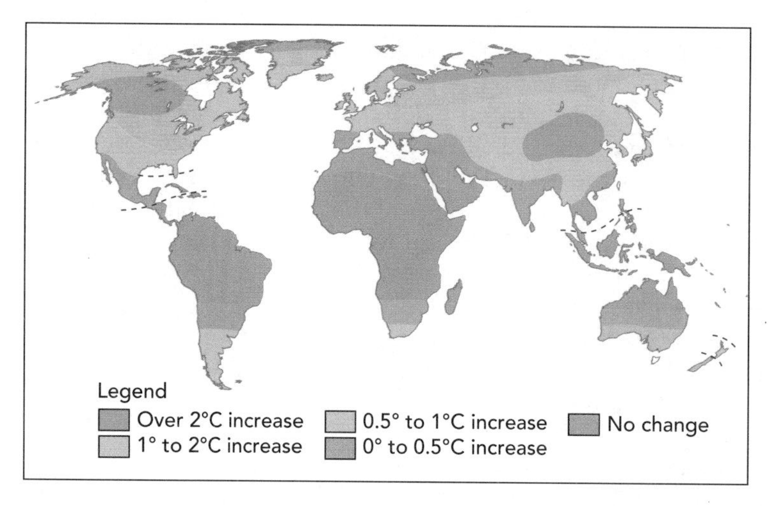

Figure 16.4
Greenhouse Effect: Possible Temperature Changes

north and south poles, and of all glaciers and snow fields, will increase. All of the models indicate that the largest temperature rises will be experienced in the mid- and high latitude areas of the world. A conservative estimate is that a one-metre rise in worldwide sea levels will occur by the middle of the next century. The flood damage of such a rise would reach into the hundreds of billions of dollars. For example, 10 000 000 people in Pakistan now live on land that is 1.5 m or less above sea level, and two thirds of the country of the Netherlands is below sea level. A 1.5 m rise in sea level would force the movement of 20 million people in Great Britain alone. Coastal cities around the world would face massive problems of relocation and redevelopment. In addition, a one-metre rise in sea levels would increase tides by almost 2 percent and also result in an increase in the flood damage caused by hurricanes and typhoons. The encroaching salt water could contaminate freshwater aquifers such as the one supplying Miami, Florida.

Of particular worry is the Ross or West Antarctic Ice Sheet, which forms part of the ice pack found on the land mass of Antarctica. At the moment, this ice sheet, which consists of over two million cubic kilometres of ice, is held in place on underwater bedrock by two partly floating ice sheets. If a warmer ocean disrupts the delicate balance of these ice sheets, the large West Antarctic Sheet could slip into the ocean, causing an almost immediate increase in world sea levels of between three and five metres.

Weather and Climatic Changes

Some of the likely climatic changes that will result from the greenhouse effect are summarized in Figures 16.4 and 16.5. Note that some areas of the world, such as India, the Ukraine, and large parts of Australia, might benefit from the changes, while other areas, most notably the "breadbasket" of central North America, will suffer. Such radical climatic changes could impact on the balance of political and economic power in the world.

Weather patterns will be severely disrupted. Increased incidences of tropical cyclones will occur as warmer ocean waters give birth to more tropical storms. In addition, some models predict that the intensity of hurricanes will increase by

Figure 16.5
Greenhouse Effect: Possible Precipitation Changes

40 to 50 percent. As winters become warmer, less precipitation will fall as snow and more as rainfall in the mid- and high latitudes. This factor alone can have severe consequences for river basins and reservoirs, as run-off from the rainfall will occur immediately, causing many areas to lose their spring and summer supply of water, which now accumulates as snowfall over the winter months. Increased incidences of heat waves and droughts will affect the middle of North America. The decade of the 1980s produced the four hottest summers experienced in Toronto, Ontario, since records have been kept.

Biological Changes

In theory, since plants use carbon dioxide to grow, increased levels of CO_2 should result in increased plant growth. Combined with warmer winters and longer growing seasons, the impact on agriculture should be a positive one. However, as Figure 16.5 indicates, many areas will become drier, necessitating large-scale irrigation projects. In addition, while the climate of areas such as northern Canada becomes more suitable for agriculture, the soils of these regions might not be able to sustain the demands of crops. Of particular concern is the "breadbasket" of North America. This area of the world currently produces the majority of the planet's surplus food supplies. All of the climatic models predict that this "breadbasket" will become much drier in the future, which would lead to a decrease in agricultural production. Hotter weather means more evaporation, less ground water, and drier soil conditions.

A further problem for agriculture lies with the possible infestation of tropical pests and parasites. As mid-latitude areas become warmer, they will become part of the ranges of a host of insects, diseases, and bacteria which are now confined to the tropics. Also, longer growing seasons will give existing insect pests more time to multiply and might result in greater agricultural damage than at present.

As the climate warms, forests in the northern hemisphere should move northwards from the existing tree line. One estimate by Environment Canada suggests that a 1°C rise in temperature will result in a northward move of 100 km in the tree line. Warmer, drier conditions will increase the risk of forest fires in many

areas, while some areas wi
in the amount and rate of f
Ecosystems across the
under stress as they atten
the changes in the world's
animal populations wil
migrate in order to m
desired living conditions.
will flourish, while other
threat of extinction. For
fish species such as lake
sensitive to water temper
water temperatures monit
northern Ontario have be
the late 1960s, putting fish

Changes in the Hydrologic Cycle

Although the amount of water on the planet will not vary, the greenhouse effect holds the potential to significantly change the "storage" reservoirs discussed in Chapter 10. The oceans will contain more of the earth's water resources as ice caps and glaciers melt and ocean levels rise. Evaporation rates will increase due to warmer temperatures, and the ability of the atmosphere to hold water vapour will increase due to the higher temperatures. Parts of the world will become drier, while other parts will become wetter. For North America, it appears as though groundwater supplies will decrease, as will the amount of water available in many of our lakes and rivers. Many computer models predict a decrease in water levels in the Great Lakes system of between 1 and 2 m over the next hundred years. At the same time, demand for the fresh water of the Great Lakes system will increase as irrigation becomes more necessary and as pressure increases to divert water to the ever drier areas of the continent. A decreased

in the areas of prevention, adjustment, and knowledge. Most scientists now believe that it cannot be reversed or stopped. It is happening now and will continue at least through the next century. The Environmental Protection Agency in the United States examined a number of possibilities which would slow down the impact of the greenhouse effect. That organization found that even such drastic action as a 300 percent tax on fossil fuels would only delay a 2°C warming of the planet by five years. A complete ban on coal would delay the 2°C change by only 15 years. If such statistics are correct, then we must learn to adapt to the changes which the greenhouse effect will bring about. In your lifetime, you will experience a climatic change greater than that which has occurred over the last 10 000 years. We need to continue to work on perfecting our models and our understanding of the processes involved in this change. Knowledge will enable us to plan and to attempt to adapt to this new physical world we are creating. International cooperation is necessary and worldwide strategies are now being developed to cope with many environmental problems. The World Commission on Environment and Development was created by the United Nations in 1983 and global conferences specifically on the atmosphere are now held annually.

Many nations have now committed themselves to reducing the amount of greenhouse gases which they release into the atmosphere. This can be achieved by encouraging and, if necessary, forcing industries to dramatically cut their emissions of greenhouse gases. If we look at carbon dioxide as an example, whenever one tonne of fossils fuels is burned, four tonnes of carbon dioxide are released into the atmosphere. It is estimated that at least five gigatonnes (five thousand million tonnes) of fossil fuels are burned worldwide each year and this sum is increasing. In the last ten years, we have added as much CO_2 into our atmosphere as we did in the previous one hundred years. Clearly, part of the solution lies with reduced emissions.

Automobile engines must be made more efficient and residential heating and insulation procedures must be improved. Alternative energy sources such as solar, wind, and geothermal will need to be further developed and energy conservation stressed.

The earth's forests play a vital role in regulating the climates of the world. Our forests must be preserved and, if possible, re-established in areas which have been deforested.

Many of the solutions suggested above will help to solve other environmental problems, such as the depletion of the ozone layer and acid rain.

It's a Fact...

In a December, 1991, Greenpeace poll of leading climatologists, almost half suggested that a "runaway" greenhouse warming is possible. This would involve a much greater and faster warming than current predictions from our models. The most recent conservative models predict the earth's temperature will increase by 3°C by the year 2100. Some models predict an increase as high as 6°C by the year 2100.

QUESTIONS

1. In point form, classify all the possible effects of the greenhouse effect under the following headings: Beneficial Effects, Harmful Effects, Effects Which Might Be Either Beneficial or Harmful.
2. a) List six ways in which you personally contribute to the greenhouse effect. List both direct ways, such as driving in the family car, and indirect ways, such as the fuel needed to produce the food you eat.
   b) Explain three actions you could take that would help solve the environmental problem of the greenhouse effect.
3. a) Classify the possible solutions to the greenhouse effect under the following headings: Prevention and Treatment, Adjustment, Research and Education.
   b) Rank the solutions you have listed from "most effective" down to "least effective". Explain and justify your rankings.

4. Why is there so much uncertainty concerning the seriousness of the "greenhouse effect"?
5. Study Figure 16.3. Write a note summarizing the changes that might occur in each of the following areas of Canada as a result of the greenhouse effect: Prince Edward Island, Southern Ontario, the coastal areas of British Columbia, the southern Prairies, the northern Prairies.

16•2 Case Study: The Depletion of the Ozone Layer

Problem

The depletion of the **ozone layer** refers to the chemical breakdown of a part of the ozone layer which is located in the stratosphere. This layer of ozone (O_3) absorbs harmful ultraviolet radiation in the very short wavelengths emitted by our sun. The ozone layer protects Planet Earth from the harmful effects of this very intense shortwave radiation. As the ozone layer thins, more ultraviolet radiation reaches the earth's surface, resulting in disruptions in plant growth and negative health impacts on humans.

The Natural System

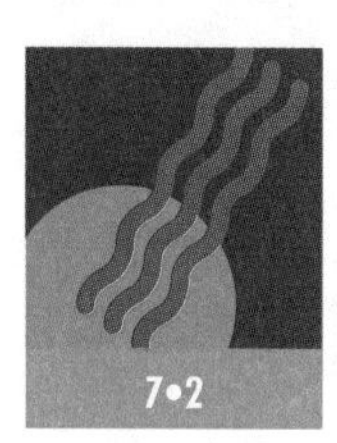

The ozone layer is located 12 to 45 km above the earth's surface (with a concentration at an altitude of 25 km) in the stratosphere. Ozone, a molecule made up of three oxygen atoms, absorbs up to 99 percent of the harmful shortwave solar radiation with wavelengths below 0.3 microns. These wavelengths can be harmful to both people and plants; thus, this natural ozone system is an important defence which Planet Earth has against the more destructive wavelengths of the sun.

The Altered System

The natural system is being altered by the introduction into the atmosphere of a group of chemicals called chlorofluorocarbons (CFCs). CFCs were originally thought to be perfect chemicals when they were developed in the 1920s. They are odourless, chemically inert, non-toxic, and non-flammable. They do not cause cancer, threaten our lakes and rivers, or become concentrated in ecological food chains. CFCs are extremely flexible in terms of their industrial uses. They are used as propellants in aerosol sprays; as refrigerants in air conditioners, freezers, and refrigerators; as microchip cleansers; and as components in foam products, insulation, and packaging. About 800 million tonnes of CFCs were manufactured in 1988 alone.

CFCs in aerosol sprays enter the atmosphere directly, while CFCs used in foam packaging and insulation materials diffuse into the atmosphere through the porous material or enter directly when these materials are burned. CFC solvents used in the electronics industry evaporate readily into the atmosphere, while discarded refrigerators and other appliances represent an additional important source of atmospheric CFCs.

CFCs are very inert and can remain in our atmosphere for decades. Scientists believe that the two most common CFCs called CFC-11 ($CFCl_3$) and CFC-12 (CF_2Cl_2) remain in the atmosphere for 80 and 120 years respectively. By contrast, gases such as sulphur dioxide and nitrous oxide are

"washed out" of the atmosphere within weeks. This characteristic of CFCs to remain in the atmosphere for a long period of time enables these gases to drift slowly upwards through the troposphere and the tropopause before entering the stratosphere. This journey might take ten years or more. Upon entering the stratosphere, the CFCs react chemically with the naturally occurring ozone layer found there.

Ozone itself is produced when an oxygen molecule, O_2, is broken down into two oxygen atoms (O + O) by the effects of solar ultraviolet radiation. Each oxygen atom which is released is then free to combine with another oxygen molecule to form a molecule of ozone (O_3). Ozone is capable of absorbing vast amounts of ultraviolet shortwave radiation. It is constantly broken apart when it absorbs this shortwave radiation, but quickly reforms as ozone to absorb more radiation. When CFCs are introduced into this oxygen-ozone system, they both hinder the formation of new ozone and destroy the existing ozone.

When ultraviolet radiation strikes a CFC molecule, a chlorine atom can be released. This chlorine atom can combine with one of the freed oxygen atoms in the stratosphere to form chlorine monoxide (ClO), or it can directly "steal" an oxygen atom from a molecule of ozone to form a chlorine molecule (ClO) and an oxygen molecule (O_2). When the ClO molecule collides with another atom of oxygen (O), the two oxygen atoms (O + O) combine to form a molecule of oxygen (O_2), freeing the atom of chlorine (C) to "attack" another ozone molecule. As this chemical process occurs over and over again, the amount of molecular ozone is decreased and less ultraviolet solar radiation is absorbed in the stratosphere, causing an increase in the amount of this dangerous shortwave radiation received at the earth's surface. (See Figure 16.6.) One chlorine atom is capable of destroying hundreds of thousands of ozone molecules if no other chemical process occurs to stop the reactions. The presence of bromine in the atmosphere, which is both a naturally occurring chemical as well as a manufactured pollutant released from fire retardants and fumigants, contributes to the breakdown of ozone by reacting with chlorine to produce free oxygen atoms. In addition, bromine can contribute to the breakdown of ozone by a direct chemical reaction.

There is uncertainty concerning the degree to which our ozone shield has been damaged. Some measurements suggest that over the Antarctic area, where the damage appears to be the greatest, the level of atmospheric ozone has decreased by as much as 40 percent since 1977. Some scientists attribute this depletion, which occurs most dramatically in springtime, to natural dynamic forces at work in the atmosphere. However, most scientists are now convinced that the chemical depletion described above is a significant factor in the creation and growth of this ozone hole over the Antarctic. This depleted area now covers an area greater than that of the United States. The effect seems to be greater in the Antarctic due to high-altitude atmospheric circulation patterns and unique chemical reactions found in the very cold air of the Antarctic polar vortex. A similar hole has now been discovered over the Arctic Ocean.

Although the ozone layer "holes" over both the Antarctic and Arctic have received the most publicity in recent years, the worldwide thinning of the ozone layer throughout the atmosphere is of most concern. NASA's global habitability project has placed the worldwide depletion at 6 percent since 1970. Ozone levels over Toronto, Ontario, have decreased by 4 percent over the last ten years. For three

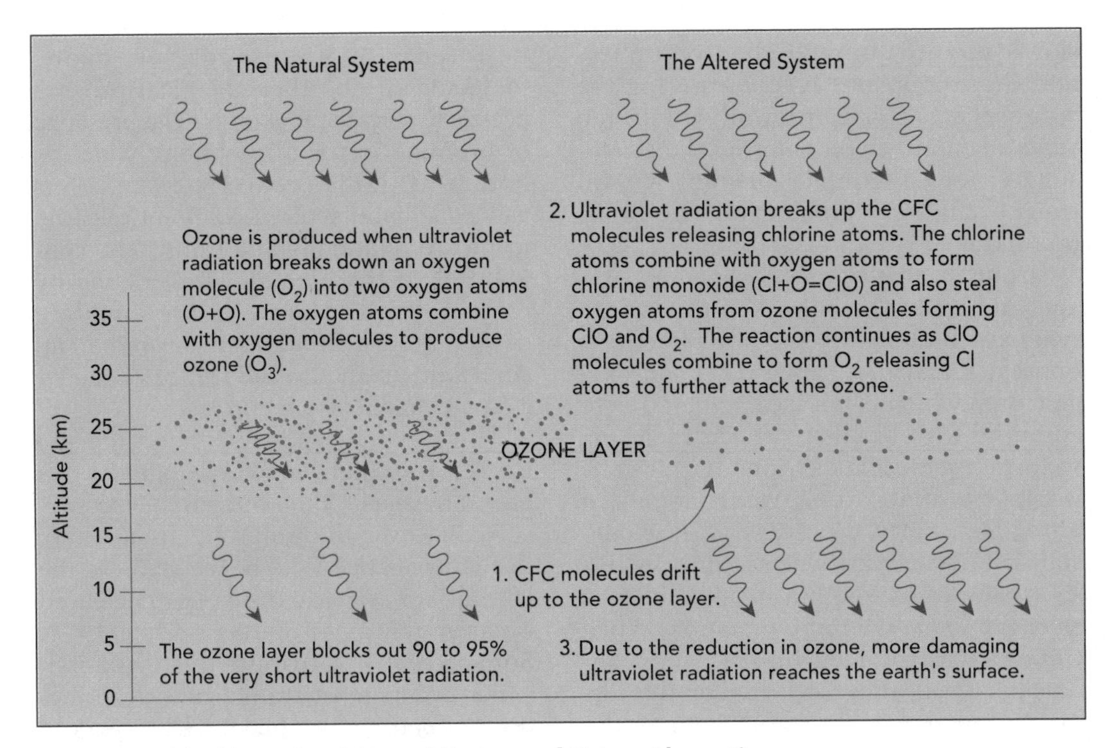

Figure 16.6 **The Natural and Altered Systems of Ozone Absorption**

weeks in December of 1987, scientists recorded a "dip" in the ozone shield over Australia of 10 percent. Another report by the National Center for Atmospheric Research in Boulder, Colorado, has suggested that a 6 percent decline over the last 15 years has occurred in the ozone layer over the northern hemisphere. Computer models of stratospheric chemistry predict that if CFC use were to grow at 3 percent annually (the rate of growth is now 6 percent), then by the year 2030, the ozone layer would be 10-20 percent depleted.

As part of a natural yearly variation, the ozone layer becomes thinner in late winter and early spring. However, a thinning of the ozone layer of 15 percent over yearly norms was experienced over eastern Canada in the spring of 1992. This compares with a 10 percent thinning in 1991 and a 6 percent thinning in 1990.

Effects

As the ozone shield is depleted, more of the sun's ultraviolet radiation with wavelengths below 3 microns reaches the earth's surface. These short wavelengths of solar radiation are known to cause skin cancers and eye damage in human beings. The United States Environmental Protection Agency (EPA) estimates that a 2.5 percent depletion in our ozone shield results in an additional 15 000 victims of melanoma per year in the United States alone. Melanoma is a deadly form of skin cancer. Over the next century, the EPA's report estimates that over 200 000 000 additional cases of skin cancer will occur in the United States if present trends continue. It is thought that a 1 percent decrease in atmospheric ozone results in a 2 percent increase in the amount of ultraviolet radiation which reaches the earth's surface, and this in turn means a 4 percent increase in skin cancer rates. In addition, an increase in ultraviolet radiation weakens the body's natural immune systems, which could lead to an increase in the number of deaths from a wide range of diseases. It is known that an increase in ultraviolet radiation interrupts the photosynthesis process on both the land and water, which could lead to reduced crop yields and damage to the oceans' plankton and algae growth. Such damage would disrupt the planet's entire food chain.

Solutions

Most scientists who believe in the chemical theories and models of ozone depletion point out that even if we stop the production of CFCs today, the damage to our ozone shield will continue well into the twenty-first century due to the long life of atmospheric CFCs and the existing quantities of these chemicals. The use of CFCs is growing at a rate of 6 percent per year at the present time; this production represents a reservoir of these chemicals that will enter and remain in the stratosphere for the next 100 years.

In September of 1987, in Montreal, forty-two nations reached agreement on a program to protect the ozone layer. The agreement, known as the Montreal Protocol, was strengthened in 1990 and will attempt to:

i) freeze the production of CFCs and halons such as bromine at 1986 levels;
ii) reduce the amount of CFCs entering the atmosphere by 50 percent by 1999;
iii) encourage the development of environmentally safe alternatives to CFCs;
iv) provide for cooperation and sharing of information amongst nations; and
v) provide for trade sanctions against countries that do not ratify the protocol.

Most of the forty-two countries that signed the protocol have been slow to implement any of its provisions, largely because effective alternatives are not readily available. One suggestion to deal with this problem would involve a substantial tax on all CFCs produced. This would encourage the search for substitutes and hopefully lead to recycling and

It's a Fact...

- A 1992 U.N. report estimates that every 1 percent depletion of the ozone layer results annually in 300 000 additional skin cancers, 2 million additional eye cataracts, and up to 150 000 more cases of blindness on a worldwide basis.

- Although ozone high in the stratosphere is beneficial to the human race, ozone at ground level is a major pollutant. It is formed by chemical reactions involving nitrogen and carbon compounds and is a major component of smog. It threatens human health and injures trees and other vegetation.

conservation programs. Many countries such as Sweden, the United States, and Canada have banned the use of aerosol sprays for over a decade for all but medical purposes; however, many countries have not, and today aerosols represent 27 percent of worldwide CFC use. On a smaller scale, many cities and industries have now banned the use of CFCs in the packaging of products such as eggs and hamburgers. Many scientists claim that a scaled reduction in the use of CFCs is not enough. They claim that the potential threat to our planet is so great that an immediate freeze on the production of all CFCs must be implemented immediately.

QUESTIONS

6. Classify the possible solutions to ozone layer depletion under the following headings: Prevention and Treatment, Adjustment, Research and Education. Based on your own knowledge and opinion, rank the solutions you have listed from "most effective" down to "least effective". Explain and justify your rankings.
7. Why is there so much uncertainty about the degree of damage to the ozone layer and the seriousness of the problem?
8. Write a letter to the manager of a local industry explaining why he or she should not be using foam containers or packaging that contain CFCs.
9. a) Explain two specific actions you personally could take to help solve this environmental problem.
   b) Explain three specific actions you could take to reduce your personal risk of receiving too much ultraviolet radiation.

16•3 Case Study: Hydro Dams — A Planned Alteration

At the beginning of this chapter, we pointed out that it is inevitable that our natural physical systems will be altered by human activity. The problems dealt with so far have been unplanned alterations. There are also many examples of the planned or deliberate alteration of physical systems. Even with planning, however, the consequences resulting from the alteration of physical systems are sometimes unexpected and are always difficult to predict. One such planned alteration of a physical system is the building of hydro-electric dams on the world's river systems.

Problem

The number and size of hydro-electric dams have been increasing at a rapid rate over the last few decades. Over 20 percent of surface run-off in both North America and Africa is now controlled or regulated by dams. In Europe and Asia, the percentage stands at 15 percent. Approximately 500 dams are added to the world total each year. The concerns over the impact of hydro dams on both natural systems and indigenous lifestyles is increasing.

The Natural System

Rivers drain the surface of the earth, channelling surface and groundwater toward the oceans. The rivers also transport sediment, material that has been eroded from the surface of the earth. The water and sediment are important to wildlife and vegetation along the course of the river. In some places, the whole cycle of human activity is determined by the rhythm of rivers.

The Altered System

The development of large lakes or catchment areas upstream from the dams is necessary in most cases to control the flow of water. Lake Mead in the United States, which was created by the building of the Hoover Dam, contains some 38 billion m^3 of water. This large lake is now dwarfed in size by more recent developments. Lake Nasser, created by the Aswan Dam in Egypt, contains 157 billion m^3 of water. In northern Quebec, as part of the first phase of the James Bay Project, the building of nine dams and 206 dikes along three rivers flooded close to 12 000 km^2 of land. Future plans for developments associated with the James Bay Project would eventually flood over 25 000 km^2 of land, an area the size of Lake Erie! (See Figure 16.7.)

Not only are large artificial lakes created by hydro dams, but also the flow of the rivers themselves is changed. In northern areas, river levels are usually lowest in the winter and highest in the spring when ice melts. This cycle is opposite, however, to the demand for hydroelectricity, which is greatest in the winter. The flow rate of rivers is adjusted accordingly by the dams to meet hydroelectric needs.

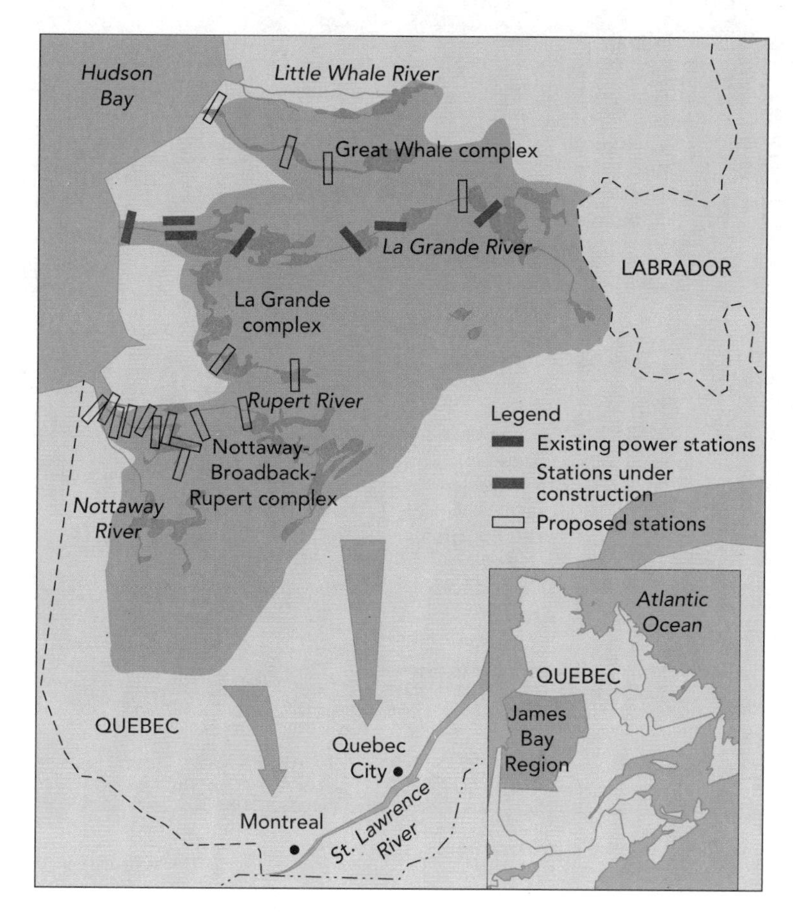

Figure 16.7
The James Bay Project: Present and Future Dams

Effects

The formation of such huge artificial bodies of water and the influence of the dams themselves give rise to new ecosystems both upstream and downstream of the dam. They affect local climates, local water chemistry, traditional flora and fauna, natural erosion and deposition rates, and human occupation. Figure 16.8 illustrates some of the effects which hydro dams have on the environment.

The 10 000 Cree and 2000 Inuit in Northern Quebec have outlined specific environmental impacts associated with the James Bay Project, impacts which they fear can destroy their traditional cultures.

- When you put more quantities of water down a river channel in winter, you flush out fish and their fertilized eggs.
- When a vast quantity of vegetation ends up under water due to the flooding of the land, it decomposes, releasing toxic methyl mercury. The mercury enters the food chain. Increased concentrations now exist in the whitefish, pickerel, pike, and lake trout upon which the Cree depend.

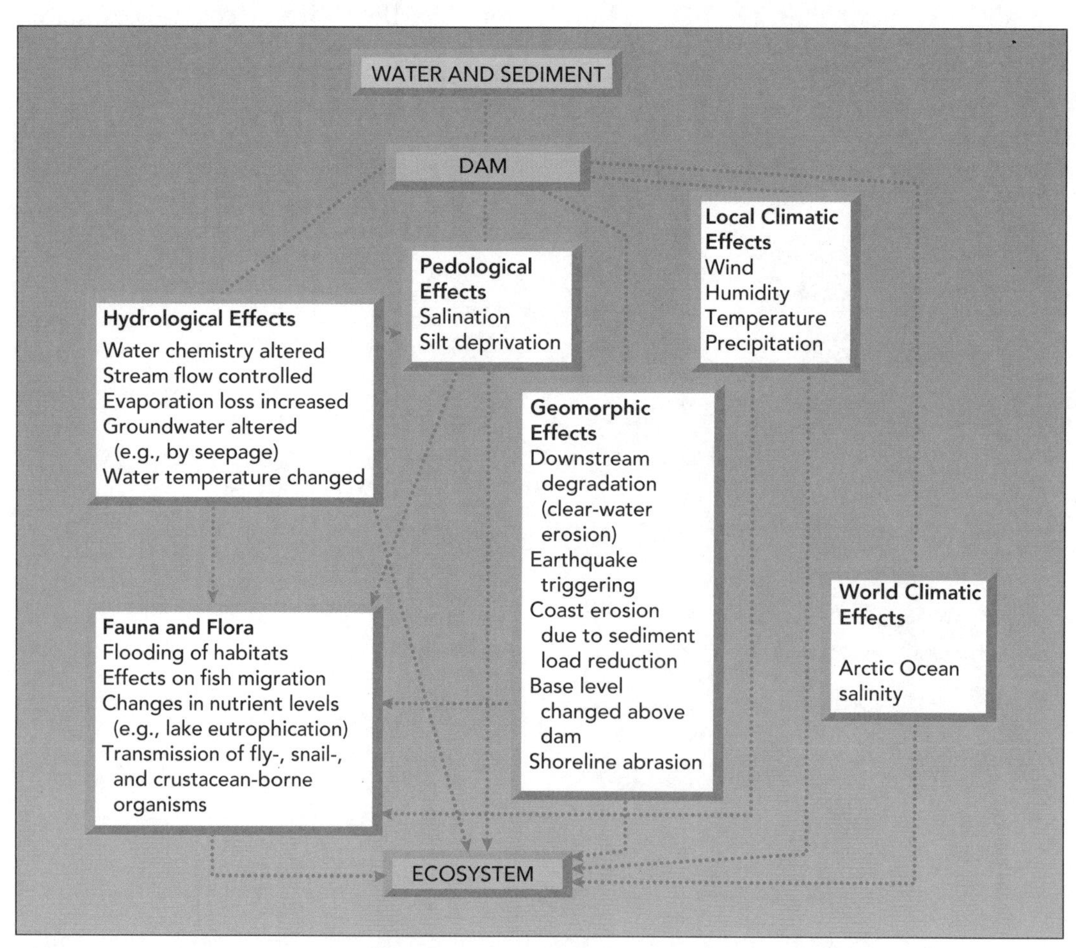

Figure 16.8 The Effects of Hydro Dam on the Environment

More than half the Cree living in the original dam areas now have concentrations of mercury in their bodies that exceed the limits set by the World Health Organization. Mercury poisoning can cause neurological diseases. As the mercury makes its way to Hudson Bay, it will contaminate that ecosystem, and animals such as the beluga whale and seal will be affected.

- When shorelines are drowned and marsh and wet lands are cleared, the traditional nesting grounds of migratory birds such as yellowlegs, geese, ducks, sandpipers, and snipes are destroyed.
- When you disrupt the flow of rivers through diversion projects, you run the risk of altering the entire hydrologic cycle. In the James Bay region, it is feared that larger amounts of fresh water will enter Hudson Bay, altering the salinity and water temperature of that saltwater ecosystem. The effects on marine life are difficult to predict.
- When you build roads and increase the flow of rivers and flood large areas of land, you disrupt the traditional migratory and breeding patterns of the caribou. The effects of this disruption are not yet known.

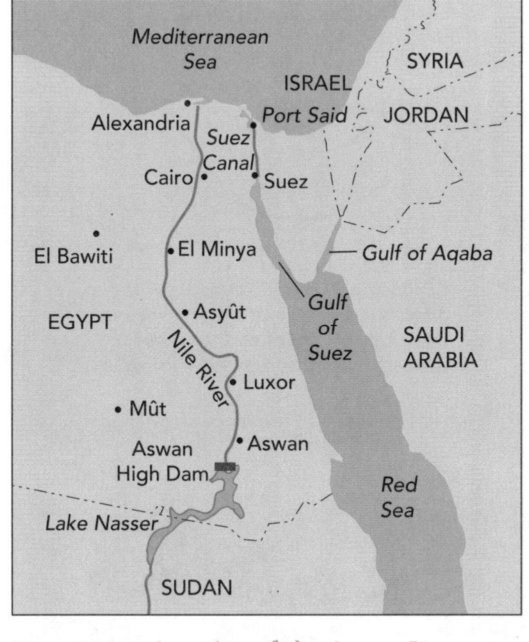

Figure 16.9a Location of the Aswan Dam

Each new dam alters the environment in unique ways, which are not always predictable. For example, examine Figure 16.9b. The Aswan Dam has acted like a filter taking much of the silt out of the Nile waters. Traditionally, this silt has fertilized the areas along the banks of the Nile whenever the river has flooded. Without this natural fertilizer, crop yields decreased and the people of Egypt have been forced to turn to expensive and environmentally damaging chemical fertilizers. The effects of this reduction in silt load have even been felt by the fish stocks of the Mediterranean Sea as the silt traditionally was a

	Jan.											Dec.
BEFORE	64	50	45	42	43	85	674	2702	2422	925	124	77
AFTER	44	47	45	50	51	49	48	45	41	43	48	47

Figure 16.9b Silt Concentrations in Parts per Million in the Nile River Measured Near Asyût Before and After the Construction of the Aswan Dam

source of nutrients for the fish. In addition, the now "cleaner" water of the Nile has caused an increase in erosion both of the Nile Delta and the river bed itself. This increase in erosion is called "clear-water erosion" and has been measured in many places in the world. However, it should be noted that some dams have had just the opposite effect. Because dams regulate the flow of the water, floods, which at times in certain areas cleared the river basin of accumulated sediments, no longer occur. The result in these areas is increased deposition of sediments and shallower river channels.

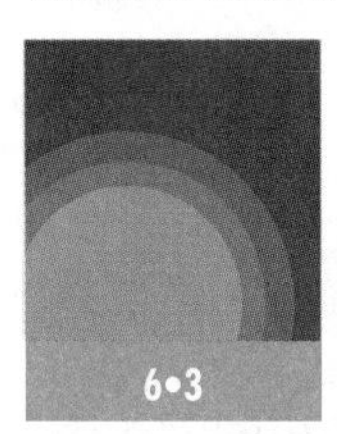

In recent years, scientists have suggested that large reservoirs created behind hydro dams seem to increase the number of earthquakes which occur in the area of the dam. The added weight of vast volumes of water combined with changes in subsurface water flows and pressures apparently have the potential to initiate faulting. Figures 16.10 and 16.11 illustrate data that support the link between seismic activity and dams.

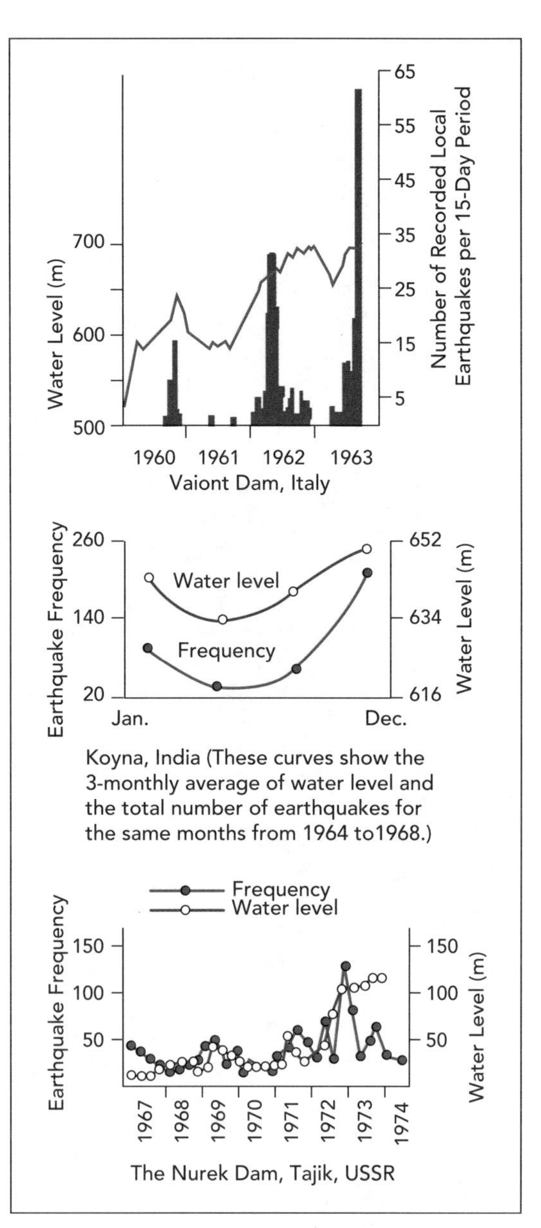

Figure 16.10 **The Link Between Seismic Activity and the Construction of Large Dams**

Solutions

Some of the impacts of hydro dams are predictable and their effects can be minimized by careful planning. Fish ladders can be built to allow for the normal upstream migration of fish during the mating season. Wildlife can be moved prior to the flooding of the land and the new artificial lakes can be stocked with supplies of fish. After construction, natural vegetation can be replanted and aquatic vegetation that is especially sensitive can be reintroduced wherever possible. Areas which have become prone to erosion due to the alteration of stream flow can be banked and reinforced. Despite the procedures noted above, it should be stressed that each new dam alters the environment in unique ways, which are not always predictable.

Dam	Year of Completion	Year of Largest Earthquake
Marathon, Greece	1929/30	1938
Boulder, USA	1936	1939
Clark Hill, USA	1952	1974
Kariba, Zimbabwe/Zambia	1959	1963
Grandval, France	1959	1963
Hsingfengkiang, China	1959	1962
Kurobe, Japan	1960	1961
Camarillas, Spain	1960	1961
Canelles, Spain	1960	1962
Vaiont, Italy	1961	1963
Monteynaard, France	1962	1963
Koyna, India	1962	1967
Benmore, New Zealand	1965	1966
Kremasta, Greece	1965	1966
Piastra, Italy	1965	1966
Countra, Switzerland	1965	1965
Najina Basta, Yugoslavia	1966	1967
Oroville, USA	1968	1975
Nurek, USSR	1969	1972
Vouglans, France	1970	1971
Talbingo, Australia	1971	1972
Jocasses, USA	1972	1975
Keban, Turkey	1973	1974
Manis 3, Canada	1975	1975

Figure 16.11
List of Dams for Which Associated Seismic Phenomena Are Reported

Hydro dams illustrate clearly that we are both controllers of our environment as well as part of the global ecosphere. One change in a physical system initiates many others. We must continue to attempt to understand our role in the disruption of fragile global environments and reduce our human impact as much as possible. To fail to do so is to court ecological disaster.

It's a Fact...

Lac Bienville on the Grand Baleine would double in size if the James Bay II Project goes ahead. Lac Bienville is an important nesting, feeding, and resting area for migratory birds. It is not known if the birds could adjust to the higher lake level or the changing shorelines, which would go up and down as the reservoir level changes.

QUESTIONS

10. Supporters of the James Bay Project point out that when completed the project will produce 26 000 megawatts of electricity. The electricity will meet the industrial and residential demands of a province in which 70 percent of homes are heated electrically. In addition, they claim that a billion dollars a year will be made by selling excess power to the United States. They claim that the project is already attracting industries and tourists to northern Quebec.
    a) Devise questions to test your fellow students on their knowledge of the James Bay Project and on their opinions concerning this project.
    b) Use your questions to poll at least 15 of your fellow students who are not enrolled in Physical Geography.
    c) Graph your results and state conclusions from your study.
11. Explain why the flooding of large areas is often necessary in the development of hydro-electric generating stations.
12. In a number of cases, developers of hydro-electric projects have found multiple uses for the reservoirs that were created. Additional uses include recreation, irrigation, flood control, water supplies for urban areas, and so on.
    a) Will such uses increase or decrease the impacts the dams have on natural systems? Explain.
    b) Why might developers of the dams look for multiple uses for the reservoirs?
13. In a chart format, compare typical conditions in a river system upstream and downstream of a hydro-electric dam. Topics you might include in your chart are: water depth, stream flow, erosion, sedimentation, fish species.

Review

- Environmental problems occur because natural physical systems have been altered by human activity.
- The solutions to environmental problems fall into the areas of prevention, treatment, knowledge, adjustment, and philosophy.
- Even the planned alterations of physical systems bring about many effects which cannot be anticipated or planned for.

- Greenhouse gases are preventing longwave radiation from escaping the atmosphere, creating a warmer climate through the greenhouse effect.
- Chemicals that have proven to be very useful for humans are destroying the ozone layer found high in the atmosphere.
- Hydro-electric dams are an example of the planned alteration of natural cascading systems.

Geographic Terms

greenhouse gases
carbon dioxide sink
chlorofluorocarbons (CFCs)
ozone layer

Explorations

1. Apply the five-stage case study format used in this chapter (problem-natural system-altered system-effects-solutions) to the investigation of another altered physical system. Possible topics might be: the draining of wet lands, water pollution in your local area, aquifer depletion, the use of lawn chemicals, cloud seeding, landfills.
2. We noted at the beginning of this chapter that some disruption of physical systems is inevitable due to the presence of human beings. How can we determine which alterations are acceptable in terms of their impact and which alterations are not acceptable? In a group, brainstorm a list of criteria or conditions that you feel should be met before you would consider that altering a natural system is justified. Rank order your criteria. Apply your list of criteria to the following and in each case state whether or not the alteration of the natural system is justified.
   a) the use of rivers for the purpose of irrigation
   b) the use of artificial weed killers on suburban lawns
   c) cloud seeding in order to generate rainfall for dry areas
   d) the James Bay Project
   e) the driving of an automobile
3. Choose either the greenhouse effect or the depletion of the ozone layer.
   a) Make up six specific questions related to your topic which would test an individual's knowledge and concern about the topic.
   b) Use your questions to poll ten adults and ten students concerning their knowledge and concern about the topic. Do not use students in your class.
   c) Graph your results in a way that will allow you to contrast the adult responses with those of the students.
   d) Analyse your graphs and summarize your findings in a paragraph.
4. Explain in your own words why each of the five types of solutions (prevention, treatment, adjustment, knowledge, philosophy) has a role to play in solving our environmental problems.

CHAPTER

17

FRAGILE ENVIRONMENTS

OBJECTIVES:

By the end of this chapter, you will be able to:

- appreciate the difficulty in protecting fragile environments from human activity;
- recognize that large-scale international cooperation is necessary to fully resolve most environmental problems;
- understand that humans make economic choices that can have far-reaching implications for the environment;
- reflect on how you as an individual have the power to take action to protect fragile environments;
- appreciate the diversity and importance of the earth's biomes.

CHAPTER 1: The Nature of Physical Geography

CHAPTER 2: Earth: Its Place in the Universe

CHAPTER 3: The Earth in Motion

CHAPTER 4: The Earth's Interior

CHAPTER 5: The Earth's Crust

CHAPTER 6: The Lithosphere in Motion: Plate Tectonics

CHAPTER 7: Solar Radiation

CHAPTER 8: Climate

CHAPTER 9: Weather

CHAPTER 10: The Hydrosphere and the Hydrologic Cycle

CHAPTER 11: Natural Vegetation and Soil Systems

CHAPTER 12: Denudation: Weathering and Mass Wasting

CHAPTER 13: Distinctive Landscapes: Humid and Arid Environments

CHAPTER 14: Distinctive Landscapes: Glacial, Periglacial, and Coastal Environments

CHAPTER 15: Natural Hazards: Disrupting Human Systems

CHAPTER 16: The Disruption of Natural Systems

CHAPTER 17: Fragile Environments

Introduction

Chapter 11 dealt with the earth's biomes, those major ecosystems that are recognizable by their vegetation types. In this chapter, we will examine how human activity has altered some of these complex and fragile systems. Components within each of these systems are dependent upon one another. As in a child's kaleidoscope, when one item, shape, or colour is altered, the entire pattern changes. Such is the case with ecosystems. Change one of the variables, for example, climate, soil, vegetation, or life form, and the entire system readjusts and changes. Indeed, the entire ecosystem of the planet is influenced.

17•1 Environment Under Siege: The Tropical Rainforest

Figure 17.1 Photo of a Rainforest, Equador

Figure 11.5 (page 219) shows the location of the world's tropical rainforests. These

remarkable forests, which now cover approximately 7 percent of the earth's surface area, provide a home for between 50 and 80 percent of the planet's plant and animal species, most of which have not yet been discovered and catalogued. Human activity is now destroying this rich ecosystem at an alarming rate. Estimates vary; however, at least 11 000 000 ha per year of tropical rainforest are being cut down. The ecosystems of a further 5 000 000 ha of tropical rainforest are severely altered due to the selective harvesting of various parts of the forests. Thirty to forty hectares are being lost each minute. This amount equals approximately 550 km^2 each day! And the rate of destruction is increasing. Our grandchildren are in danger of living in a world without tropical forests. We are losing these remarkable areas even before we have gained an understanding of their genetic diversity or of their role in regulating the earth's climate.

Causes of Destruction

The forests are being cut down for a number of reasons. Hardwood varieties, such as teak and mahogany, are in great demand in the developed world and are a source of income for the countries of the developing world. The natural vegetation of the rainforest is replaced in many areas by cash crops, such as palm oil, nuts, and coffee, which are then sold to the developed world. In Central America and Brazil, millions of hectares have been cleared for cattle ranches. Again, much of the beef produced is exported to the developed world in order to satisfy the demand for hamburgers and steak. The building of new roads and highways into the interior of the rainforests has encouraged the settlement of farmers who clear the land and attempt to grow crops in order to support their families.

In much of the developing world, wood remains the most important source of energy for cooking and heating. As population increases, more and more tropical forest is cut down to satisfy these basic human needs. In addition to the above reasons, as the developing world attempts to industrialize, added pressure will be put on the rainforests. Massive hydro-electric projects flood millions of hectares of forest, while mineral exploration and extraction further disrupt these important ecosystems.

Figure 17.2 Photo of a Cleared Section of Rainforest, Amazonia, Brazil

Why Should We Be Concerned?

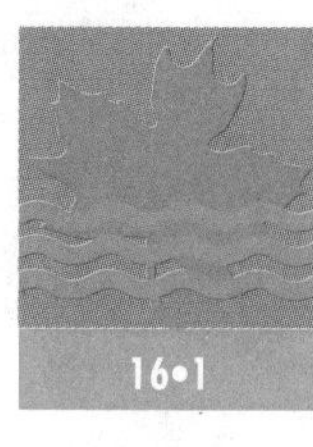

The rainforests are important regulators and components of the solar radiation balance, the hydrologic cycle, and the carbon cycle. Alter the rainforest and you alter all of these important natural cascading systems.

The cutting down of the rainforest contributes to the greenhouse effect. The earth's tropical forests are a carbon dioxide "sink", removing carbon dioxide from the atmosphere and replacing it with oxygen through the process of photosynthesis. As the forests are cut down, they can no longer perform this function. In addition, large areas of the forests are being burned to make space for farming and pasture lands. The burning of the trees directly adds to the carbon dioxide content of the atmosphere. As much as 20 percent of the global increase in atmospheric carbon dioxide can be attributed to the destruction of the rainforests. The remaining 80 percent is due to the burning of fossil fuels. This alteration of the carbon cycle, and consequently of the solar radiation balance, will have profound effects on the human race in the twenty-first century.

As the tropical forests are cut down, the hydrologic cycle is severely altered. The forests regulate the flows of matter and energy within the cycle. Close to 75 percent of the water that falls on a tropical forest is returned to the atmosphere through evaporation and transpiration. The remaining 25 percent is held by the root systems of the trees and slowly released into the ground and rivers. Without the trees, soil erosion increases as the rainfall runs off the land taking the soil with it. Leaching, the movement of nutrients downwards, increases because there are no longer any root systems to capture the nutrients.

Climates in both the immediate area and in distant locations change as less water is supplied to the atmosphere through evaporation and transpiration. These processes play an important role in cooling the equatorial areas as they consume heat. Also, clouds, which are produced by the addition of water vapour to the atmosphere, reflect solar radiation and add to the cooling effect. Much of this water vapour is transported to higher latitudes where it condenses and falls as rain, bringing moisture and heat to areas far distant from the tropical forests. Without the rainforests, the equatorial areas will become hotter and drier. The risk of desertification is increasing in many areas of the world.

We noted earlier that 50 to 80 percent of the earth's species of plants and animals are found in the rainforests. Most of these life forms are at risk of becoming extinct before we have had a chance to even catalogue them. Although biologists have identified over 1 400 000 species of plants and animals, estimates are that the earth probably contains between 10 000 000 and 40 000 000 different species. This great diversity is necessary if ecosystems are to survive. For example, if the peccary, an Amazonian rodent, is destroyed, a number of other species that depend on the mud pools it creates also disappear.

Genetic diversity allows for the adaptation and evolution of species and ensures a stability for life on the planet. A significant reduction in the number of species would disrupt the natural evolution of life on earth and could threaten the earth's ability to rejuvenate itself. It is estimated that well over 90 percent of all species that have ever lived have disappeared. As human activity accelerates and adds to the natural extinction of species, the ability of the earth to compensate comes into question. At the moment, human activity is destroying perhaps 1000 species per year, more than two a day. Many medicines in use today have been derived or developed from tropical plants. The cure for diseases such as cancer and AIDS might lie in the diversity of plants found in the rainforest. Most of our food crops today are the result of the genetic engineering of various species of wild plants. We are endangering our own survival by our wanton devastation of species other than ourselves.

Figure 17.3 **Photo of Replanted Area of Rainforest in Southern India**

Solutions

To say that the solution to the destruction of the rainforest biome lies in stopping the cutting down of the trees is overly simple. The developing countries use their forests as sources of income to help them solve economic and social problems. To expect countries in Africa, South and Central America, and Asia to simply give up this economic resource without having it replaced is unrealistic. Countries burdened with debt, preoccupied with housing, medical, and educational problems, and faced with exploding populations will continue to seek economic solutions that rely on the exploitation of their most valuable resources.

A major reason for deforestation is to satisfy the demands of the wealthier countries of the world. In Brazil, for example, beef cattle graze in deforested areas to supply meat to richer countries. The rainforest also supplies the developed world with a variety of hardwoods for use in furniture manufacturing and construction. A change in our own lifestyles, combined with massive amounts of support for the developing countries, will be necessary if the rainforests are to be saved. We must cooperate on a world scale to allow the developing countries to share in the richness of the planet.

As we await worldwide policies that will protect the rainforests, some methods can be used to ensure that the development of the forests proceeds in ways that are as ecologically sound as possible. Ideally, we would want to restore deforested land to its natural condition. However, **restoration** is extremely expensive and efforts have met with limited success. One of the problems is soil erosion. The soils of the rainforest are shallow and easily eroded and leached of nutrients when the anchoring root systems of the trees and plants are removed. Once an area's soil is carried away or extensively leached, it can take decades for nature to restore it.

An alternative approach is **agroforestry**, the planting of fruit, nut, or selected hardwood trees to replace the original vegetation. In Puerto Rico, in the Luquillo Experimental Forest, it was found that this policy of **replacement**, for example, the replacement of a natural forest with a mahogany plantation, can result in the survival of approximately 60 percent of the species found in the original forest. A policy of **rehabilitation** goes beyond replacement in the sense that a varied and complex number of new and traditional species are introduced into an area that had been deforested.

Brazil has now developed a policy that ensures at least 50 percent of the traditional rainforest is preserved in areas that are being developed. Dr. Thomas Lovejoy of the World Wildlife Fund, along with other scientists, is attempting to determine the minimal size that a section of forest

It's a Fact...

- In the time it has taken you to read this section on the rainforests, another 100 ha of forest have disappeared.
- Seventy-five percent of the destruction of rainforests is occurring in Brazil because Brazil contains most of the world's rainforests. From a percentage standpoint, Costa Rica, Thailand, and India are losing their rainforests at the greatest rate.

can be in order to ensure the survival of various species of plants and animals. Such research increases our knowledge of the ecosystem and will hopefully allow us to plan for its survival.

QUESTIONS

1. In what ways could the destruction of the world's rainforests influence your life?
2. Create a poster or diagram to summarize the effects that the earth will experience if the rainforests continue to be destroyed.
3. Explain why the destruction of the rainforests is an environmental, political, and economic issue.
4. Pretend that you are in charge of presenting a plan of action to the United Nations for reducing the destruction of the rainforests. Include in your plan a goal statement, a list of problems that will have to be solved, resources that will help solve the problems, and strategies that you wish to implement.

17·2 Environment Under Stress: Boreal Forests

A vast coniferous forest encircles the globe in the northern hemisphere of the planet. Stretching across the continents of North America, Europe, and Asia, this huge biome is home to a vast diversity of plant and animal life. Increasingly, this biome, upon which many northern nations have built their wealth, is coming under stress. The source of this stress is a process that involves the increase in the **acidification** of this environment. The term **acid rain** has been often used to describe the source of this acidification. However, this is somewhat misleading because the problem relates to the increasing acidity of all atmospheric deposits, both wet and dry. These acidic deposits increase the acidity of our lakes, rivers, soils, and of the air we breathe.

The acidity of a solution is dependent upon the amount of hydrogen ions ($H+$) that are present. Acidity is measured by the **pH scale**, which ranges from 0 to 14. A value of 7 represents a solution that is neutral, while values below 7 represent increasing acidity, and values above 7 represent increasing alkalinity. Because the scale is logarithmic, a value of 5 represents ten times the acidity of a value of 6. Figure 17.5 illustrates a pH scale.

Causes of the Stress

Normal rain water has a pH value of 5.6 or higher. Our ecosystems have adjusted to an input of water with this level of

Figure 17.4 **Photo of a Boreal Forest in Yoho National Park, British Columbia**

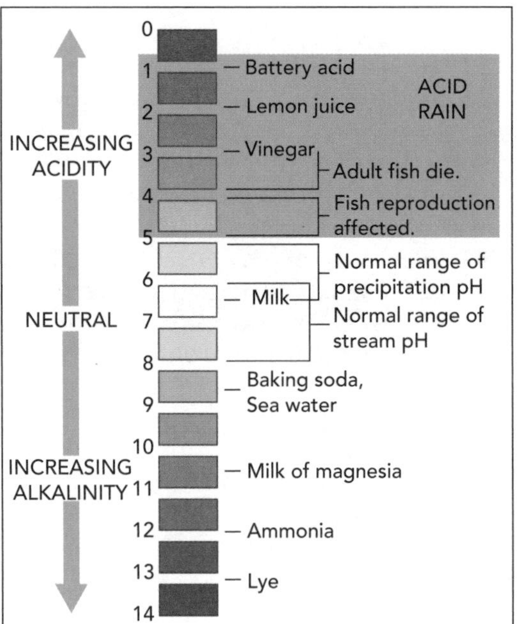

The acidity of a water sample is measured on a pH scale. This scale ranges from 0 (maximum acidity) to 14 (maximum alkalinity). The middle of the scale, 7, represents the neutral point. The acidity increases from neutral toward 0.

Because the scale is logarithmic, a difference of one pH unit represents a tenfold change. For example, the acidity of a sample with a pH of 5 is ten times greater than that of a sample with a pH of 6. A difference of 2 units, from 6 to 4, would mean that the acidity is one hundred times greater, and so on.

Normal rain has a pH of 5.6 — slightly acidic because of the carbon dioxide picked up in the earth's atmosphere by the rain.

Figure 17.5 **The pH Scale**

acidity. Analysis of droplets of water from pre-industrial times indicate that this pH value of 5.6 has been the norm for at least the last few thousand years.

The burning of fossil fuels and the smelting of various ores are the two major processes that lead to acidic deposits. These processes produce gases that react chemically with the water vapour present in the atmosphere to produce acids. The major gases contributing to the acidification of our environment are sulphur dioxide, nitric oxide, and nitrogen dioxide. The gases can also react with other elements in the air, such as sodium, calcium, and magnesium, to form various sulphates and nitrates which fall to the earth as dry deposition. These dry deposits then combine with the water in our soil, lakes, and rivers to form acids. Some studies have suggested that in certain areas of northeastern North America dry deposition contributes ten times the amount of acid to the earth's surface than does wet deposition.

Sulphur dioxide is released into the atmosphere from many natural sources, such as volcanoes and decaying organic material in swamps. However, since the start of the Industrial Revolution, human activities have greatly added to the amount of sulphur dioxide that is placed into our atmosphere. Sulphur dioxide (SO_2) combines with water vapour (H_2O) to form sulphuric acid (H_2SO_4). Rain "water" with a pH as low as 2.4 has been recorded. This rain is really diluted sulphuric acid with an acidity close to that of vinegar.

Nitric oxide (NO) and nitrogen dioxide (NO_2) are produced in certain industrial processes, such as the burning of coal.

However, its main source is the internal combustion engine. Every time we drive cars, trucks, and buses, or use gas lawnmowers, snowblowers, or outboard motors, we add nitric oxide to the atmosphere. Nitric oxide combines with oxygen in the atmosphere to form nitrogen dioxide (NO_2), which combines with more oxygen to form a nitrate (NO_3). Nitrates are more soluble in water than are sulphates and are, therefore, more likely to be dissolved in the atmosphere and fall as wet deposition, acid rain. The nitrates (NO_3) combine with water (H_2O) to form nitric acid (HNO_3) and a hydroxyl ion (OH-).

The wet form of acid deposition is easier to measure than dry deposition and therefore more is known about this form. The ratio of sulphuric acid to nitric acid varies with the location and the time of year. In general, eastern North America experiences twice as much sulphuric acid deposition than nitric acid deposition, while in western North America the ratio is much closer to being even.

Why Should We Be Concerned?

Over a long period of time, our ecosystems have evolved to be in balance with a natural level of acidification. As the acidity of our environments becomes greater, all ecosystems are at risk.

As lakes become more acidic, fish, insects, and other water life begin to disappear. At a pH level of 5.0, only a limited number of fish species can survive. As the pH drops below 5.0, frogs, snails, and most other aquatic life disappear. No species of fish can survive acidity levels of 4.5 or lower. At the moment, some 15 000 lakes in

Location of Major Source Areas of Sulphur Dioxide Emissions

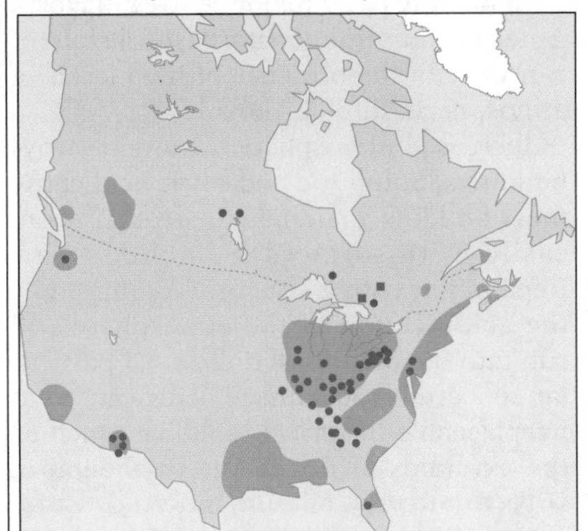

The dots indicate the sources with SO_2 emissions between 100 and 500 kilotonnes per year. The squares are sources of more than 500 kilotonnes. Smaller sources in shaded areas also account for a significant portion of total emissions.

Location of Major Source Areas of Nitrogen Oxide Emissions

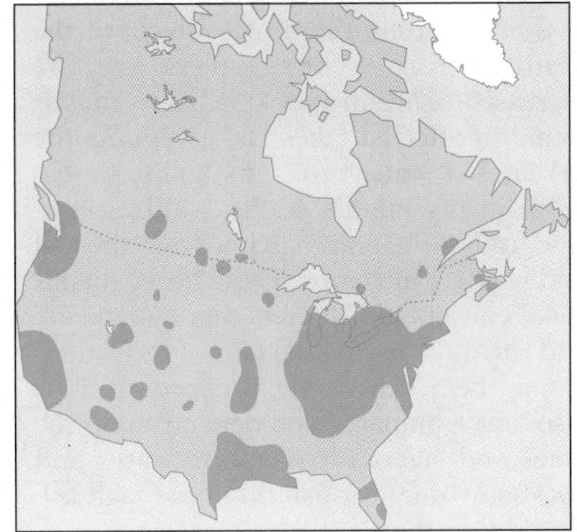

Approximate pictorial representation

Figure 17.6 Acid Rain Sources

Canada are considered "dead" and another 40 000 are at risk. Some 600 000 lakes in Canada are vulnerable to acidification.

In northern areas, **acid shock** occurs in the springtime. This refers to the sudden increase in the acidity levels of the lakes and rivers due to the melting of snow. Spring meltwater has been measured to be 100 times the acidity level of the lakes and rivers into which it is flowing. This sudden increase in acidity disrupts the breeding and reproduction cycles of many aquatic species.

In addition to damaging traditional wildlife species, the presence of acid waters leads to an increase in acid-loving organisms which are not traditionally found in our lakes and rivers. An acid-loving moss called sphagnum is threatening many lakes in northern Canada, while mongeotia algae, which float in thick masses on the surface of lakes with pH levels approaching 5, are becoming more and more common in affected waters.

Compounding the direct effects of the acidification of our waters are the increased concentrations of heavy metals found in affected lakes and rivers. As the pH level of water drops, its ability to dissolve heavy metals such as aluminum, lead, cadmium, and mercury from the soil and bedrock increases. These heavy metals cause cancers in fish and other aquatic life and can directly clog the gills of fish, suffocating them. Heavy metal concentrations also affect humans who depend upon the lakes and rivers for drinking water and who sometimes eat fish that have high levels of heavy metals in their tissues.

The increased acidification of the environment is also having disastrous effects on the world's forests. As acidity increases, plant growth becomes stunted because the absorption of nutrients is affected and root growth is interfered with. Some European forests have already been decimated by environmental acidification. In some parts of North America, 80 percent of the sugar maples have been damaged, threatening the entire maple sugar industry in parts of Canada and the northeastern United States.

The effect of environmental acidity on agriculture is less well documented than is its effects on forests. One study from the University of Guelph in Ontario has suggested that corn yields have been reduced by 10 percent by the increased acidity of rainfall.

There are also direct health effects on humans. Acidic air pollution increases the health risks for individuals suffering from respiratory diseases such as emphysema and chronic bronchitis. Some scientists have reported that children can develop decreased lung function and can suffer from more colds and allergies if they live in areas of high acidic air pollution. A study of 29 major centres in North America has linked increased levels of cancer with the presence of high levels of atmospheric sulphur dioxide.

Increased atmospheric acidity destroys buildings. Sulphuric and nitric acid cause most building materials to weather more easily, with surfaces crumbling away. Repairs to existing structures damaged by the acidification of the atmosphere will run into the billions of dollars. Of particular concern is the damage being caused to irreplaceable historical buildings, such as the pyramids of Egypt, the Parthenon in Greece, and the Coliseum in Rome.

Finally, the acidification of our environment has international repercussions that can strain the relationship between countries. Sulphur dioxide produced in Great Britain falls as acid rain in Sweden, and large quantities of sulphur dioxide, nitrogen dioxide, and nitric oxide drift

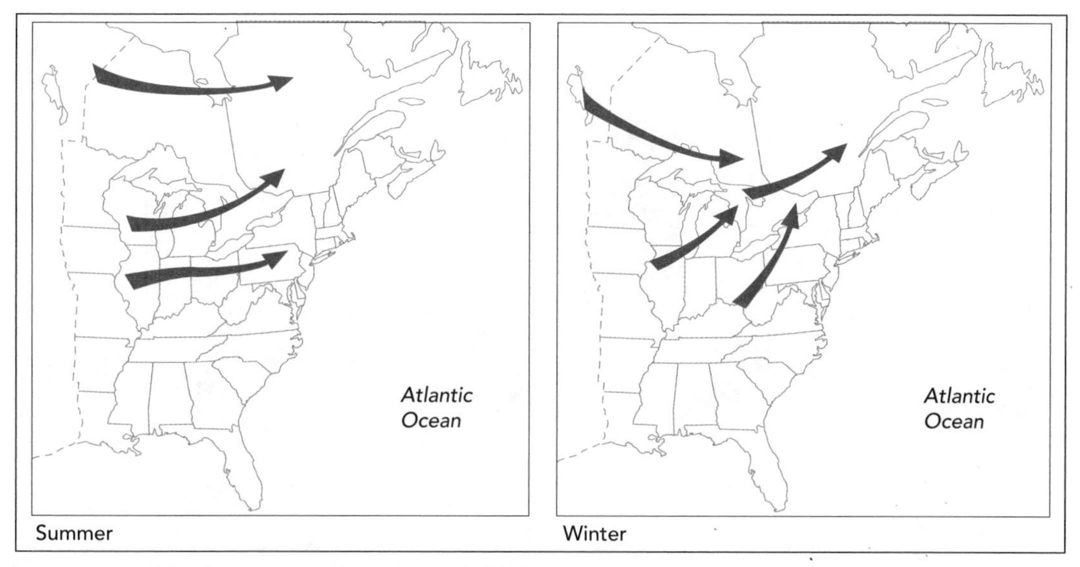

Figure 17.7 **Air Movements That Affect Acid Rain**

across the border from the United States into Canada. International commitment and understanding will be necessary in order to prevent strain in future relationships between countries.

Solutions

As with most environmental problems, the solution to the acidification of our environment can be broken down into five areas — prevention, treatment, knowledge, adjustment, and philosophy. With acidification, prevention is of prime importance. The industrialized world must drastically cut its emission of sulphur dioxide and nitrogen oxide.

Each year, the world pours more than 100 million tonnes of both sulphur dioxide and nitric oxide into our atmosphere. There is a limit to what our environment can bear, and many scientists feel that we have reached that limit. Immediate reductions in SO_2, NO, and NO_2 emissions are necessary. Industries must develop the technology to switch to low sulphur coal, scrubbers of the type shown in Figure 17.8 must be placed on smokestacks, and emission controls must be used on all automobiles. The expense of these prevention methods will be great; however, the costs of not taking these preventive steps are greater. Human health, ecosystems, and food chains are being threatened and we must take action before our natural systems become so damaged that they cannot be saved. In 1984, a group of industrial nations signed an agreement of intent to cut their emissions of acid rain by 30 percent by 1993. Some of these countries, including Canada, have since increased their target reductions to 50 percent, and a number of other countries have signed the agreement.

In terms of treatment, some countries, such as Sweden, have experimented with the liming of lakes. This process is called **buffering**. Lime is a powerful alkali or base and will cause the pH level to rise. Natural buffers in the form of sedimentary rocks and alkaline soils negate the

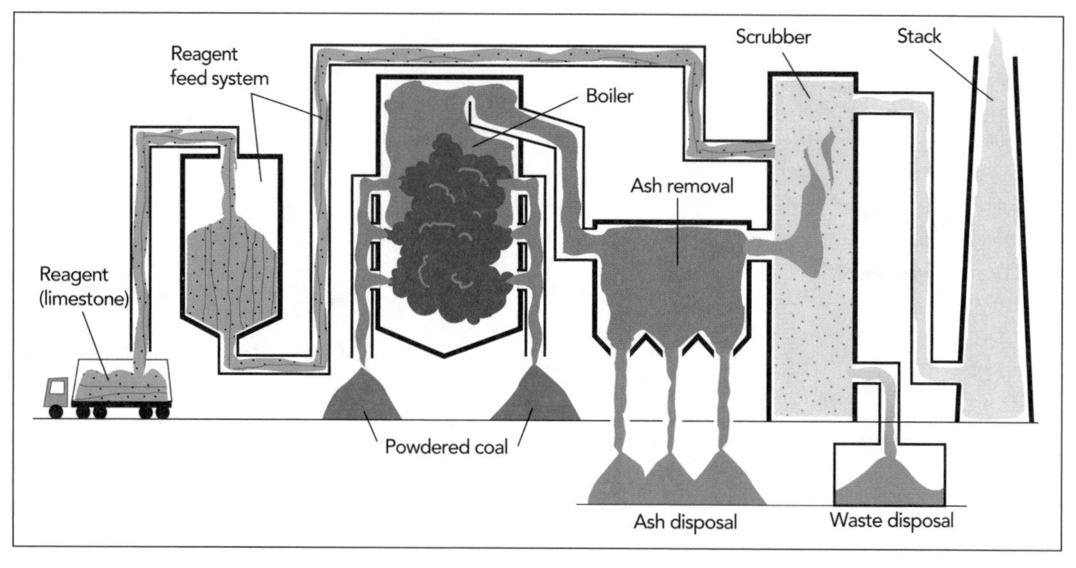

Figure 17.8 How Scrubbers Work Scrubbers remove sulphur oxides from the flue gases given off by coal-fired generating stations. They do this by mixing the flue gases with an alkaline substance, such as finely crushed limestone, that reacts with the sulphur oxides to form solid compounds. These compounds are then collected and disposed of safely. Despite their ordinary sounding name, scrubbers are massive structures that are expensive to build and run.

damaging effects of acidification. Much of the damage due to acidification occurs in igneous rock areas, which have shallow soils. In these areas, there are no natural buffers to offset the effect of increasing acidity. By adding lime to the lakes in these areas, we can temporarily prevent the pH level from falling to dangerous levels. The problem is that this process is expensive. To "lime" the 50 000 endangered lakes in Ontario and Quebec would cost approximately a billion dollars a year. Also, this liming would not offset the damage being done to forests, buildings, and human health.

QUESTIONS

5. List three specific ways in which you can change your lifestyle in order to decrease the degree of acidification of our boreal ecosystems.
6. In a visual way, summarize the effects of acidification on the coniferous biome.
7. Explain why the increasing acidification of the northern forests is a political, environmental, and economic issue for Canadians.
8. Write a letter to your member of parliament or a newspaper expressing your concerns about the acidification of our boreal biome.

It's a Fact...

New burners can reduce the amount of nitrogen dioxide and nitric oxide emissions produced in coal generating plants by 35 percent.

17·3 Environment Destroyed: Desert Margins

All biomes have limits on the amount of human activity that they can tolerate. Nowhere do these limits become clearer than on those transitional lands located on the borders of deserts. Rainfall in these areas is sparse and there are prolonged dry periods. The vegetation is typically grasslands, with scattered trees such as the acacia. These areas are constantly at risk of developing into deserts themselves, a process called **desertification**.

Thirty-five percent of the earth's surface falls into the category of land that is at risk of desertification. Close to a billion people inhabit these marginal lands. Over the last century, semi-arid areas the size of Canada have been converted by misuse and overexploitation from productive drylands into deserts. In Northern Africa, the Sahara Desert has been enlarged by an area greater than that of Ontario. At least 6 million hectares of formerly productive land is converted into desert each year. Figure 17.9 illustrates the areas of the world which are threatened by desertification.

Most of the land that has undergone desertification was previously used as rangeland for the grazing of goats and cattle; however, croplands fed both by irrigation and natural rainfall have also suffered severe desertification.

The major cause of desertification is human misuse of the land. As population increases, so does the demand for pasture land and cropland. Fragile ecosystems which were once the home to a wide variety of wild animals as well as to nomadic human societies are asked to house much denser human populations. In turn, these human populations attempt to graze

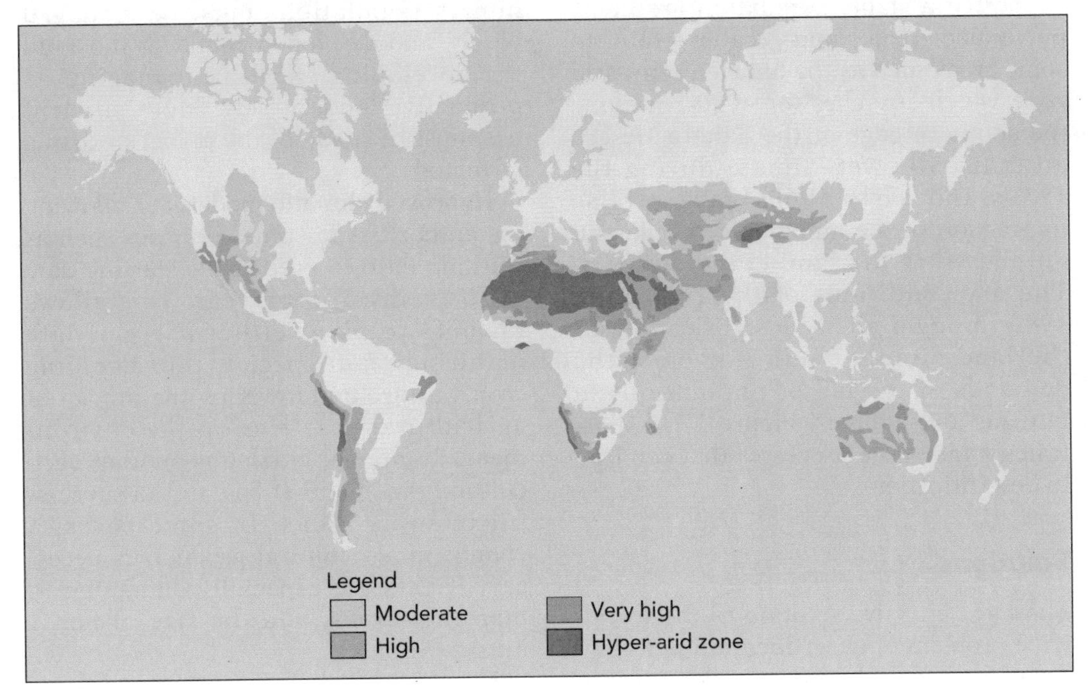

Figure 17.9 Areas of the World Threatened by Desertification

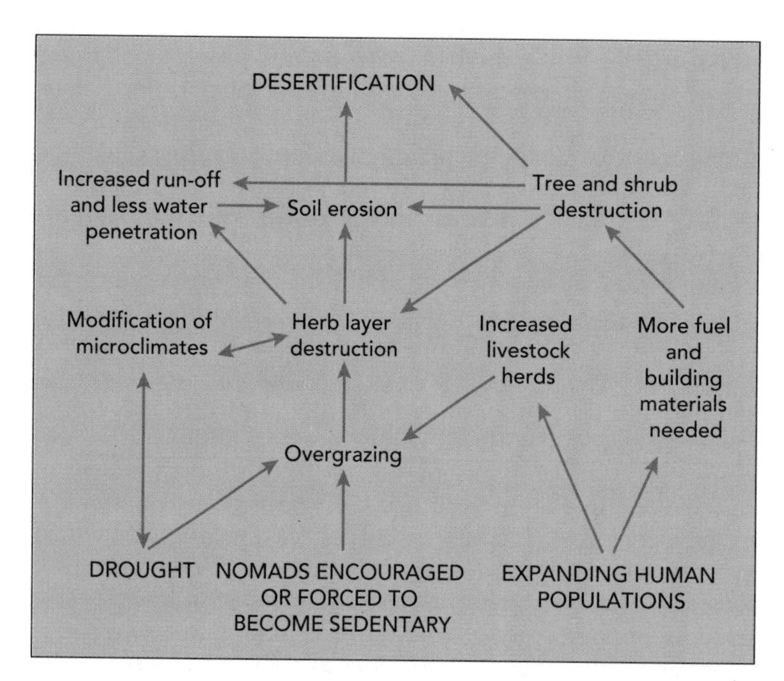

Figure 17.10
Factors Leading to Desertification

domestic animals and to grow a wide variety of crops. Areas which for centuries supported a stable population are being put under greater and greater strain. In some areas such as the Sahel, a semi-arid zone which stretches for 4000 km along the southern edge of the Sahara Desert, an unusually wet climate during the 1950s and early 1960s gave a false impression of the amount of people and animals that the land could support. The dry conditions of the 1970s and 1980s brought about desertification, as the land simply could not meet the demands of the larger populations of humans and animals. Figure 17.10 illustrates some of the processes that can lead to desertification.

Solutions

The most effective solution to the world-wide problem of desertification is to halt the movement of human populations into areas that simply cannot sustain them over a long period of time. In order to accomplish this, land that can support human populations must be managed wisely, and developed countries must aid the developing world in managing its productive lands wisely. More efficient use must be made of the planet's existing farmland.

Improved climatic models are allowing the authorities in some countries such as Burkina Faso to predict the starting date of the rains in each year. This allows farmers to plant either early or late maturing plant species, thus avoiding crop wastage and lowering the impact of agricultural practices on the environment. Rotational grazing is another agricultural practice that has proved successful in some areas. In some cases, a change in agricultural products is necessary. Goats are especially damaging to marginal areas, and the size of herds must be limited.

In areas already suffering from desertification, suitable vegetation covers such

It's a Fact...

Some estimates state that 20 000 000 km^2 are now in danger of turning into deserts due to both climatic change and human misuse. This is an area greater than the combined sizes of Canada and the United States!

as clover and alfalfa must be planted in order to prevent further erosion of the soil and to restore soil fertility. Casuarina trees grow fast in sand and can provide windbreaks, while leucaena trees add to soil fertility and grow very rapidly. Knowledge and education remain important components of the solution to all environmental problems. A reforestation program in northern Burkina Faso failed because authorities planted eucalyptus trees, which were not part of the traditional ecosystem. The eucalyptus trees did not return the necessary nutrients to the soil. In addition, the local population was not educated regarding the goals of the program and cut the trees for firewood.

QUESTIONS

9. Why should the process of desertification occuring in the Sahel be of concern to Canadians?
10. Produce a series of sketches to show the process of desertification as it might occur on the edges of deserts.
11. Suggest the ways in which the process of desertification is linked to the destruction of the rainforests and the greenhouse effect.
12. Desertification most often occurs in countries that do not have the economic means to deal with it. What steps can be taken by developed countries to aid the developing countries in dealing with this global problem?

17·4 Environment at the Threshold: The Great Lakes

Ten thousand years ago, retreating ice sheets left behind five gigantic basins in eastern North America. These basins filled with glacial meltwater and became the largest, most impressive system of lakes in the world. The five Great Lakes of North America cover an area of 244 000 km^2 and stretch for over 1300 km. They contain 23 000 km^3 of water, which represents 18 percent of all the surface fresh water on earth. Figures 17.11 and 17.12 illustrate the characteristics of the five lakes as they exist today.

Because of the size of the watershed, many complex ecosystems exist in the Great Lakes region. Colder northern areas of the watershed are dominated by the Precambrian rock of the Canadian Shield, thin layers of acidic soil, and coniferous forests. In the warmer south, glacial deposits dominate the landscape. Sedimentary rock is overlain by thick deposits of clay, sand, silt, and gravel. Mixed forests and deciduous vegetation dominate the southern regions of the watershed.

The impact of the 40 000 000 people who live in the Great Lakes watershed has placed the lakes at a threshold. Some scientists believe that we have one more decade to begin to effectively deal with the environmental problems created by human activities before an ecological disaster takes place.

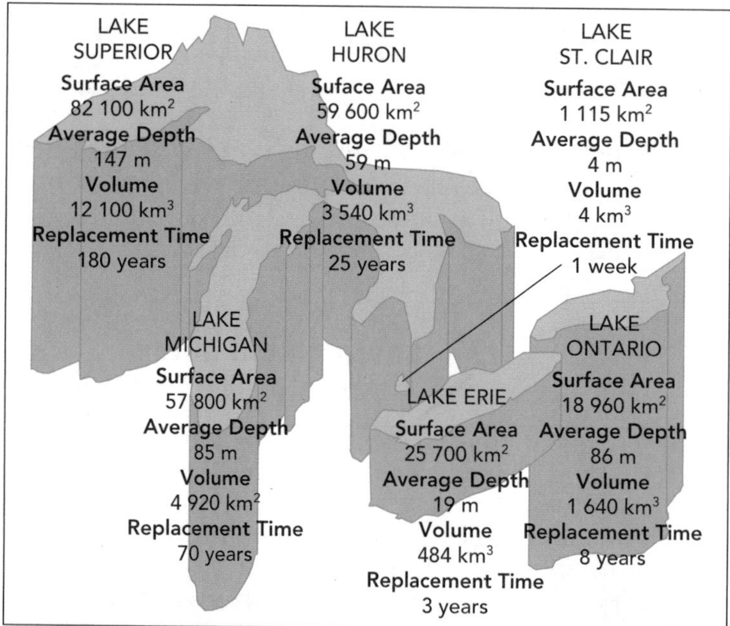

Figure 17.11 **Characteristics of the Great Lakes**

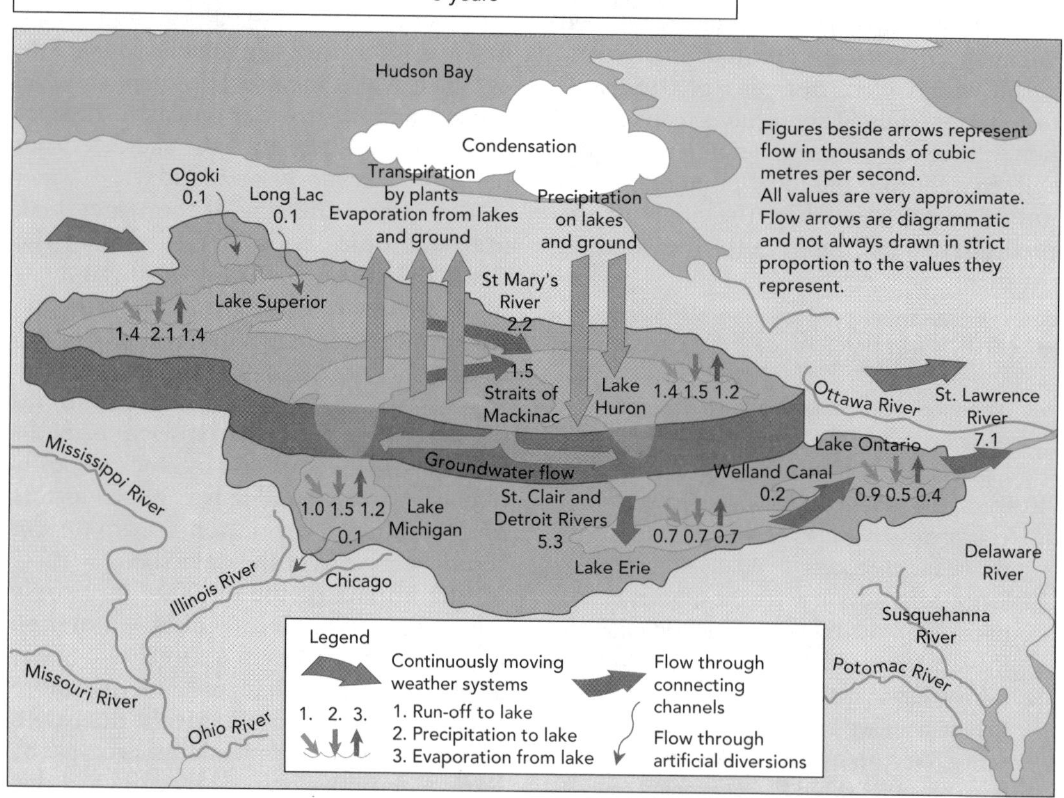

Figure 17.12 **The Great Lakes Water System**

Causes of Disruption

The natural ecosystems of the lakes have been seriously altered by chemicals. Sources of pollution include sewage and waste from urban areas, discharges from industrial areas, leachate from disposal sites, and chemical run-off from agricultural lands. In addition, the large surface area of the lakes makes them susceptible to atmospheric pollutants that fall with rain or snow or as dust. All these sources have contributed to the chemical soup of today's lakes. Although pollution of the lakes was a concern as early as 1918 when the first International Joint Commission labelled their status " gross and foul", it was not until the mid-1960s that the lakes' condition prompted a major international response. By the mid-1960s, Lake Erie was, in some journals, pronounced "dead". Phosphorous pollution from detergents and industrial sewers had reached levels which produced algae in such quantities that oxygen levels had dropped to zero in some bottom areas of the lake. The algae swallowed up oxygen, which resulted in the death of most other forms of marine life. This situation prompted governments on both sides of the border to improve sewage treatment plants and to encourage consumers to use phosphorous-free detergents. Lake Erie, at least in terms of its oxygen levels, sprung back to "life". The problem today is much more complex than elevated phosphorous levels.

Over 60 000 chemicals are produced and used by North American society. Of this amount, over 1000 chemicals have been detected in the Great Lakes watershed. Of these 1000 chemicals, approximately 300 are **toxic**, that is, they are damaging to plants, animals, or humans. Only 30 of these toxic substances are now being measured and regulated. The effects of many of these chemicals are unknown; however, eleven have been identified as critical pollutants which potentially can have negative health effects on humans. These eleven chemicals are outlined in Figure 17.13.

Forty-two "hotspots" have been identified as extreme sources of chemical pollution in the Great Lakes watershed (Figure 17.14).

Some chemicals, such as fluoride, end up in the water on purpose. Fluoride is added to the drinking water of many communities to help prevent tooth decay. Chlorine is added at treatment plants in order to kill bacteria. Some scientists think too much chlorine is used and that it can produce cancer-causing by-products. Chlorine is necessary because a harmful type of bacteria called coliform is present in the water. The major source of coliform is residential sewage. Many existing sewage plants cannot handle the volume of waste that they are now receiving and, in some communities, raw sewage is still dumped directly into the lakes. High levels of coliform are the major reason why many beaches along the Great Lakes must be closed in the summer.

Pollution is not the only reason for the alteration of the natural ecosystems of the Great Lakes. Humans have inadvertently allowed foreign life forms to enter the ecosystems. For example, shipping activities and the building of canals allowed the lamprey eel to appear in the lakes in the 1920s. It nearly wiped out many local fish populations and altered the food chain in the lakes. More recently, the zebra mussel has appeared. It is a miniature mollusk that is native to the Caspian Sea. It first appeared in the Great Lakes watershed in Lake St. Clair in 1988. The suspicion is that zebra mussels were deposited in the Great Lakes through the dumping of ballast water from an ocean-going cargo ship. Since 1988, this mussel has spread rapidly. Female mussels lay 40 000 eggs a year and the larvae can drift for hundreds of kilometres. Concentrations of up to 700 000 per square metre have been found, and the zebra mussels are now clogging intake pipes, weighing down buoys, and damaging the spawning grounds of walleye and lake trout. The mussels

Chemical	Use and Source	Effects
Polychlorinated Biphenyls (PCBs)	Formerly used for insulating fluid in electrical equipment and in various industrial activities. They enter lakes through leaks and spills and by travelling through the atmosphere as by-products of incineration. They were banned in 1980 but continue to exist in electrical equipment which predates 1980.	They cause cancers and reproductive failure in mammals.
DDT	DDT was a commonly used insecticide during the 1950s and 1960s. Although banned in 1969, it remains in sediments in our environment today.	It causes reproductive failure. Its effects were particularly devastating on the populations of various birds.
Dieldrin	An insecticide which enters the ecosystem through spraying.	It is a suspected carcinogen. (A carcinogen is a substance that causes cancer.)
Toxaphene	An insecticide usually used on cotton crops. Its appearance in Lake Superior indicates the importance of the atmospheric transport of toxic substances.	It is a carcinogen.
Dioxin and Benzofuran	Both substances are by-products produced during the manufacture of pesticides. They are also released during the incineration of some chemicals.	These substances are extremely toxic and are powerful carcinogens.
Mirex	A pesticide manufactured in New York State. It has been found in Lake Ontario as a result of leakage from chemical dump sites along the Niagara River.	It causes cancer and reproductive problems.
Mercury	Mercury is produced by pulp and paper mills as well as by a wide variety of industrial processes. It is also carried by atmospheric processes.	It affects the central nervous system causing tremors, mental disease, and death.
Alkylated Lead	Traditionally, the main source has been leaded gasoline used in automobiles. This is now banned in many states and provinces. Lead also enters the environment through various industrial processes.	Lead concentrates in the skeleton leading to lead poisoning. It is especially damaging to young children.
Benzo(a) Pyrene	Released during the steel and aluminum making processes. (It is also concentrated in cigarette smoke.)	It is a carcinogen and has been linked to tumors in fish.
Hexacholobenzene	A by-product of the chemical industry created during the production of solvents and pesticides.	It is a carcinogen.

Figure 17.13 Toxic Chemicals Found in the Waters of the Great Lakes

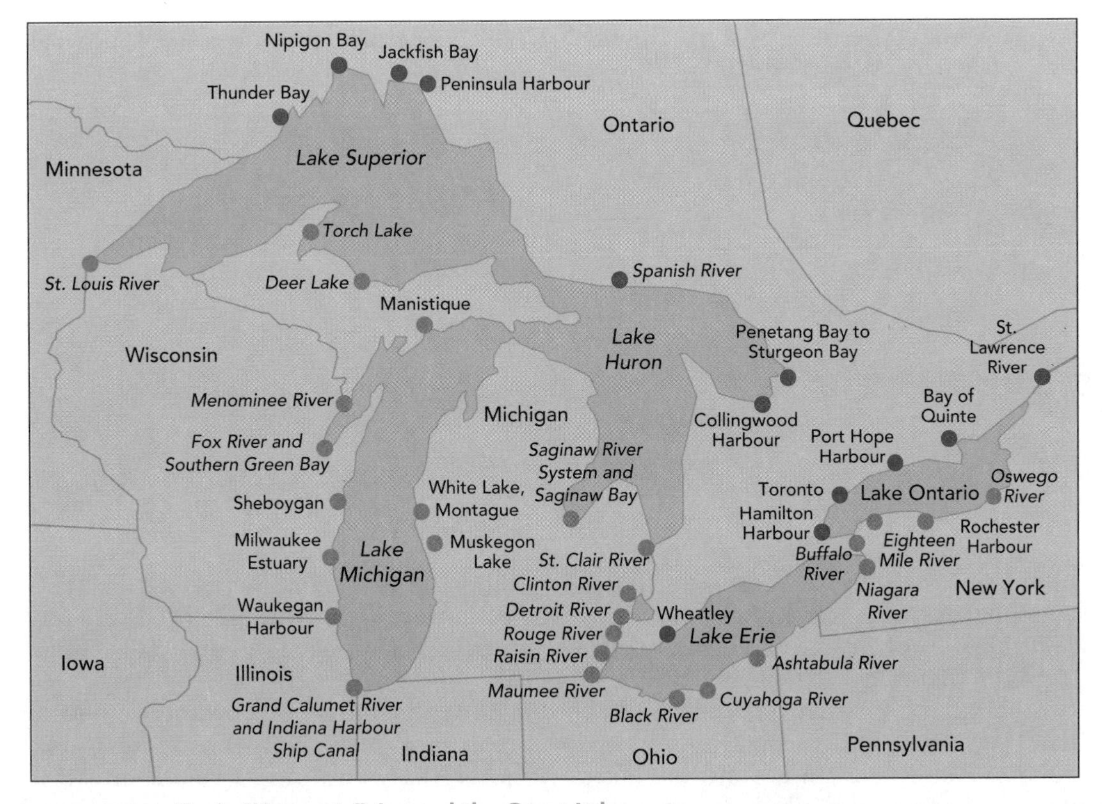

Figure 17.14 Toxic "Hotspots" Around the Great Lakes

attach themselves to any hard surface. They have the potential to seriously disrupt the natural ecosystems of the lakes.

The building of dams and locks, the dredging of channels, the diverting of rivers, and the construction of canals have all led to additional changes in the natural ecosystems of these lakes.

Why Should We Be Concerned?

Of immediate concern are the effects on human health caused by the pollution of the lakes. Some studies claim increases in cancer and infertility rates amongst people living in the watershed. People have already been warned not to eat large quantities of fish taken from the lakes. Figure 17.15 illustrates how one chemical, PCB, is concentrated on its journey through the food chain, a process termed **biomagnification**. Although humans have been warned about eating the fish and, at times, swimming in the water, authorities insist that the drinking water is still safe.

The disruption of the natural ecosystems has economic implications as well. Five billion dollars a year are now being spent on the zebra mussel problem alone. A large commercial fishing industry on the lakes might also be at risk if measures are not taken to clean up the system.

Solutions

In 1909, the Boundary Water Treaty was signed by the United States and Canada. It created the International Joint Commission

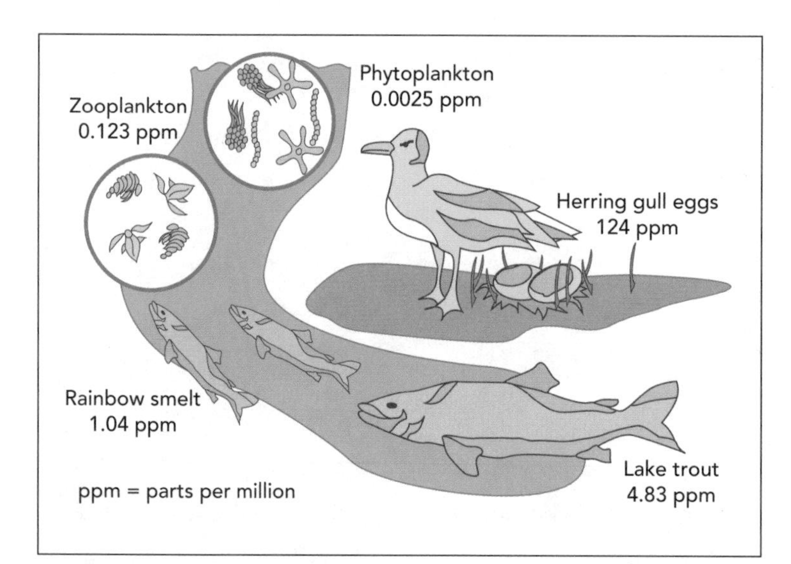

Figure 17.15
Concentration of PCBs Through the Food Chain

to monitor the lakes and resolve disputes over the use of water resources. Since 1972, a series of Great Lakes Water Quality Agreements has been signed by the two countries. A great deal of research has been done, and the examination of herring gull eggs from around the lakes over the last two decades indicates some improvement in certain chemical contaminations, such as DDT and PCBs. The difficulty lies with the number of different sources of pollutants, the increasing human use of the lakes, and the amount of money needed to address certain problems. Along the banks of the Niagara River alone there are over 200 toxic waste dumps, at least 22 of which are known to be leaking into the river. The Niagara River is just one of the 42 hotspots illustrated in Figure 17.14. For each identified hotspot, a Remedial Action Plan, or RAP, is being developed. The hotspot approach does not take into account atmospheric deposition nor run-off from agricultural sources. In addition, each RAP involves the expenditure of millions of dollars. Estimates for the clean-up of Hamilton Harbour alone range from 100 to 200 million dollars. It is increasingly apparent, however, that we cannot afford not to spend the money to clean up the Great Lakes system.

Zero discharge is a procedure which involves industry reusing waste water rather than dumping it back into the lakes. This procedure is becoming increasingly widespread in industry and promises to reduce the toxins entering the water. Mobile PCB decontamination labs are becoming available to municipalities. These labs break down the PCBs into less harmful chemical compounds. The technology now exists to burn toxic chemicals in high temperature kilns. Controversy still exists, however, as to the effectiveness of such procedures. In addition, the locating of such kilns is a problem as no community wishes to have these toxic substances transported and burned near it. Sewage treatment plants must be upgraded and new ones built. Storm sewers must be enlarged to ensure that untreated sewage is not swept out into the lakes during times of heavy rainfall. Ecologically sound substitutes for pesticides and insecticides must be developed and employed. Existing dump sites must be contained and cleaned up.

The environmental problems of the Great Lakes watershed require immediate economic and political action. The numerous state, provincial, and federal governments involved must coordinate plans and commit billions of dollars over the next decade if the Great Lakes are to be saved.

QUESTIONS

13. List all the ways, both direct and indirect, in which people make use of the Great Lakes.
14. Examine a map of the Great Lakes region. Identify and count all the states and provinces that border the Great Lakes. Add to your total number the two federal governments. In what ways does your total of governments indicate a difficulty in solving the environmental problems associated with the Great Lakes?
15. Ask 15 people to name as many of the toxic chemicals found in the waters of the Great Lakes as they can. (See Figure 17.13.) State two conclusions from the results of your survey.
16. The Jesuits reported that the native peoples referred to the Great Lakes as "sweetwater seas". Do you feel that this name will ever again be appropriate for the lakes? Explain.

It's a Fact..

It is estimated that 80 percent of Lake Superior's toxins are from windblown atmospheric deposits. The figure for Lake Michigan is 50 percent.

Review

- The tropical rainforests are being cut down for a variety of reasons, which include demand for hardwoods, land for settlers, cattle ranching, and mineral exploration.
- The destruction of the rainforests has global implications.
- Protection of the rainforests is the responsibility of all the world community, not just the countries in which the forests are located.
- Acid rain is produced by human activities, particularly the burning of fossil fuels and the use of automobiles.
- Acidification of the environment affects the vegetation, life forms, and people of the boreal biome.
- Acid rain is an international problem which knows no political boundaries.
- Desertification occurs when an arid ecosystem has been used beyond its capacity.
- Desertification is an international problem that requires worldwide cooperation.
- The ecosystems of the Great Lakes watershed are being altered by human interference on a massive scale.
- The chemical pollution of the Great Lakes threatens the health of more than 40 million people.
- International cooperation and great sums of money will be necessary to restore the Great Lakes to health.
- Environmental problems are also political and economic in nature.

Geographic Terms

restoration
agroforestry
replacement
rehabilitation
acidification
acid rain
pH scale
acid shock
buffering
desertification
toxic
biomagnification
zero discharge

Explorations

1. Construct a chart to compare the four fragile environments discussed in this chapter. Your chart might include these headings: ecosystem affected, location, problem, causes, consequences, solutions.
2. The oceans are fragile environments that suffer from human misuse. Research one of the following topics and summarize your findings in a two- to three-page report. The topics are:

– the impact of petroleum exploration in Arctic waters
– oil spills in the oceans
– mining on the ocean floors
– managing the fish stocks of the Atlantic Ocean
– whaling: past, present, and future

3. Copy the chart below into your notebook and fill in the statistics using atlases, yearbooks, or other statistical material available to you in the school. Be sure to fill in the correct units for each of the variables in the chart.

Country	Gross National Product Per Capita	Energy Consumption Per Capita	Average Life Expectancy Male/Female	Birth Rate
Brazil				
Canada				
India				
Sudan				
Sweden				
United States				

a) What do the statistics in your chart suggest to you concerning the ability of different countries to solve their environmental problems? Explain.
b) What do the statistics suggest concerning the necessity of international cooperation in terms of the protection of fragile environments? Explain.
c) The concept of "Spaceship Earth" suggests that our planet is like a spaceship with limited resources. Some people on the spaceship are living very well, other people are not. How is the concept of Spaceship Earth related to the solution of many of the world's environmental problems?

4. Choose a fragile environment not specifically dealt with in this chapter. Prepare an oral or visual report on the stresses that your chosen environment is being subjected to. Some possible environments to deal with are: Antarctica, the Great Barrier Reef, the Florida Everglades, wetlands.

Related Careers

- environmental planner
- environmental economist
- waste reduction and recycling manager
- landfill specialist
- wildlife manager
- environmental consultant
- ecologist
- forester
- water manager
- mine rehabilitation specialist
- parks supervisor
- environmental health officer

GLOSSARY

abrasion: the wearing away of bedrock caused by the rubbing, scouring, or scraping action of rock fragments or particles carried by streams, ice, wind, or waves.

absolute humidity: the actual amount of water vapour in a given volume of air; usually expressed in g/m^3.

abyssal plain: the huge bottom regions of the oceans. Most of the abyssal plain is found at depths between 4000 m and 6000 m.

accretion: the process of growth through accumulation; often used to refer to the growth of the planets through the gravitational attraction of plantesimals, asteroids, and meteoroids: the growth of continents or cratons by the addition of new rocks along their edges through mountain-building activity or collisions with other blocks of continental crust.

acid rain: wet form of acid deposition. Any rainwater that is more acidic than normal can be called acid rain.

acid shock: the sudden increase in acidity experienced by lakes and rivers when acidic snow melts in the spring.

acidification: the process that is increasing the acidity of our environments.

active layer: the layer of moisture-saturated material overlying permafrost that melts during the short summer season of high latitude regions.

adiabatic lapse rate: the rate at which temperatures in an air mass change due to expansion or compaction of that air mass as a result of changes in air pressure. In dry air, the adiabatic lapse rate equals 10°C per 1000 m, while in saturated air the rate equals 3°C per 1000 m.

aeolian erosion: see *eolian erosion.*

aftershock: a lower-magnitude earthquake that often occurs after a major earthquake.

aggradation: the processes that build up the land surface through the deposition of rock materials.

agroforestry: the planting of alternative trees with economic value in an area that has been deforested.

albedo: the amount of radiation that is reflected off a surface.

alluvial fan: a cone-shaped deposit of sediment formed where a fast-moving stream carrying large amounts of sediment flows abruptly onto a level plain, valley floor, or basin; common in high latitude and arid mountain ranges.

alluvium: stream-deposited sediments commonly found in stream beds, flood plains, deltas, and alluvial fans.

alpine or mountain glacier: a glacier that forms in an upland or mountain area, occupying valleys and basins; usually much smaller than continental glaciers.

andesitic magma: highly viscous, silica-rich, gaseous, molten rock formed by the melting of subducting oceanic plates in the asthenosphere that produces violent volcanic activity.

Antarctic bottom water: a major deep sea current of cold, dense water that flows at a depth of approximately 4 km north from its source in the Antarctic.

anticline: rock beds that have been folded or bent upwards to form a hill or mountain.

anticyclone: an area of high atmospheric pressure.

aphelion: the point on the earth's elliptical orbit where it is at the greatest distance from the sun.

aquatic biosphere: the life forms of the earth that live primarily in the seas and oceans.

aquifer: areas beneath the earth's surface in which fresh water is stored in layers of porous and permeable rock.

arch: a natural opening cut through a rocky headland by wave action; often an enlarged cave.

arête: a knife-edged ridge formed between the steep walls of two or more adjacent glacial cirques.

asteroid: a small body of rock with a diameter of less than 800 km orbiting the sun between Mars and Jupiter.

asthenosphere: the plastic (part solid, part liquid) layer of the upper mantle directly below the lithosphere that can flow slowly when put under constant pressure.

atmosphere: the mixture of gases, mainly hydrogen and oxygen, found above the earth's surface.

atmospheric hazard: a violent or powerful force resulting from processes operating in the atmosphere.

attrition: the breakdown of rock particles caused by collision with one another as they are carried by water, wind, or waves.

autotrophic: life forms, such as plants, that are able to produce their own foods (glucose).

avalanche: the rapid flow of snow, ice, and rock materials, common in mid- and high latitude mountains.

axis: an imaginary line joining the north and south poles around which the earth rotates.

azonal soil: an immature soil that has not had time to develop horizons because of constant erosion and deposition of materials, as on an alluvial flood plain, sand dune, or steep slope subject to landslides or slumping.

badlands: a landscape of gullies, carved hillsides, and ravines cut by the erosive action of water into relatively soft rocks in a semi-arid climate.

bahada: a gently sloping surface of alluvial materials stretching from the foot of a mountain to an inland desert basin formed by the growing together of a series of alluvial fans.

baroclinic wave theory of cylcones: the theory put forth by Charney to explain the development of mid-latitude cyclones.

barrier island (offshore bar): a long deposit of sand which forms parallel to the coastline and is separated from it by a lagoon.

basal slippage: the sliding of a glacier over bedrock, aided by the lubricating effect of water, melted by the heat of friction, along the surface between the glacier and its bed.

base level: the lowest level to which a land surface can be eroded (i.e., sea level).

batholith: a massive, often bottomless intrusion of magma that cools beneath the earth's surface to form igneous intrusive rock.

bayhead beach: a deposit of sand formed at the head of a bay through wave action, usually along an irregular coastline.

baymouth sand bar: a deposit of sand forming a narrow band linking two headlands across the mouth of a bay on an irregular coastline.

bedding plane: a natural, parallel surface separating different layers or strata of sedimentary rocks.

bedrock plain: a level to rolling area where glacial erosion has stripped most of the soil and rock fragments to expose the underlying bedrock.

biochemical cycle: a cycle that moves the chemical elements necessary for the survival of life through the ecosphere; powered by both energy from above and below.

biomagnification: the increase in toxic chemicals as they work their way through the food chain.

biome: a group of ecosystems with similar plant and animal species that exist under a similar climate.

biosphere: the thin layer below, on, and above the earth's surface where life forms exist.

block mountain: see *horst.*

boreal: the biome consisting of coniferous forest and animal species adapted to subarctic climates with short, cool summers and long, cold winters; common trees include spruce, pine, cedar, fir, and larch.

braided stream: a stream made up of a series of small, shallow, interlaced channels, separated by sand bars; typical of desert and glacial meltwater streams with great seasonal variations in water flow.

breaker: an oversteepened wave that falls over on itself, spending its energy against the shoreline.

bridal veil fall: a waterfall created where a stream flows from a hanging valley in the main valley in a glaciated mountain area.

buffering: the process of adding an alkali or base such as lime to a lake in order to reduce the acidity.

butte: an isolated, flat-topped hill with steep slopes that is smaller than a mesa.

caliche: a layer of calcium carbonate or other soluble salts formed near the surface of soils developed under arid and semi-arid climates.

cap rock: an impermeable and non-porous rock that does not allow oil or natural gas to pass through it.

capacity: the quantity of sediment of a given size that can be carried by a river, wave, wind, or glacier.

carbon dioxide sink: items in the physical world that have the ability to absorb carbon dioxide from the atmosphere. Tropical rainforests and the oceans are carbon dioxide sinks.

carnivore: an animal that preys on herbivores as its primary source of food.

catastrophic event: a sudden event that alters the physical world over a relatively short period of time.

cation: a positively charged ion that is attracted and held by positively charged particles of clay and humus in a soil.

cave: a hollow or cavity cut into weaker rock beds of a headland by wave action.

cavern system: an interconnected series of underground cavities or caves formed by the dissolving of limestone rocks, and often occupied by an underground river system.

change: physical geography sees the earth as dynamic and changing. Change is constantly occurring over both space and time.

chemical weathering: the chemical decay or alteration of rocks by a change in their chemical composition.

chlorofluorocarbons: a very stable group of gases that are contributing to both the greenhouse effect and the depletion of the ozone layer. Known also as CFCs.

cinder cone: a steep-sided volcanic peak, with a large summit crater, composed mainly of volcanic ash and rock spewed out of the vent during explosive eruptions.

cirque: a bowl-shaped depression cut into a mountain by a glacier as it advances from an icefield or snowfield.

clastic: loose sediments produced by the weathering or breakdown of rocks.

clay plain: a level to rolling plain formed by the deposition of clay in the calm water of a glacial lake.

climate: the long-term characteristics of the atmosphere. Climate is determined by averaging and totalling weather statistics over many years.

climograph: a combination graph that displays average monthly temperatures as a line graph and total monthly precipitation in the form of a bar graph.

collision zone: the zone where two plates containing continents are crashing into one another at a converging plate boundary.

competence: the largest size and weight of rock and soil particles that can be moved by an erosive agent such as water, wind, waves, or ice.

composite volcano: a smooth-sloped volcanic peak with a summit crater made up of alternating layers of ash and lava, formed from andesitic magma at subduction zones.

condensation: the change of state from vapour to liquid.

coniferous: cone-bearing, needle-leafed, evergreen trees adapted to survive in cold subarctic climates and infertile, acidic soils of high latitudes.

continental glacier: a single ice sheet that covers all, or a large portion, of a continent.

continental volcanic arc: an arc-like chain of volcanic mountains formed on a continent bordering a subduction zone at a converging plate boundary.

continuously habitable zone (CHZ): the zone within the solar system where a planet could, theoretically, have temperatures between -80°C and 100°C over a long enough time for life to form, according to the evolutionary theory.

convection current: a current that forms in liquids or gases when warmer, less dense materials rise while cooler, more dense materials sink, setting up a cycle of movement.

converging plate boundary: the boundary between two plates that are moving towards one another.

core: the innermost layer of the earth's interior, believed to be composed of a mixture of iron, nickel, and traces of other heavy metals, which is divided into a solid inner core and a liquid outer core.

Coriolis force: a force that deflects air and water currents to the right of their direction of travel in the northern hemisphere and to the left in the southern hemisphere.

corrosion: the dissolving of soluble minerals by water in streams or waves; common in humid areas underlain by limestone rocks.

craton: the most stable part of a continent, often made up of ancient rocks, that experiences few earthquakes or volcanic eruptions; often known as shields.

crust: the thin, solid, outer layer of the earth that is chemically different from the mantle rocks below.

cuesta: a ridge with one steep side and one gently sloping side formed by the differential weathering of gently dipping sedimentary rock beds.

cyclone: an area of low atmospheric pressure.

cyclonic disturbance: an area of low atmospheric pressure. Other names include cyclone or depression.

debris slide: the rapid downhill slide of loose rock and soil materials down a steep slope, triggered by an earthquake or undercutting by river or glacial erosion.

deflation hollow: a large hollow formed in desert areas by the removal of fine-grained mineral particles by wind action.

degradation: the process of lowering the earth's surface by the processes of weathering, erosion, and transportation.

delta: a level area of alluvial deposits formed at the mouth of a river, where it enters a shallow and/or calm water body.

denudation: the range of agents that work constantly to lower the earth's surface to a common level.

deposition: the dropping of rock materials that occurs when a drop in energy slows a transporting agent, such as running water, wind, or waves.

depression: an area of low atmospheric pressure. Other names include cyclone or cyclonic disturbance.

desert pavement: an extensive area of closely spaced pebbles and stones covering the desert floor, formed by the removal of finer soil particles by wind action.

desert varnish: a black-coloured iron or manganese oxide crust formed on exposed rock surfaces in desert areas.

desertification: the process by which desert conditions extend into formerly non-desert areas, largely by human actions but also due to climatic changes.

differentiation: the process by which compounds of different densities separated into distinct layers within the molten interior of the earth to form the core, mantle, and crust.

disaster: an extreme natural event that affects human beings and their creations in a negative way.

distributary: a stream channel that carries part of a river's flow across a delta.

diverging plate boundary: the boundary between two plates that are moving apart or separating, at a mid-ocean ridge.

doline: a large, enclosed, cone- or bowl-shaped hollow created by the dissolving of soluble limestone rock, often forming the entrance to underground caverns.

drainage basin: the area of land that is drained by a single river system.

drought: an extended period of low rainfall during which the growth of food and other crops is drastically reduced or even completely halted.

drumlin: a streamlined, tear-shaped hill with a wide, round front end, or stoss, and a longer, tapering tail formed when a glacier re-advances over a previously existing moraine.

dry valley: a valley without a river; commonly in karst landscapes where a river sinks through cracks in the limestone and flows completely below ground level, through cavern systems.

dyke: an intrusion of magma that cuts across the original rock beds and cools beneath the earth's surface to form igneous intrusive rock.

earthquake: a sudden shaking of the ground caused by waves generated by the movement of blocks of the earth's crust; earthquake intensity is often measured by the Richter scale.

earthquake focus: the centre of movement of the rocks of the crust along a fault.

eccentricity: the change in the orbit of the earth from a more circular to a more elliptical shape over a period of approximately 100 000 years.

ecosphere: the zone on earth in which all the major processes that affect life operate; includes both the biosphere and hydrosphere and extends into the atmosphere and the lithosphere.

ecosystem: a community of plants and animals that interact with one another and with the environment in which they live.

El Niño event: the disruption of the usual upwelling of cold, deep, ocean water that normally occurs at five locations in the oceans. It is believed that this disruption can lead to major changes in the earth's weather patterns.

electromagnetic wave: any object possessing heat gives off electromagnetic energy, which is usually visualized as travelling in waves at the speed of light.

element: a material that cannot be subdivided or broken down by ordinary chemical means into simpler materials.

energy balance: the equation that explains how the net radiation available at the earth's surface is utilized or consumed by the various physical and biological systems operating on the planet.

environmental lapse rate: the rate at which air temperatures decrease with altitude. In stable air, the lapse rate is 6.4°C per 1000 m.

eolian erosion: the action of wind in eroding, transporting, and depositing sediments in arid and semi-arid climates.

ephemeral stream: a stream that flows only rarely; common in desert areas where precipitation is sparse and very irregular.

epicentre: the point on the earth's surface directly above an earthquake focus.

equinox: the two times during the year (March 21 and September 21) when the sun is directly overhead at noon at the equator and the length of day and night is approximately equal throughout the world.

erg: the name used in the desert areas of North Africa and the Middle East to describe a sea of sand covering thousands of square kilometres.

erosion: the removal and movement of rock debris and associated organic matter from one part of the earth's surface to another through agents such as running water, ice, waves, and wind.

escarpment: a steep slope or cliff marking the edge of a plateau or any other level upland surface.

esker: a long, snake-like ridge of sorted sands, gravels, and boulders left behind by a meltwater stream that flowed on, inside, or beneath a glacier.

evaporation: the change of state from liquid to vapour.

evapotranspiration: the loss of water to the atmosphere through the combined processes of evaporation and transpiration.

exfoliation: the expansion, cracking, and peeling of rock layers off a newly exposed rock surface resulting from the release of pressure due to the erosion of overlying bedrock.

extreme natural event: a violent or powerful force caused by a natural process.

extrusive: an igneous rock that cooled quickly from magma erupted onto the earth's surface.

fault scarp: the often straight, continuous cliff created by the uplift of the earth's crust along a fault line.

faulting: the process by which rocks move past one another along a fracture or crack in the earth's crust, usually occurring where plates are separating, sliding past one another, or colliding.

fiord: a glacially eroded, U-shaped valley flooded by the sea when the glacier has melted.

firn: granular snow or compacted ice crystals that are being transformed into solid glacial ice by periodic melting and freezing and the pressure of overlying snow layers.

flood: an inundation of any land area not normally covered with water; usually caused by rivers overflowing their banks and inundating their flood plains or by high water levels associated with storm surges along exposed coastlines.

flood plain: a level, gently sloping alluvial plain, found on one or both sides of a stream channel, that is subject to periodic flooding.

folding: the process that bends and twists rocks through compression or squeezing.

foliated: a banded structure within metamorphic rocks caused by the lining up of different minerals into parallel layers.

frontal wave theory of cyclonic development: the theory put forth by Bjerknes to explain the development and effects of mid-latitude cyclones.

food chain: the general flow of energy and nutrients in an ecosystem.

fossil fuel: all fuels formed from the remains of formerly living organisms, including peat, lignite, coal, oil, and natural gas.

free oxygen: oxygen (O_2) in the atmosphere that has not combined with any other element(s).

frost action: a process of physical weathering carried out by the constant freezing and thawing of water.

frost shattering: the splitting apart of rocks by the freezing and expansion of water in fractures and joints; common in high altitude and high latitude regions where the temperatures frequently rise and fall about the freezing point.

Gaia Hypothesis: an hypothesis which emphasizes the interrelationships between the physical and biological worlds and suggests that all of Planet Earth can be thought of as a living entity.

gas giant planet: the larger, outer planets of the solar system (Jupiter, Saturn, Uranus, and Neptune) that are composed largely of lighter elements in the form of gases and ices.

geologic time: the division of the earth's history into eras and periods dating back to approximately 5 billion years.

geological hazard: a violent or powerful natural event resulting from tectonic or geologic processes.

geomorphology: the scientific study of the landforms that make up the earth's surface.

gibber: an Australian term referring to large areas covered with desert pavements.

glacial erratic: a rock carried from its area of origin and deposited at a distant location by a glacier.

glacial ice: a solid, impermeable mass of ice formed from firn, or granular snow, by the pressure of overlying firn and snow layers and by melting and refreezing.

glacial plucking: a process by which sections of rock, frozen to the bottom of a glacier, are pulled out of place and carried away as the ice advances.

glacial till: ice-deposited material made up of an unsorted jumble of clay, sand, silt, gravel, and boulders.

graben: another word for rift valley, formed where a block of the earth's crust drops down between two parallel fault lines, often where the crust is moving apart or creating tension.

graded stream: a stream that has adjusted its channel so that the amount of sediment it can carry matches its ability to carry such materials; a stream with a smooth cross-sectional profile from its source to its mouth.

greenhouse gases: those gases in the atmosphere that have the ability to "trap" outgoing longwave radiation. These gases lead to an increase in the net radiation available at the earth's surface.

grike: a deep crack, from 15 to 60 cm in width, created by the dissolving of soluble limestone along fractures or weaknesses in the rock.

ground moraine: an undulating surface of glacial till laid down beneath a glacier.

groundwater: water that is found beneath the surface of the earth.

hanging valley: a U-shaped valley cut by a smaller tributary glacier that lies at a higher elevation than the deeper U-shaped valley eroded by the main glacier.

headland: a point of land that extends outwards into a body of water from a coastline; often having a steep cliff face.

headward erosion: the action of a river in cutting its channel upstream, lengthening its valley.

heat island: the increased temperatures found in the centre of large urban areas compared with surrounding rural or suburban areas.

herbivore: an animal that consumes plants as its direct source of food.

hetertrophic: life forms, such as animals, that are unable to manufacture their own foods.

holistic: physical geography examines "the big picture" and studies the interrelationships among all the various phenomena and energy flows on the earth.

horizon: a distinct horizontal layer found in mature soils, created by the movement of water, upward and/or downward, through a soil profile.

horn peak: a rectangular, sharp-pointed peak formed where several glacial cirques erode back into a single mountain.

horst: a steep-sided mountain formed where a block of the earth's crust has been squeezed upward between two parallel fault lines; also known as a block mountain.

hot spot: a point on the earth's surface where strong upward convection currents or plumes of hot magma in the upper mantle push up below the plates of the lithosphere causing volcanic activity.

humus: the black, partly decomposed organic matter found within the upper layer of a soil.

hurricane: an intense large cyclone, usually found in tropical areas, in which wind speeds exceed 117 km/h; also known as a typhoon.

hydraulic pressure: the breakdown of solid rocks caused by the sheer weight and high pressure exerted by water in fast flowing streams or strong waves.

hydrologic cycle: the movement of water through the hydrosphere. The hydrologic cycle is a closed cascading system.

hydrolysis: a form of chemical weathering in which water combines with minerals to form hydroxides that expand and eventually cause the disintegration of the rock.

hydrosphere: the part of our planet where water, in its varied forms, is found.

hygroscopic particle: any solid particle in the atmosphere around which water vapour can condense.

hythergraph: a scatter graph in which each of the twelve points represents both the average monthly temperature and the total monthly precipitation for a location.

ice wedging: see *frost shattering.*

ice-dammed lake: a lake formed when a glacier blocks the downslope movement of glacial meltwaters.

icefield (snowfield): a large area where the accumulation of snow has sufficient depth to form glacial ice.

igneous: a rock formed from the solidification of molten material or magma; from the Latin word for fire.

integrative: physical geography combines information from a number of other disciplines.

interlobate moraine: a rolling line of hills, composed of glacial till, deposited where two different lobes of ice meet.

intermittent stream: a stream that flows only part of a year, during the wet season.

intertropical convergence zone: the zone where the northeast and southeast trade winds meet.

intrazonal soil: a soil that has not developed distinct horizons and characteristics because of local factors such as unusual parent materials or poor drainage.

intrusive: an igneous rock that cooled slowly below the earth's surface.

island arc: an arc-shaped chain of volcanic mountains, often rising above sea level as islands, formed on the ocean floor at a subduction zone at a converging plate boundary.

isobar: a line on a weather map that joins points of equal atmospheric pressure.

isostasy: the state of balance maintained by the earth's crust as it "floats" on the plastic layer of the upper mantle, based on the principle of displacement discovered by Archimedes.

jet stream: the high altitude winds that flow around the planet in both hemispheres at heights of between 9 and 12 km. Their path is wavy or meandering in nature and they mark the division between cold polar air and warm tropical air.

joint: a fracture or crack in a rock created by tectonic processes such as folding or faulting or by the release of pressure when overlying rock layers are eroded.

kame: a small, conical hill composed of sorted sands and gravels deposited where a glacial meltwater stream entered a pond or lake forming a delta.

karst landscape: a land surface created under a humid climate by the dissolving of limestone rock and characterized by hollows and depressions, an underground rather than surface drainage system, and limestone caverns.

kettle: a lake formed when a block of ice, trapped beneath glacial deposits, melts and leaves behind a small depression.

L (long) wave: slow moving earthquake wave that passes along the earth's surface, moving the ground sideways as it passes; also known as Love wave.

laccolith: an intrusion that forces apart the local rock beds to form an enlarged, dome-shaped chamber of magma that cools to form igneous intrusive rock.

lagoon: the coastal water body separating a barrier island or offshore bar from the mainland.

lateral erosion: the widening of a river valley by sideways erosion, usually brought about by the undercutting of the banks on the outside curves of meanders.

lateral moraine: a rolling line of hills deposited along the sides of a glacier and composed of glacial till.

lava: magma that has erupted onto the earth's surface.

Law of Uniformitarianism: assumes that the physical laws and processes that operate on the earth today operated in similar ways in the past.

leaching: the process by which percolating rain water removes soluble substances from one soil horizon to another.

levee: a ridge of alluvial material found on both sides of a river that stands above the level of the flood plain.

light year: the distance travelled by light in one year (9.46 trillion kilometres).

lithification: the process that turns sediments into sedimentary rock, usually through cementation or compaction, or a combination of the two.

lithosphere: the solid outer layer of the earth where the rocks are less dense and more rigid than those of the asthenosphere below; includes the top part of the mantle and all of the crust.

loam: an easily ploughed soil composed of sand, silt, and clay; combines many of the individual properties of each of these soil textures.

longshore drift: a current that moves down a coastline within the shallow breaker zone next to the shoreline.

longwave radiation: any electromagnetic energy whose wavelengths are greater than 3 microns.

magma: molten rock that exists under great pressure below the surface of the earth.

magnetic field: the magnetic lines of force created by movements of the molten iron and nickel layer of the earth's outer core.

magnetic reversal: a periodic change in the polarity of the earth's magnetic field from normal (north) to reversed (south), or vice versa.

mantle: the layer of the earth's interior differing in chemical composition from the crust above and the core below and having a density ranging from 3.0 to 3.3 g/cc.

mass wasting: the downhill movement of weathered materials caused by the pull of gravity.

meander: a broad, curving bend in a river flowing over an alluvial flood plain.

medial moraine: a ridge of glacial till dropped where two glaciers flow together to form a single river of ice.

mesa: a tableland or isolated, flat-topped hill with steep cliffs or escarpments marking its outer edges.

mesosphere: the largely solid layer of the earth's interior located between the core and the asthenosphere.

metamorphic rock: a rock that was changed by great heat and pressure from its original state.

meteorite: matter that has fallen to the earth's surface from outer space.

meteoroid: a very small body of rock within the solar system moving in orbit about the sun.

mid-ocean ridge: the ridge that marks the boundary between two or more separating plates.

mineral: the material formed when two or more elements combine in a crystalline structure.

mineral matter: inorganic material produced by the weathering of rocks.

monsoon: a wind system that develops as a result of the existence of high pressure systems over large land masses in the winter and low pressure systems over large land masses in the summer. Monsoon winds usually result in distinctive wet and dry seasons.

moraine: a deposit of glacial till transported and deposited by a glacier.

natural hazard: a violent or powerful natural event that can potentially affect human beings and their creations in negative ways.

neap tide: the lower tide caused when the sun and moon are at right angles to one another, offsetting their gravitational attraction.

nebula: a cloud of dust particles and gas found in outer space.

nebular hypothesis: a hypothesis suggesting the solar system formed from a large, disc-shaped cloud of gases and dust.

net radiation (Rn): the amount of electromagnetic energy available for use by the earth's physical systems at the earth's surface.

non-clastic: sediments produced by chemical or organic processes.

normal fault: a fault resulting from the upward movement on one side of a fault line and/or the downward movement on the other, forming a cliff or fault scarp.

North Atlantic Deep Water Current: a major deep sea current of cold dense water that flows south from its source in the Arctic.

nutrient cycle: the processes that move the chemical elements needed for life through the ecosphere.

nutrients: the important chemical elements and compounds needed to maintain life, including carbon, oxygen, and nitrogen.

obsidian: an extrusive igneous or volcanic rock that cools quickly to form volcanic glass.

omnivore: an animal that includes both animal flesh and plant foods within its diet.

organic matter: the remains of plants and animals living on, or in, the soil.

outwash plain: a gently sloping plain composed of sand, gravel, and silt laid down by streams flowing out from the front of a melting ice sheet.

oxbow lake: a crescent-shaped lake on a river flood plain formed when a stream cuts through the narrow neck of a meander.

oxidation: a form of chemical weathering in which oxygen combines with certain minerals in the rock to form oxides and hydroxides and leading to the disintegration of the rock.

ozone layer: a layer of ozone (O_3) concentrated at a height of about 25 km in the stratosphere. Ozone absorbs the harmful, very short wavelengths of solar radiation, preventing this radiation from reaching the earth's surface.

P (primary) wave: the fastest moving compressional earthquake wave that passes through both the solid and liquid layers of the earth's interior.

parent material: the weathered rock fragments that provide the mineral matter from which soils develop.

particulate matter: solid particles in the atmosphere.

patterned ground: a ground surface broken into regular polygon shapes by intense frost action of high latitude periglacial landscapes.

ped: a larger unit or block formed when the individual grains of a soil group together.

pedestal rock: a large boulder of more resistant rock perched on a narrow, wind eroded base of softer rock.

pediment: a bedrock plain that slopes down gently from the base of a steep mountain range into the centre of a desert basin.

perennial stream: a stream that contains water for all, or most, of the year.

periglacial landscape: a poorly drained, bog- and pond-strewn landscape formed in high latitude regions that are underlain by permafrost and subject to constant freezing and thawing.

perihelion: the point on the earth's elliptical orbit where it is closest to the sun.

permafrost: ground that has been frozen for more than two years; found in high latitude regions.

pH scale: a measure of the concentration of hydrogen ions found within a soil that determines the acidity or alkalinity of a soil; a reading of 3 to 7 is acidic and 7 to 11 is alkaline.

photosynthesis: the process by which sunlight converts carbon dioxide and water into glucose (sugar) and oxygen.

physical weathering: the mechanical disintegration or fragmentation of rocks into smaller particles with little change in their chemical composition.

pillow lava: a pillow-shaped igneous rock formed by the rapid surface cooling of magma in direct contact with cold ocean water, often at mid-ocean ridges.

pingo: a hill with a core of ice, water, and mud that rises above the land surface of a periglacial landscape.

pitted outwash plain: a gently sloping outwash plain composed of sand, gravel, and silt that is dotted with kettle lakes and depressions formed by the melting of ice blocks buried under the outwash sediments.

planetesimal: a tiny planet formed from dust and gas that eventually was pulled together by gravitational attraction to form the larger planets of the solar system.

plastic deformation: the imperceptible flow of a solid, such as ice, without any cracking or rupturing, through the slippage and rotation of individual crystals put under constant pressure or stress.

plate: a rigid slab of solid lithosphere rock that has defined boundaries and floats on the denser rocks of the asthenosphere.

plate tectonics: the study of the movement of the earth's plates and the effects they have on the surface features of the lithosphere.

plateau: an extensive elevated area of level land, usually bounded on at least one side by a steep escarpment.

playa: a dry lake bed made up of alluvium crusted with soluble salts formed in the centre of a desert basin; occasionally filled with water after infrequent desert storms.

plutonic rock: another name for an intrusive igneous rock.

point bar: a deposit of sandy materials on the inside bend of a meander.

polar front: the boundary zone between cold, dry, polar air and warmer, more humid, subtropical air. It is along the polar front that most of the weather systems affecting the mid-latitude zones develop.

polje: a large basin or valley formed in limestone regions by the collapse of an underground cavern system.

precession of the equinoxes: the slow change in the time of year when the earth is at its aphelion and its perihelion; the aphelion currently occurs on July 4 and the perihelion on January 3; a cycle that takes approximately 21 000 to 22 000 years to complete.

pressure gradient: the change in air pressure as one moves from one location on the earth to another.

rainpit: a small hollow created by raindrops dissolving soluble limestone on an otherwise level rock surface.

recessional moraine: a rolling line of hills, composed of glacial till, deposited where a retreating ice sheet stalled or had a slight re-advance.

recumbent fold: a fold that has been compressed so severely that it falls over or is overturned.

reg: an Algerian term referring to large areas covered with desert pavements.

rehabilitation: the process of introducing a wide variety of plant and animal species into an area that has been deforested.

relative humidity: the ratio obtained by comparing the amount of water vapour in a given volume of air with the maximum amount of water vapour which that volume of air could hold at a given temperature.

remote-sensing: the process of gathering information by use of electronic or other sensing devices mounted on satellites or placed on the earth's surface.

replacement: the process of planting different species of trees in an area that has been deforested.

reservoir rock: a porous and permeable rock, usually sandstone or a coral reef formation, that holds oil and natural gas within a trap or reservoir.

respiration: the process that oxidizes glucose, releasing energy and carbon dioxide; the opposite of photosynthesis.

restoration: the process of returning an ecosystem, such as a rainforest, back to its original state.

reverse fault: a fault where one block of the earth's crust moves upward against another.

revolution: the motion of the solar system in its orbit around the centre of the Milky Way galaxy; the motion of the earth in its orbit around the sun.

rift valley: a steep-sided valley formed when a block of the earth's crust falls down between two parallel fault lines; also known as a graben.

rock: a consolidated mixture of one or more minerals.

rock basin lake: a lake occupying a basin gouged out of less resistant bedrock by glacial erosion.

rock cycle: the linked series of processes that form and alter the rocks of the earth.

rock glacier: the slow, downhill movement of large volumes of rock fragments, especially in mid- and high latitude mountains where frost shattering, or ice wedging, is common.

rock knob topography: a land surface consisting of rounded rock hills and rock basin lakes formed by glacial erosion.

rock slide: the rapid slide of large sections of rock in steeply sloping mountain areas.

Rossby wave: a wave or meandering in the jet stream.

rotation: the west to east spinning of the earth on its axis.

run-off: the water that flows over the ground surface, usually during a heavy rainstorm.

S (secondary) wave: the slower, transverse or shear earthquake wave that can pass only through the solid layers of the earth's interior.

salinity: the total weight of dissolved salts in a fluid. The average salinity of sea water is 3.5 percent or 35 000 parts per million.

salt dome: a large mass of salt that forces its way upward through overlying sedimentary rock layers, often bending or distorting these layers as it rises.

saltation: the movement of rock fragments by bouncing or hopping along the earth's surface by running water or wind.

salt: a chemical compound formed when the hydrogen in an acid is replaced in whole or in part by a metal or an electropositive radical. Dissolved salts, such as sodium chloride, magnesium chloride, and sodium sulphate, result in the characteristic salinity of sea water.

sand plain: a level to rolling plain formed by the deposition of sands and gravels by glacial streams as they enter a glacial lake.

sand sea: an extensive desert area covered with sand dunes.

schlerophyll: the species of evergreen shrubs and trees with tough, leathery leaves that are adapted to climates with long, dry summers, such as those bordering the Mediterranean Sea; trees such as oak, olive, and pine make up schlerophyll forests.

scree: see *talus*.

sea stack: an isolated spire or pillar of rock rising above water level along a coastline; often formed where a sea arch has collapsed as a result of wave action.

sea-floor spreading: the process that creates new sea floor as plates spread apart or separate at mid-ocean ridges.

sediment: rock particles of various sizes that have been transported by erosive agents.

sedimentary rock: a rock formed by chemical precipitation or cementing together of mineral grains deposited by erosive agents.

seismic tomography: a computer technique that analyses earthquake waves recorded at hundreds of seismic stations to produce maps showing variations in the density and temperature of the earth's mantle.

seismology: the scientific study of earthquakes, the seismic waves they generate, and the passage of these waves through the earth's interior.

sensible heat flux: the movement of heat in the atmosphere by convection currents in the air.

serir: a Libyan term referring to large areas covered with desert pavements.

shearing: the process that occurs when the ice crystals within a glacier slip over, past, and beside one another along miniature faults within the ice mass.

sheet flooding: a continuous film of water flowing over level, gently sloping ground surfaces during heavy rainstorms; common on alluvial fans in desert regions where there is little vegetation cover.

sheeting: see *exfoliation*.

shield volcano: a gently rising, smooth-sloped volcanic dome formed from very fluid (low viscosity) basaltic lava typical of mid-ocean ridges and hot spots.

shortwave radiation: any electromagnetic energy whose wavelengths are less than 3 microns. Most solar radiation is shortwave energy.

sial: the lighter (density of 2.8 g/cm^3) granitic rocks of the continents that are largely composed of the elements silicon and aluminum.

silicate: a relatively light chemical compound containing the elements silica and oxygen.

sill: an intrusion of magma that follows the layers of the original rock beds and cools beneath the earth's surface to form igneous intrusive rock.

sima: the denser (density of 3.0 g/cm^3) basaltic rocks of the ocean basins that are largely composed of the elements silicon and magnesium.

slump: the rotational slip of a block of soil and rock along a concave surface; common in wet, clay soils.

snout: the lowest or advancing front of a glacier.

snow belt: the increase in snow amounts found in the lee of many large lakes.

soil: a complex mixture of animal, mineral, and organic materials, capable of supporting plant life, and differentiated into horizons or layers.

soil air: the air that exists between the mineral and organic particles that make up the soil; provides oxygen for the survival and growth of organisms living within the soil.

soil creep: the slow and unspectacular downhill movement of loose rock and soil, often shown by the downslope tilt of fences and walls.

soil water: the water that exists within a soil; provides moisture for the survival and growth of plant and animal organisms living within the soil.

solar constant: the amount of solar radiation received on a surface perpendicular to the sun's rays at the outer limit of the earth's atmosphere. It is equal to about 1400 watts per square metre (1400 W/m^2).

solar radiation: the energy given off by the sun.

solar system: the nine planets and asteroids that revolve in almost circular orbits about the sun.

solifluction: the slow, downhill movement of water-saturated rock and soil materials; commonly occurs in summer in high latitude climates where the soil is underlain by permafrost.

solstice: the two times during the year when the midday sun is directly overhead at noon at one of the tropics; usually June 21 over the Tropic of Cancer (summer solstice) and December 21 over the Tropic of Capricorn (winter solstice).

solution: the dissolving and removal of soluble minerals from rocks by flowing water or waves.

solution furrow (rill): a small channel, ranging in length from a few millimetres to several metres, formed by the dissolving of limestone rock by water.

source rock: a rock from which oil and natural gas remains are squeezed by the pressure of overlying rocks.

spatial: geography is spatial in that it deals with the location, distribution, and pattern of phenomena.

spillway: a wide, shallow valley eroded by a stream carrying water from a melting glacier.

spit: a sand bar, formed by wave action, extending out into a body of water that is attached to land at only one end.

spring tide: the higher tide caused by the greater gravitational attraction that occurs when the sun, moon, and earth line up.

stalactite: a narrow calcite tube or cone deposited by water seeping downward from the roof of an underground cave or cavern in limestone rock.

stalagmite: a wide pedestal of calcium carbonate, or calcite, deposited by water falling to the floor of an underground cave or cavern in limestone rock.

stone circle: a dome-shaped area of fine soil surrounded by a circular ring of stones formed by intense frost action in a periglacial landscape.

storm surge: a sudden rise in water levels along a coastline caused by high winds and low atmospheric pressure; often associated with tropical cyclones or hurricanes.

striation: a scratch or groove cut into the surfaces of bedrock by boulders and pebbles frozen into the ice along the bottom of a glacier.

strike-slip fault: a fault where two sections of the earth's crust move almost horizontally past each other.

subduction: the downward movement and eventual melting of an oceanic plate as it sinks into the asthenosphere along converging plate boundaries.

subduction zone: the zone where an oceanic plate is sinking below a plate containing continents at a converging plate boundary.

subpolar lows: the zones of ascending air found at approximately 60 degrees north and south of the equator. The wind systems known as the polar easterlies and the westerlies blow into the subpolar lows.

subtropical highs: the zones of descending air found approximately 30 degrees north and south of the equator. The wind systems known as the westerlies and the trade winds blow out of the subtropical high pressure zones.

sunspot: a cooler, darker region of the surface of the sun created by a magnetic storm and often giving rise to solar flares.

supercontinent: a single, massive continent made up of all of the world's continents; Pangea is the last supercontinent that began to break up about 100 million years ago to form the smaller, more numerous continents we know today.

suspension: the process by which light rock fragments are held and carried within water and wind currents by turbulent flow.

syncline: rock beds that have been folded or bent downwards to form a valley.

systems oriented: all the complex topics in physical geography —flows of matter and energy, spatial phenomena, and human interventions— can be linked and studied as parts of one of four types of systems.

taiga: the Russian term referring to the boreal or short coniferous forest biome developed under subarctic climates of high latitudes.

talus: a fan-shaped pile of rock fragments that accumulates at the base of steep slopes in mountain regions where frost shattering is common.

tarn: a lake occupying the bottom of a cirque eroded by a glacier that has since completely melted.

tectonics: the processes that deform the earth's lithosphere and the rock structures and surface features created by these processes.

terminal moraine: a rolling line of hills deposited at the furthest extent of glacial advance and composed of glacial till.

terrestrial biosphere: the life forms of the earth that live primarily on land.

terrestrial planets: the smaller, inner planets of the solar system (Mercury, Venus, Earth, and Mars) which are largely composed of heavier elements in the form of solid rock materials.

texture: the sizes of the mineral particles that make up a soil.

thermal expansion: the disintegration of rocks caused by the differential heating and cooling of different minerals within them; common in granitic rocks in desert environments.

thermocline: the boundary zone between the deep, cold, dense bottom waters of the ocean and the warmer, less dense surface waters. The thermocline is usually found at a depth of about one kilometre.

tide: a regular rise and fall of sea level produced by the gravitational attraction of the moon and the centrifugal force of the earth's rotation.

tilted block mountain: a mountain formed where a block of the earth's crust moves upward at an angle between two parallel fault lines.

time zone: a north-south division of the earth, approximately 15° of latitude in width, with a time one hour ahead of the zone to the west and one hour behind the zone to the east.

tombolo: a sea stack or small island attached to the mainland by a sandbar.

topsoil: the material contained in the upper or A-horizon of a soil profile.

tornado: a rapidly spiralling funnel of air developed around an intense low pressure centre associated with massive thunderstorms that develop along fronts between warm tropical and cool polar air masses.

toxic: chemical compounds that are damaging to plants, animals, and/or humans.

traction: the rolling or tumbling of rock fragments along the earth's surface by running water or wind.

transform fault: a fault formed by the horizontal movement of the earth's crust, occurring where two plates are sliding past one another.

transform plate boundary: the boundary between two plates that are slipping or sliding past one another.

transpiration: the loss of water vapour through the leaves of plants into the atmosphere.

transportation: the movement of rock debris from one part of the earth's surface to another by running water, ice, wind, or wave action.

trap (reservoir): a rock structure in which oil and/or natural gas collects, due to overlying impermeable and non-porous rock.

tropical cyclone: a very intense storm with high winds, thick clouds, and high precipitation originating over warm tropical seas; also known as hurricanes (North America), typhoons (East and South Asia), and willy willys (Australia).

trough: a weak low pressure front found in tropical areas. These fronts seldom develop through the stages associated with mid-latitude cyclones.

tsunami: an ocean wave generated by a large displacement of water caused by an undersea earthquake or volcanic eruption.

tundra: the biome with an absence of trees and made up of hardy plants and animals adapted to life under cold, high latitude climates and boggy, permafrost soil conditions; common plants include mosses, lichens, grasses, and annual flowering plants.

typhoon: an intense, large cyclone, usually found in tropical areas, in which wind speeds exceed 117 km/h; also called a hurricane.

U-shaped valley: a wide, deep valley with a U-shaped cross-section, formed by glacial erosion in a mountainous region.

uvala: a large, flat-bottomed, steep-sided depression, several kilometres in diameter, found in limestone regions and created where several dolines have grown together.

vertical erosion: the downward cutting action of a river that deepens its bed, often forming a V-shaped valley.

vertical zonation: changes in the natural vegetation that occur on the slopes of mountains as a result of changes in altitude, climate, soils, and exposure.

viscosity: the resistance to flow of a liquid, such as magma or molten rock.

volcanic neck: a block of hard rock left standing above the landscape when an extinct composite volcano is eroded away over millions of years.

volcanic rock: another name for an extrusive igneous rock.

vulcanism: the movement of molten rock, or magma, beneath or above the earth's surface.

wadi: a steep-sided, flat-floored river valley, typical of desert regions, that seldom contains flowing water, except after infrequent storms; frequently subject to flash floods.

water table: the level beneath the earth's surface below which the soil and rock are saturated with groundwater.

wave-cut platform: a level terrace cut by wave action into a cliff, usually along a steeply sloping coastline.

weather: the short-term characteristics of the atmosphere. The components of weather include temperature, precipitation, wind, humidity, cloud cover, visibility, and air pressure.

weathering: the disintegration or decomposition of rocks in place on or near the earth's surface.

xerophyte: a plant species, such as the cactus, that has adapted to survive in extremely dry climates.

zero discharge: the process of reusing or containing all waste materials from industry in order to prevent them from entering the ecosystem.

zonal soil: a soil whose distinctive horizons and characteristics show the influence of the climate and vegetation under which it has formed.

INDEX

A
Abrasion, 259, 280, 293
Abyssal plain, 198
Accretion, 30, 51, 111
Acid rain, 349
Acid shock, 352
Acidification, 349
Adiabatic lapse rates, 153, 154
Aftershocks, 315
Aggradation, 242
Agroforestry, 348
Air masses, 169–71. *See also* Wind systems
Albedo, 125, 139, 140, 159, 277, 278
Alluvial fans, 264, 270
Alluvium, 260
Alpine glaciers, 280
Altered systems, 9
Altitude, effect on temperature, 152, 153
Anemometer, 170
Antarctic Bottom Water, 200
Anticyclones, 143, 173
Aphelion, 39
Aquatic biosphere, 211
Aquifers, 202–204
Arch, 294
Arêtes, 285
Arid landscapes, 268–72
Asteroids, 23, 58
Asthenosphere, 55, 63, 87, 92, 109
Atmosphere, 3, 4, 33, 205–207
Atmospheric Environment Service (AES), 189
Auroras, 29
Autotrophic life forms, 216
Avalanches, 250, 291

B
Badlands, 271
Bahadas, 264
Barometer, 170
Barrier islands, 296
Basal slippage, 279
Base level, 241, 260
Bayhead beaches, 295
Baymouth sandbar, 296
Bedding planes, 248
Bedrock plains, 281
Big Bang Hypothesis, 21, 38–39
Biochemical cycle, 212–13
Biome, 218
Biosphere, 4, 211
Boreal forests, 220, 349–54
Breaker waves, 293
Bridal veil falls, 287
Buffering, 353
Buttes, 271

C
Cambrian period, 211
Cap rock, 270
Carbon cycle, 214
Carbon dioxide, 126, 162–63
Carbon dioxide sink, 326
Carnivore, 216
Cascading systems, 8, 196, 255, 279
Catastrophic events, 13
Cations, 228
Cave, 294
Cavern systems, 267
Chlorofluorocarbons (CFCs), 326, 332–33, 335–36
Cirques, 285
Climate, 132
    change, 162–63
    classification systems, 133–36
    controls, 133, 137–61
Climograph, 137
Coastal landscapes, 291–93
Cold accretion theory, 51
Comfort Index, 171
Condensation, 205, 206
Condensation point, 154, 178
Conduction, 128
Coniferous forests, 220
Continental glaciers, 280
Continental shelf, 201
Continents, 112–14
Continuously habitable zone (CHZ), 34
Convection, 128
Convection currents, 88, 92, 149
Convectional rainfall, 177, 178, 179, 268
Coriolis force, 46, 142, 143, 174, 178
Corrosion, 259, 293
Cosmic egg, 21
Craters, 28
Cratons, 111
Cyclones, 143, 173, 174, 309–310
Cyclonic rainfall, 178, 179

D
Debris slide, 250
Deciduous forests, 221–22, 224
Deflation hollows, 272
Degradation, 242
Delta, 264, 291
Denudation, 241–42
Deposition, 242
Depression, 174
Desert margins, 355–57
Desert pavements, 271
Desert varnish, 272
Desertification, 306, 355–56
Deserts, 222–23
Differentiation, 30, 33, 53, 54
Disaster, 303–304
    natural, 304–305
Distributaries, 264
Dolines, 267
Drainage basin, 258
Drought, 306
Drumlins, 283
Dry valleys, 267

E
Earth
    age of, 10
    arrival of life, 33
    atmosphere, characteristics of, 123
    axis, 41–45
    core, 54, 56
    crust, 54, 67
    elements making up, 53–56
    formation of, 51
    interior of, 56–57, 63–64
    internal heat sources, 51–52
    internal temperature of, 51–52, 56
    magnetic field, 63
    mantle, 54, 63, 68
    orbit of, 39–40, 44–45
    origin of, 30–33
    place in solar system, 34
    rotation of, 46
Earthquake, 96, 109–10, 313
    epicentre, 61
    focus of, 58
    forecasting, 101
Eccentricity, 40
Ecosphere, 4, 211–13
Ecosystems, 9, 201, 217
Eccentricity, 40
El Niño, 190–92
Elements, 67–68
Energy balance, 128–29
Environmental lapse rate, 152, 154
Equinoxes, vernal and autumnal, 41, 43
Ergs, 272
Erosion, 242
    eolian, 271
    headward, 264
    lateral, 261–62, 264
    vertical, 264, 265
Escarpments, 270
Eskers, 284
Evaporation, 198, 205

Evapotranspiration, 159, 231
Exfoliation, 244

F
Falls, 250–51
Fiords, 286
Firn, 278
Floods, 269–70, 307–308
Flood plains, 262–63, 265, 307
Flows, 249–50
Foliated, 79
Food chain, 217
Fossil fuels, 76–77
Free oxygen, 33
Frontal wave theory of cyclonic development, 173
Frost shattering, 243, 248

G
Gaia Hypothesis, 13
Galaxies, 18, 19
- elliptical, 22
- moving, 38–39

Gamma rays, 120
Gas giant planets, 24
Geologic time scale, 10, 11, 86–87
Geomorphology, 255
Gibbers, 272
Glacial erratics, 282
Glacial ice, 278
Glacial plucking, 280
Glacial till, 282
Glaciation
- alpine, 285–88
- continental, 280–84
- theory of, 276–77

Glaciers, 201–202, 249, 278–80, 291
Glucose, 215, 216
Granite, 71
Grasslands, 221–22
Gravitational energy, 30, 243
Great Lakes, 357–63
Greenhouse effect, 123, 128, 214, 310, 324–31
Greenhouse gases, 162–63
Grikes, 267
Groundwater, 202–204
Gulf Stream, 147, 206

H
Hanging valleys, 287
Hazards
- atmospheric, 306–12
- geological, 313–18
- natural, 303, 318–19

Headlands, 293
Heat island, 159
Herbivore, 216
Heterotrophic life forms, 216
Highland vegetation, complex, 224–25
Horn peak, 285
Humid landscapes, 257–63
Humidity, 206
Humus, 228
Hurricanes, 178–82
Hydraulic pressure, 259, 294
Hydro dams, 336–41
Hydrogen, 119
Hydrologic cycle, 196–98, 212, 213
Hydrolysis, 245–46
Hydrosphere, 3–4, 56, 67, 69, 196
Hygroscopic particles, 207
Hytherograph, 137, 138

I
Ice age, 44
Ice-dammed lakes, 287
Ice sheets, 201–202
Ice shelf, 201
Ice wedging, 243
Icefields, 285
Intermittent streams, 258
International Date Line, 46
Intertropical convergence zone, 178
Isobars, 174
Isostasy, 81–83

J
Jet streams, 143, 176
Joints, 248

K
Kames, 284
Karst landforms, 267–68
Karst landscapes, 266–68
Kettles, 283
Kinetic energy, 52

L
Lagoons, 296
Lakes, 204
Landform classification, 256–57
Lava, 71
Law of Uniformitarianism, 12–13
Levees, 262
Light year, 18
Lithification, 75
Lithosphere, 3, 4, 54, 55, 86, 214
Lithosphere plates, 88–89
Loam, 226
Long (L/Love) waves, 60
Longshore drift, 293
Longwave radiation, 126, 159

M
Magma, 70, 71
Magnetic reversals, 93–94
Mass wasting, 249–51
- creep, 249
- factors affecting, 251
- slides, 250

Megalopolises, 159
Mesas, 271
Mesosphere, 54
Mesozoic Era, 211
Meteorites, 23, 51, 52, 57–58
- chondrite, 30
- impacts, 28

Meteoroids, 23
Milankovitch Cycles, 127, 162
Milky Way, 18, 39
Mineral matter, 226
Minerals, 67–68
Monsoons, 150, 151, 152
Moraines, 282–83, 287
Morphological systems, 8, 255
Mountain glaciers, 280
Mountains, effect on climate, 153–54

N
Nebula, 25
Nebular hypothesis, 25, 28
Net radiation (Rn), 122, 126–28
Nitrogen cycle, 215
North Atlantic Deep Water Current, 200
North Atlantic drift, 148
Northern lights, 64
Nuclear fusion, 29
Nutrient cycles, 213–15
Nutrients, 213

O
Obsidian, 71
Ocean currents, 147–48
Oceans, 33, 80–83, 198–201
Offshore bars, 296
Omnivore, 216
Organic matter, 28, 226
Orographic rainfall, 178, 179
Oxbow lake, 262, 265
Oxidation, 245
Oxygen cycle, 214–15
Ozone, 125, 128
Ozone layer, 4, 332–35

P
Pangea, 113–14
Parent material, 226
Particulate matter pollution, 159
Patterned ground, 290
Pedestal rocks, 271
Pediments, 270
Peds, 227
Periglacial landscapes, 288–91
Perihelion, 39
Permafrost, 288
Permeability, 203
pH scale, 228, 349
Photosynthesis, 159, 213–16
Physical geography
- characteristics of, 6–7

framework for studying, 3–5
systems in, 8–9
theories in, 10, 12–13
Pillow lavas, 95
Pingos, 290
Plains, types of, 284
Planetesimals, 28, 51, 58
Planets, 24, 34
Plastic deformation, 279
Plate tectonics, 88–114, 162–63
boundaries, 89–91
faulting, 89
folding, 89, 108
inactive zones, 111–14
mid-ocean ridge, 92, 94–96
sea-floor spreading, 92–94, 96
subduction, 103–104
transform faults, 97–101
vulcanism, 89, 106–107
Plateau, 270
Playas, 270
Pleistocene, 277, 278
Point bar, 261
Polar front, 172–73
Poljes, 267
Porosity, 203
Precession of the equinoxes, 40, 43
Precipitation, 207
Predators, 217
Primary (P) waves, 60–61
Primitive atmospheres, 28
Protosun, 28

R

Radiation balance, 127
Radioactive decay, 10, 30, 52
Radiosonde balloons, 186–87
Rainforest, tropical, 223–24, 345–49
Rainpits, 267
Regs, 272
Rehabilitation, 348
Remote-sensing, 318–19
Replacement, 348
Respiration, 215, 216
Restoration, 348
Revolution, 39
Richter scale, 313
Rift valley, 96
Rills, 267
Rivers, 204, 264–65
Rock basin lakes, 281
Rock cycle, 69, 212, 213
Rock glaciers, 249–50
Rock knob topography, 281
Rock slide, 250
Rocks, 68–70
igneous, 70–72
metallic mineral deposits, 72
metamorphic, 78–80
sedimentary, 72–77
Ross ice shelf, 201–202
Rossby waves, 176
Run-off, 257, 258, 263, 269

S

Salinity, 199
Saltation, 260
Salts, 199
Saltwater, 199–201
San Andreas Fault, 100–101, 314–15
Sand plains, 284
Sand seas, 272
Satellite photos, 187–90
Schlerophyll forests, 222
Scree, 251, 291
Sea stack, 294
Seasons, 40, 41
Secondary (S) waves, 60–61
Seismic tomography, 57, 61–62
Seismic waves, 58, 60, 91
Seismology, 58–61
Sensible heat flux, 152
Serir, 272
Shearing, 279
Sheet flooding, 269–70
Sheeting, 244
Shortwave radiation, 122, 124, 125
Sial, 81
Silicates, 54
Sima, 81
Slump, 250
Snout, 279
Snow belt, 161
Snowfields, 285
Soil, 226–29, 236
classification and distribution, 233–34
creep, 249
Soil formation, 230–32
leaching, 230–31
Soil orders, 235–36
Soil systems, 226–29
Solar constant, 121, 122
Solar energy, 243
Solar flares, 29
Solar radiation, 119, 132, 149, 152, 205
balance, 122–26
reflection/scattering, 124
variations in, and climate, 137–40
Solar system, 20
components of, 23–24
five zones of, 24
origin of, 25–29
Solar wind, 28, 30
Solifluction, 249, 289, 291
Solution, 260
Solution furrows, 267
Spillways, 284
Spit, 296
Stalactites, 268
Stalagmites, 268
Stars, 18
Steady-State Hypothesis, 21–22
Stone circles, 290
Storm surges, 309
Stream(s)
braided, 263, 288, 291
capacity, 260
competence, 259
deposition, 260
ephemeral, 258
erosion, 259
frequency and velocity of flow, 258–59
graded, 260–61
intermittent, 258
meanders, 261–62
perennial, 258
transportation, 259
Striations, 280
Summer solstice, 41
Sun, 20, 29, 119, 120, 162
Sunspots, 29
Supercontinents, 113–14
Suspension, 260
Systems, natural, 4

T

T. Tauri Wind, 28
Taiga, 220
Talus, 251, 291
Tarns, 285
Terrestrial biosphere, 211
Terrestrial planets, 24
Thermal equator, 145
Thermal expansion, 243–44
Thermocline, 200
Thermocouple, 170
Tides, 297–98
Time zones, 46
Tombolo, 294
Topsoil, 229
Tornadoes, 183–86, 311–12
Traction, 260
Trade winds, 143, 177, 191
Transportation, 242, 259
Tropical rainforests, 223–24
Troughs, 177
Tsunamis, 307, 309
Tundra, 218
Typhoons, 178–82

U

Universe, 21–22, 39
USDA Soil Classification System, 234–35
U-shaped valley, 286

V

Vertical zonation, 225
Visible spectrum, 124
Volcanoes, 94, 95, 96, 127, 162, 316–18
Voyager 2 space probe, 18, 24

W
Wadis, 269
Walker Cell, 191
Water, 196, 198–207, 226
Water cycle, 8
Water table, 202
Wave-cut platform, 295
Weather, 132–33
  equatorial, 177–86
  of mid-/high latitudes, 172–76
  predicting, 186–90
  satellite photos, 187–90
Weathering, 242, 243–46
  and gravity, 248–51
  chemical, 245–46, 248
  physical, 243–44, 248
  variations in, 247–48
Weather radar, 187
Wind chill index, 171, 172
Wind systems, global, 141–46
  pressure gradient, 142
  subpolar lows, 143
  subtropical highs, 143
  westerly winds, 143
Wind systems, regional, 150
  chinook, 154
  fohn, 154
Wind tunnels, 160
Winter solstice, 41

X
Xerophyte, 222

Z
Zero discharge, 362

PHOTO CREDITS

Cover: Imagebank; Chapter 1 Opener pp. 2-3, Figs. 1.4a-d, Figs. 1.7a-c: Gary Birchall; Chapter 2 Opener pp.16-17, Fig. 2.9 (Landsat Image): Courtesy of the Canada Centre for Remote Sensing, Energy, Mines & Resources; Chapter 3 Opener pp. 37-38: NASA; Chapter 4 Opener pp. 50-55, Figs. 4.3, 4.5: Gary Birchall; Figs. 4.6, 4.7: R. Herd/Geological Survey of Canada; Fig. 4.12: Dr. A. M. Dzeiwonski; Chapter 5 Opener pp. 66-67: Gary Birchall; Fig. 5.6: Victor Last; Figs. 5.7, 5.8, 5.12a-d: Gary Birchall; Figs. 5.15a, b, d: Ontario Geological Society; Fig. 5.15c: Alex/Arlene Penner; Chapter 6 Opener pp. 85-86: J. Fuste Raga/Masterfile; Fig. 6.10: from *Geochemical and Structural Studies of the Lamont Seamounts: Earth and Planet.* Sci. Letts., 89, 63-83, by D.J. Fornari, M.R. Perfit, J. F. Allan, R. Batiza, R. Haymon, A. Barone, W. B. F. Ryan, T. Smith, T. Simkin; Fig. 6.11: Jane Takahashi/ Volcano Observatory USGS; Figs. 6.13, 6.16: Dr. John Shelton; Fig. 6.22: J. Fuste Raga/ Masterfile; Fig. 6.23: Gary Birchall; Fig. 6.25: Earth Satellite Corp.; Chapter 7 Opener pp. 118-119: Ministry of Natural Resources; Chapter 8 Opener pp. 131-132: Gary Birchall; Chapter 9 Opener pp. 168-169, Figs. 9.1a, b, c: Bill Kiely/Atmospheric Environment Service; Fig. 9.7 John Wiley & Sons, Inc. Photo Library; Figs. 9.15, 9.16, 9.17: Environment Canada; Chapter 10 Opener pp.195-196: Arlene Penner; Chapter 11 Opener pp. 210-211, Figs.11.8, 11.9, 11.10: Gary Birchall; Fig.11.11: Victor Last; Figs. 11.12, 11.13, 11.14: Gary Birchall; Fig.11.15: Y. Shopov; Fig.11.16: Victor Last; Chapter 12 Opener pp. 240-241: Gary Birchall; Fig.12.1: Madhu Ranadive; Figs. 12.2, 12.3, 12.4, 12.5, 12.6, 12.7, 12.8: Gary Birchall; Fig.12.12: Y. Shopov; Fig. 12.14: John Wiley & Sons, Inc. Photo Library; Chapter 13 Opener pp. 254-255: Gary Birchall; Fig. 13.9: Y. Shopov; Fig.13.11: Victor Last; Fig.13.12: Gary Birchall; Fig.13.15: Gary Birchall; Fig.13.17: Victor Last; Chapter 14 Opener pp. 275-276, Figs.14.5a-e, 14.7, 14.8: Gary Birchall; Fig.14.9: Laura Radburn/Geological Survey of Canada; Fig.14.11: Denis St. Onge/Geological Survey of Canada; Chapter 15 Opener pp. 301-302, Fig. 15.1: John Wiley & Sons, Inc. Photo Library; Fig. 15.7: Canapress; Fig. 15.10: Associated Press; Chapter 16 Opener pp. 322-323: Christine Hannell; Chapter 17 Opener pp. 344-345: Masterfile; Fig.17.1: Jonathan Dunhill; Fig. 17.2 Masterfile; Fig.17.3: from *Save the Earth* by David Suzuki; Fig.17.4: Alex/Arlene Penner